Advances in Flexible and Printed Electronics

Materials, fabrication, and applications

Online at: https://doi.org/10.1088/978-0-7503-5492-9

Advances in Flexible and Printed Electronics

Materials, fabrication, and applications

Edited by
Shanmuga Sundar Dhanabalan
RMIT University, Melbourne, Victoria, Australia

Arun Thirumurugan
Sede Vallenar, Universidad de Atacama, Vallenar, Chile

IOP Publishing, Bristol, UK

ISBN 978-0-7503-5492-9 (ebook)
ISBN 978-0-7503-5490-5 (print)
ISBN 978-0-7503-5493-6 (myPrint)
ISBN 978-0-7503-5491-2 (mobi)

DOI 10.1088/978-0-7503-5492-9

Version: 20231201

IOP ebooks

British Library Cataloguing-in-Publication Data: A catalogue record for this book is available from the British Library.

Published by IOP Publishing, wholly owned by The Institute of Physics, London

IOP Publishing, No.2 The Distillery, Glassfields, Avon Street, Bristol, BS2 0GR, UK

US Office: IOP Publishing, Inc., 190 North Independence Mall West, Suite 601, Philadelphia, PA 19106, USA

This book is dedicated to the pioneers of innovation, the visionaries of tomorrow, and the champions of flexibility in thought and deed. It is a tribute to those who tirelessly explore the frontiers of materials, fabrication, and applications in the world of flexible and printed electronics.

Contents

Part II Fabrication—the art of realization

Preface

The world of electronics is undergoing a profound evolution. It is a time of remarkable transformation, where flexibility and printability redefine the boundaries of innovation. Within these leaves, we present *Advances in Flexible and Printed Electronics: Materials, Fabrication, and Applications*, a comprehensive exploration of the dynamic field that is shaping the future of electronics. Our book is thoughtfully organized into three distinct sections, each offering unique insights and perspectives:

Part I: Materials—the foundation of innovation

Our journey begins with a profound examination of materials with the four chapters 'Polymeric materials for flexible and printed electronics,' 'New generation polymer nanocomposites for flexible electronics,' 'Flexible thin films for electronics,' and 'A comparative study of carbon-based nanoribbons and MoS2-based nanoribbons for spintronics-based devices,' which introduces us to the versatile world of polymers, which form the backbone of flexible and printed electronic systems. We venture into the realm of nanoscale engineering, where polymers and nanoparticles converge to unlock new frontiers in electronic materials. These chapters invite us to explore the fascinating world of bendable, stretchable, and conformal films that challenge conventional boundaries.

Part II: Fabrication—the art of realization

The second section brings us into the realm of fabrication techniques with the two chapters 'Fabrication techniques for printed and wearable electronics' and 'Laser patterning in fabrication of flexible perovskite solar cells,' which offer a panoramic view of the techniques that breathe life into printed and wearable electronic devices, where imagination becomes reality.

Part III: Applications—pioneering new horizons

Our expedition culminates in the realm of applications. This section contains the following chapters: 'Flexible electrochemical sensors for biomedical applications,' 'Flexible conformal textile antennas,' 'Recent developments in flexible and printed reconfigurable antennas for medical and Internet of Things applications,' 'Advancement in flexible screen printing electrodes for medical and environmental applications,' 'The flexible and printed energy storage devices for foldable portable electronic devices applications,' and 'Flexible piezoelectric, triboelectric and hybrid nanogenerators.' These chapters reveal the power of real-time health monitoring and environmental sensing; establish a connection between the digital and physical worlds, paving the way for seamless connectivity; and guide us into the realms of medicine and the Internet of Things, redefining connectivity. This section also showcases the profound impact of technology on healthcare and environmental stewardship. Finally, we conclude our odyssey at the intersection of human motion and sustainability. In this captivating closing chapter, we delve into a remarkable world where the kinetic energy generated by human movement transforms into a valuable source of sustainable power. This innovative concept represents a harmonious fusion of technology and human activity, offering a promising solution to address the pressing challenges of energy generation and conservation in our dynamic world.

This book is a collective effort, bringing together the insights and expertise of leading researchers and innovators in the field of flexible and printed electronics. Our hope is that these chapters will inspire, inform, and ignite the curiosity of readers, guiding them into the dynamic world of flexibility and printability. As editors, we are humbled to present this compendium, believing it will serve as a guiding light for researchers, engineers, and enthusiasts navigating the ever-expanding horizons of flexible and printed electronics. May it kindle your imagination, spur innovation, and lead you toward new vistas of discovery and creation. With immense enthusiasm, we invite you to embark on this extraordinary odyssey through *Advances in Flexible and Printed Electronics*.

Acknowledgments

The creation of this edited book, *Advances in Flexible and Printed Electronics: Materials, Fabrication, and Applications*, has been a collaborative journey marked by dedication, expertise, and the unwavering commitment of numerous individuals. We are deeply grateful to the contributing authors whose outstanding contributions have breathed life into this comprehensive volume.

To each author, we express our gratitude for your expertise and insights that have illuminated the pages of this book, providing invaluable knowledge to our readers, and pushing the boundaries of flexible and printed electronics. Your meticulous research, innovative ideas, and dedication to your respective fields have enriched this publication beyond measure. We extend our sincere appreciation to the authors who meticulously crafted each chapter, sharing their expertise, discoveries, and passion with the world. Your collective efforts have shaped this book into a valuable resource for researchers, engineers, and enthusiasts alike, offering a treasure trove of knowledge and inspiration.

Furthermore, we would like to acknowledge the unwavering support and encouragement from our colleagues, mentors, and institutions that have made this endeavor possible. Specifically, Shanmuga Sundar Dhanabalan would like to express his sincere thanks to Prof. Sivanantha Raja Avaninathan (Alagappa Chettiar Government College of Engineering and Technology, Karaikudi, Tamilnadu, India), Prof. Marcos Flores Carrasco (Facultad de Ciencias Físicas y Matemáticas, University of Chile, Chile), and Prof. Sharath Sriram and Prof. Madhu Bhaskaran (Functional Materials and Microsystem Research Group, RMIT University, Australia) for their continuous support, guidance, and encouragement. Dhanabalan also extends his gratitude to Mrs Preethi Chidambaram and his family for their support. Arun Thirumurugan would like to express his gratitude to Dr Justin Joseyphus (National Institute of Technology, Tiruchirappalli, India) for his valuable guidance and mentorship; Prof. P V Satyam (Institute of Physics, Bhubaneswar, India) for his support and encouragement; Dr Ali Akbari-Fakhrabadi (Facultad de Ciencias Físicas y Matemáticas, University of Chile, Chile) for his guidance and collaboration; Prof. R V Mangala Raja (University of Adolfo Ibanez, Santiago, Chile) for his mentorship and support; Dr R Udaya Bhaskar and Mauricio J Morel (University of Atacama, Chile) for their support and collaboration; and Carolina Venegas, Yerko Reyes, Felipe Pizarro Barraza, and Juan Campos (Sede Vallenar, University of Atacama, Chile) for their assistance and collaboration. Furthermore, he acknowledges the financial support provided by the Agencia Nacional de Investigación y Desarrollo de Chile (ANID) through the SA 77210070 project and the continuous support of the University of Atacama.

Lastly, we express our gratitude to the readers who will embark on this journey through the world of flexible and printed electronics. It is our hope that the knowledge within these pages will inspire new discoveries, innovations, and a brighter future. Thank you all for your dedication and contributions to *Advances in Flexible and Printed Electronics*. Together, we continue to push the boundaries of what is possible in this dynamic field.

Editor biographies

Shanmuga Sundar Dhanabalan

Dr Shanmuga Sundar Dhanabalan is an accomplished researcher with a proven track record in designing, developing, and translating micro- and nano-scale devices. His primary objective is to create next-generation products that enhance quality of life and well-being, making a significant contribution to society. He is currently leading a team working on `wearable and connected sensors' at RMIT University, with a focus on materials, flexible and stretchable devices, wearables, optics, and photonics.

He graduated with a PhD in flexible electronics in June 2017 and secured a competitive postdoctoral fellowship from the Chilean government from 2018 to 2021. His studies have led to publications in referred international journals, book chapters, and books in progress as editor. He has presented plenary/keynote speeches, invited talks and guest lectures, oral and poster presentations at scientific meeting at various universities world-wide. Several outcomes have been highlighted by scientific websites (such as Photonics Media, USA).

His research work has led to securing grants from Australian government research schemes, such as the Cooperative Research Centres Projects, the ARC Research Hub for Connected Sensors for Health, Victorian Medical Research Acceleration Fund, and the Advanced Manufacturing Growth Centre's Commercialisation Fund. He has collaborations with universities from various countries, including India, Australia, Chile, Mexico, and Bangladesh.

He has served as a reviewer for over 20 prestigious specialist journals. He also served as a topical editor for highly reputed journals including IEEE Transactions on Industrial Informatics, IEEE Instrumentation and Measurement Magazine, IEEE, Energies, Computer and Electrical Engineering. He is a part of the Editorial Board of American Journal of Optics and Photonics. In addition, he served as a session chair and technical committee member in various international conferences.

Arun Thirumurugan

Dr. Arun Thirumurugan is an assistant professor at the University of ATACAMA, Sede Vallenar, Vallenar, Chile. He has completed his PhD (2010-2015) from the National Institute of Technology (NIT), Tiruchirappalli, India. He has worked as a postdoctoral fellow (2015-2017) at the Institute of Physics, Bhubaneswar, India, and then worked as a FONDECYT postdoctoral fellow (2017-2020) at the University of Chile, Santiago, Chile. His research interests are in synthesizing and surface modification of nanomaterials for the potential applications in energy and environmental applications.

List of contributors

Radhamanohar Aepuru
Departamento de Mecánica, Facultad de Ciencias Físicas y Matemáticas, Universidad de Chile, Santiago, Chile

Ali Akbari-Fakhrabadi
Departamento de Mecánica, Facultad de Ciencias Físicas y Matemáticas, Universidad de Chile, Santiago, Chile

M Alagappan
Department of Electronics and Communication Engineering, PSG College of Technology, Coimbatore, India

M S Anoop
Department of Automobile Engineering, SCMS School of Engineering and Technology, Cochin, India

R Ashok Kumar
Department of Mechanical Engineering, Madurai Institute Engineering and Technology, Sivaganga, India

Sahaya Dennish Babu
Department of Physics, Chettinadu College of Engineering and Technology, Karur, India

Samuel Paul David
Department of Physics, School of Advanced Sciences, Vellore Institute of Technology, Vellore, India

A Taksala Devapriya
Department of Electronic and Communication Engineering, Mount Zion College of Engineering and Technology, Pudukkottai, Tamil Nadu, India

Preeti Gupta
Leibniz Institute of Solid State and Material Research, Dresden, Germany

M Harikrishnan
Department of Electronics and Communication Engineering, Sri Manakula Vinayagar Engineering College, Madagadipet, Puducherry, India

Kashmira Harpale
Department of Physics and Department of Electronic and Instrumentation Science, Savitribai Phule Pune University, Pune, Maharashtra, India

Shweta Jagtap
Department of Electronic and Instrumentation Science, Savitribai Phule Pune University, Pune, Maharashtra, India

P C Jayadevan
Department of Automobile Engineering, SCMS School of Engineering and Technology, Cochin, India

D Jeyakumar
CSIR-Central Electro Chemical Research Institute, Karaikudi, Tamilnadu, India

P Justin
Department of Chemistry, Rajiv Gandhi University of Knowledge Technologies —R K Valley Campus, Kadapa, Andhra Pradesh, India

Madurakavi Karthikeyan
School of Electronics Engineering, Vellore Institute of Technology, Vellore, Tamilnadu, India

Sandeep Kumar
Department of Physics and Astronomical Science, Central University of Himachal Pradesh, Kangra, Himachal Pradesh, India

Viviana Meruane
Departamento de Mecánica, Facultad de Ciencias Físicas y Matemáticas, Universidad de Chile, Santiago, Chile

Naveen Mishra
School of Electronics Engineering, Vellore Institute of Technology, Vellore, Tamilnadu, India

D Muthukrishnan
Department of Mechanical Engineering, KLN College of Engineering, Sivaganga, India

P Nagaraja
Department of Chemistry, Rajiv Gandhi University of Knowledge Technologies— R K Valley Campus, Kadapa, Andhra Pradesh, India

K J Nagarajan
Department of Mechanical Engineering, Thiagarajar College of Engineering, Madurai, India

Durga Prasad Pabba
Department of Electricity Facultad de Ingeniería, Universidad Tecnológica Metropolitana, Santiago, Chile

J Shahitha Parveen
Department of Polymer Engineering, School of Mechanical Sciences, B S Abdur Rahman Crescent Institute of Science and Technology, Chennai, India

N Prabu
Srinivasan College of Arts and Science, Perambalur, Tamilnadu, India

J Pradeep
Department of Electronics and Communication Engineering, Sri Manakula Vinayagar Engineering College, Madagadipet, Puducherry, India

Surender Pratap
Department of Physics and Astronomical Science, Central University of Himachal Pradesh, Kangra, Himachal Pradesh, India

G R Raghav
Department of Mechanical Engineering, SCMS School of Engineering and Technology, Cochin, India

M Rajesh
Department of Computer Engineering, Sanjivani College of Engineering, India

G Ramesh
Department of Metallurgical and Materials Engineering, Rajiv Gandhi University of Knowledge Technologies—R K Valley Campus, Kadapa, Andhra Pradesh, India

G Ranga Rao
Department of Chemistry, Indian Institute of Technology Madras, Chennai, Tamilnadu, India

S Robinson
Department of Electronic and Communication Engineering, Mount Zion College of Engineering and Technology, Pudukkottai, Tamil Nadu, India

Chandrashekhar Rout
Centre for Nano and Material Sciences, Jain (Deemed-to-be University), Jain Global Campus, Bangalore, Karnataka, India

Hari Prasad Sampatirao
Department of Physics, Aditya Institute of Technology and Management (AITAM), Tekkali, India

Mani Satthiyaraju
Department of Mechanical Engineering, Kathir College of Engineering, Coimbatore, India

H Seshagiri Rao
Department of Chemistry, Rajiv Gandhi University of Knowledge Technologies—R K Valley Campus, Kadapa, Andhra Pradesh, India

R Sitharthan
School of Electrical Engineering, Vellore Institute of Technology, Chennai Campus, Tamilnadu, India

Thamizharasan Sivanesan
School of Computer Science and Engineering, Vellore Institute of Technology, Vellore, Tamilnadu, India

Ananthakumar Soosaimanickam
R&D Division, Intercomet S.L., Madrid, Spain

T Sridarshini
Department of Electronics and Communication Engineering, College of Engineering Guindy Campus, Anna University, Chennai, India

P S Sowbarnika
Department of Electronics and Communication Engineering, PSG College of Technology, Coimbatore, India

Arun Thirumurugan
Sede Vallenar, Universidad de Atacama, Vallenar, Chile

M Thirumurugan
Department of Mechanical Engineering, School of Mechanical Sciences, B S Abdur Rahman Crescent Institute of Science and Technology, Chennai, India

Part I

Materials—the foundation of innovation

IOP Publishing

Advances in Flexible and Printed Electronics
Materials, fabrication, and applications
Shanmuga Sundar Dhanabalan and Arun Thirumurugan

Chapter 1

Polymeric materials for flexible and printed electronics

Preeti Gupta

In the rapidly evolving landscape of printable electronics, the emergence of solution-processed flexible transparent electrodes represents a groundbreaking advancement with far-reaching implications. This innovative technology, poised to revolutionize various industries, offers a multitude of benefits that have the potential to reshape the way we interact with and integrate electronics into our daily lives. This chapter is a compelling exploration into this remarkable development.

1.1 Introduction

Flexible and printed electronics has revolutionized the digital industry due to the advancement in the field of electronics manufacturing and design. The boom of flexible and printed electronics has ushered in a new era of electronic device manufacturing, enabling greater design flexibility, cost efficiency, and a diverse range of applications across industries. Unlike traditional rigid circuit boards and components used in conventional consumer electronics, flexible and printed electronics involve the use of functional materials and techniques that allow electronic devices to be printed or fabricated on flexible substrates that are ultra-thin and lightweight, such as plastics, paper, and flexible glasses to name a few. This has led to a wide range of applications and has sparked a revolution in realm of industries. As technology continues to advance, the integration of flexible and printed electronics is likely to become more prevalent, leading to further innovations and transformative changes in the way we interact with electronics. At the forefront of this technological revolution are polymeric materials, which have been instrumental in unlocking the full potential of flexible electronics due to their versatility, flexibility, low-cost polymer processing, chemical structure of polymers, and more interestingly, they are biocompatible and friendly to the human body. Lately,

doi:10.1088/978-0-7503-5492-9ch1

polymers have become an integral part of flexible and printed electronics owing to vast application in consumer electronics.

This chapter aims to provide a comprehensive analysis of the role and significance of polymeric materials in enabling the development and commercialization of flexible and printed electronic devices. The most important aspect of fully developed sensory system includes functional polymers and non-functional polymers. The chapter is divided into four sections as shown in figure 1.1.

This chapter begins by highlighting the key properties and advantages of polymeric materials in the context of flexible and printed electronics. The inherent flexibility, lightweight nature, and cost-effectiveness of polymers make them highly desirable for applications where traditional rigid substrates are unsuitable. The ability of polymers to be molded into various shapes and sizes without compromising their electronic performance opens up new possibilities in wearable devices, conformal sensors, and smart textiles. The second section delves into the importance of conductive polymers in flexible electronics. These materials offer the unique capability of conducting electricity while maintaining their flexibility, which is essential for producing flexible electrodes, interconnects, and transistors. This

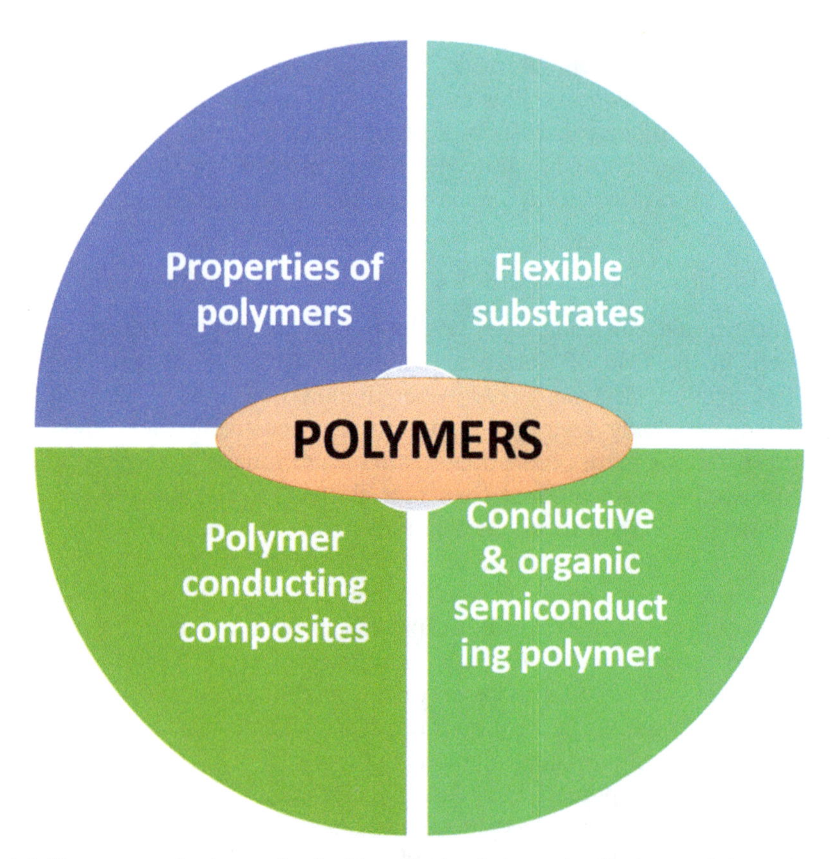

Figure 1.1. Key aspects of polymers for flexible and printed electronics (1) characteristics and properties, (2) flexible substrate, (3) conductive polymers, and (4) different types of flexible substrate.

section explores the different types of conductive polymers and their diverse applications, ranging from flexible displays to energy harvesting devices. The use of flexible substrates and encapsulants is critical to ensuring the integrity and durability of flexible electronic devices. The third section evaluates various polymeric substrates, such as polyimides (PIs) and polyethylene terephthalate (PET), and encapsulation materials that provide protection against environmental factors and mechanical stresses. The challenges and advancements in achieving high-performance flexible substrates are discussed, along with novel approaches to enhance their thermal stability and barrier properties. An essential aspect of flexible electronics is the scalable manufacturing process, which demands efficient and cost-effective printing techniques. From healthcare monitoring devices and smart wearables to Internet of Things applications and electronic textiles, the potential of polymeric materials seems limitless. The chapter also discusses ongoing research and future prospects, including the development of biodegradable and environmentally friendly polymeric materials to address sustainability concerns.

1.1.1 Properties of polymers

The word polymer is derived from Greek and can be split into two: (1) ‘poly’ means many, and (2) ‘mers’ means units or parts. Monomers are the smallest repeating unit of polymers. Polymer material plays a crucial role by offering a wide range of characteristics, such as being lightweight, thin, bendable, flexible, and scalable, as well as having low-cost mass production, that are essential for realization of printed and flexible electronics [1]. Depending upon the characteristics, it has a wide variety of application in the field of wearable, flexible sensors, flexible display, and flexible solar cells to name a few [1, 2]. Polymers are generally divided into several groups based on the monomers and how they are organized within the polymer at a single chain level. The properties of the polymer, both physical and chemical, are significantly influenced by the polymer’s size [3]. The functional groups attached to the polymer’s long chain facilitate ionic and hydrogen bonding among their own chains, giving high tensile strength and crystalline melting points [3]. Polymer composite with nanomaterial and dielectric material has attracted significant interest due to different mechanical, electrical, thermal, and optical properties.

Polymers exhibit a wide range of properties, making it suitable for a wide range of applications. Some key parameters of polymers are shown in figure 1.2.

1.1.1.1 Physical properties

The density of polymers plays an important role in determining the physical characteristics and behavior of polymers. It is influenced by various factors such as molecular weight, chemical composition, and arrangement of its polymer chains [3]. The density is a key factor in determining the mechanical, thermal, and optical properties. The higher number of branching points leads to low density polymers like polyethylene, and more linear structure gives higher packing density such as PET. Due to light weight and flexibility of low-density polymers, they are significantly used in the packaging industry. The high-density polymers are favored

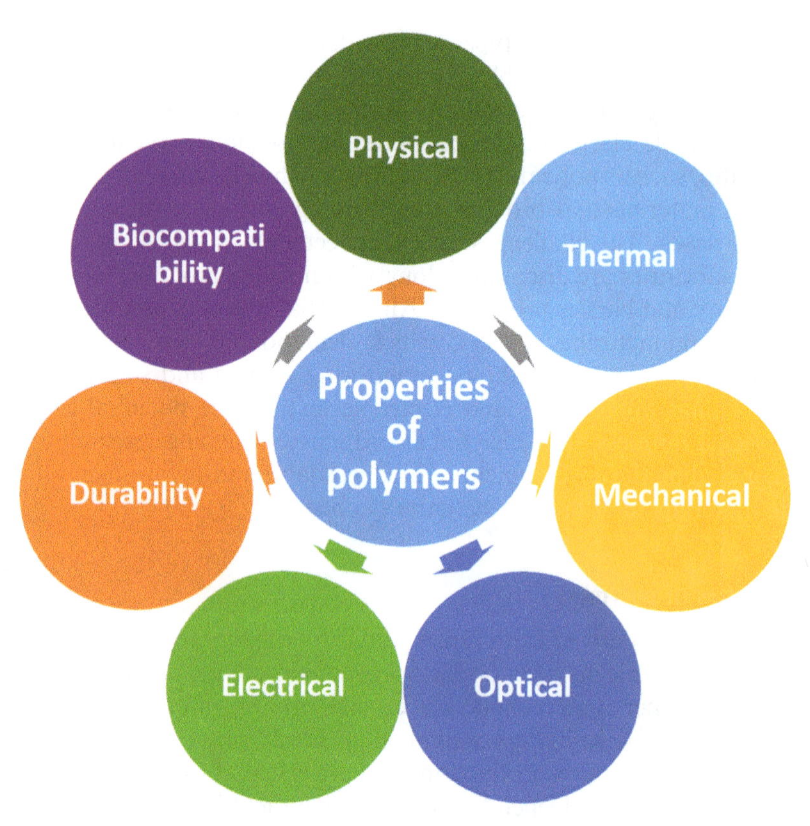

Figure 1.2. Key properties of polymers.

when rigidity and durability are required, as seen in water bottles and mechanical parts [4]. The density ranges from about 2.3 g cm^{-3} to about 6.3 g cm^{-3} [4]. The heaviest optically viable polymer has a density of only about 1.4 g cm^{-3}, whereas the lightest of these materials can float with a density of 0.83 g cm^{-3} [5]. Thus, comprehending polymer density is crucial for choosing the appropriate material for particular uses and designing new polymer composites. The density of common polymers [6] used is tabulated in table 1.1 and may vary from 0.01 to 2.30 g cm^{-3} depending upon the manufacturing process and type of polymer [5].

Quantifying the hardness of a polymer presents challenges due to its reliance on both the material and the manufacturing process. This property is influenced by various factors, including the type of polymer, its molecular weight, cross-linking, crystallinity, and the presence of additives. Polymers exhibit a wide range of hardness levels, from very soft and rubbery elastomers to rigid and tough thermoplastics, and can be adjusted using diverse techniques [3]. These include cross-linking, blending with alternative polymers or additives, and managing the crystallinity of the material. Thus, the appropriate equilibrium is pivotal between hardness and mechanical characteristics in enhancing the performance of the polymer for applications spanning from automotive to medical tools and consumer electronics [4].

Table 1.1. Density of common polymers ranging from 0.01 to 2.30 g cm^{-3}.

S. No.	Abbreviation	Polymer	Density (g cm^{-3})
1	PS	Polystyrene	0.01–1.06
2	PP	Polypropylene	0.85–0.92
3	LDPE	Low-density polyethylene	0.89–0.93
4	HPDE	High-density polyethylene	0.94–0.98
5	PA, PA 6,6	Polyamide, nylon 6,6	1.12–1.15
6	PU	Polyurethane	1.20–1.26
7	PET	Polyethene terephthalate	1.38–1.41
8	PVC	Polyvinyl chloride	1.38–1.41
9	PTFE	Polytetrafluoroethylene	2.10–2.30

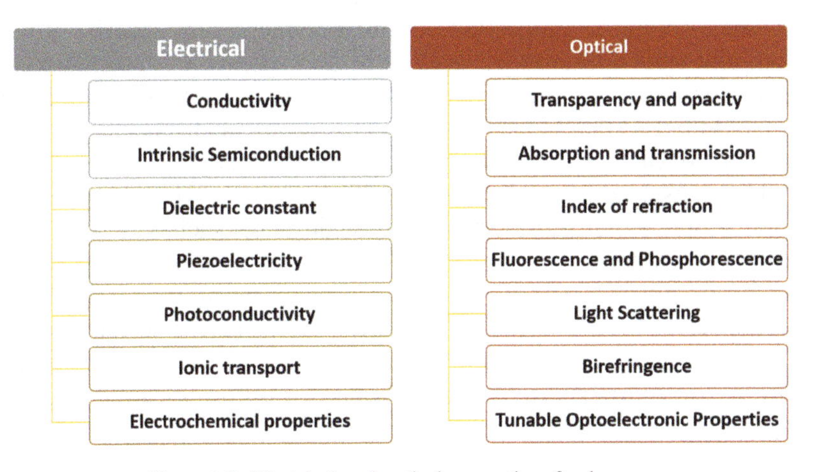

Figure 1.3. Electrical and optical properties of polymers.

1.1.1.2 Electrical properties

The electrical properties of polymers encompass a range of characteristics that describe how polymers interact with and conduct electricity.

The diverse range of electrical properties of polymers make them valuable for various applications in electronics [1], energy storage, displays, actuators, energy harvesting devices and many more [7, 8]. Some important electrical properties of polymers are shown in figure 1.3 and determine the behavior of polymers in terms of conducting, storing, and responding to electrical signals. These properties are influenced by the polymer's molecular structure, composition, and processing methods. Depending on the chemical structure, polymers can be insulators, semiconductors, or conductors of electricity [4]. Polymers mixed with solid conductive fillers such as metal particles, carbon black, graphite, and carbon nanotubes (CNTs) give polymers electrical conductivity and, hence, the term conductive polymer composite [9]. The dielectric constant determines the polymer's ability to store the electrical energy, and high dielectric constant polymers are used in energy storage devices. Many polymers are electrical insulators and prove to be beneficial in consumer applications, which include cable insulation, electrical components, and protective coating to name a few. The electroactive polymers [10] such as polypyrrole (PPy) can change shape, size, or mechanical properties in response to an electric field. PPy-based electroactive polymers have been employed in applications such as artificial muscles, soft robotics, and medical devices, showcasing their potential in creating versatile and adaptable systems [9 10]. The two conjugated conducting polymers, polyaniline (PANI) [11] and PPy, have garnered significant global attention that is attributed to their outstanding physical, thermal, and chemical stability, as well as notable conductivity [10].

1.1.1.3 Optical properties

Polymers exhibit a wide range of optical properties depending on the interaction of the light, as shown in figure 1.3. These attributes play a vital role in appearance, behavior, and potential application. Some polymers can absorb light at one wavelength and emit it at another, leading to fluorescence or phosphorescence [3]. These properties are used in sensors and luminescent materials. Polymers can disperse light into its different colors, leading to chromatic dispersion. This property is relevant in optical fibers. Polymers can have different refractive indices along different directions, leading to birefringence. This property is essential in LCDs [12]. The optical properties of polymers are crucial in fields such as optics, photonics, displays, packaging, and photography. They impact how polymers interact with light sources and influence their visual appearance, transparency, and color [12, 13]. By manipulation of these attributes, researchers have created customized materials suitable for precise applications, ranging from light-guiding elements to practical surfaces. Polymers showcase a varied collection of optical characteristics intrinsically tied to their interactions with light.

1.1.2 Flexible substrates: polymers

Substrates are an integral part of any flexible and printed device [1]. The two most common and widely used substrates are made of plastic and elastomers. The plastic substrate are polymers with good insulation, high strength, excellent bending property, good transparency, and temperature resistance [1]. The most promising plastic substrates are PI [13], PET [14], polyethylene naphthalate [15] are mainly used in the fields of flexible solar cells and flexible displays [1]. Due to recent significant advances in the digital era and wearable sensors being one of the key elements in our day-to-day life, be it smart watches, smart glasses, e-skins, or biomedical devices for regular monitoring of health, elastomers play a crucial role possessing high stretchablity and elasticity unlike plastic substrates [16–18]. The freedom of movement offered by elastomers at par with plastic substrates make them an ideal candidate for stretchable and bendable substrates without affecting the device performance. The typical elastomers are polydimethylsiloxane (PDMS) [19] and thermoplastic polyurethane (TPU) [20].

1.1.2.1 PET substrates

Polyethylene terephthalate (PET, PETE, or the obsolete PETP or PET-P) is the most common thermoplastic polymer resin of the polyester family. PET consists of repeating (C10H8O4) units and is commonly recycled to reuse and avoid environmental damage [21, 22]. The thermal and processing history of PET determines its characteristic as amorphous, which is transparent, and semi-crystalline polymer appears opaque or white depending upon the crystal structure and particle size [23]. Considering strength, durability, and transparency of PET and its vast applicability in flexible solar cells, flexible displays, and many other fields, it is one of the most commercialized and recyclable semi-crystalline thermoplastic polymer with good toughness, high transmittance in the visible light range, relatively low gas and water vapor permeability, and acceptable chemical resistance [14, 24, 25]. The high strength of PET makes it resistant to stress and deformation, making it suitable for structural integrity application. The low moisture absorption and chemical resistance makes it an ideal packaging material [23]. PET has a relatively high melting point offering high thermal stability suitable for microvan safe plastic containers [22]. Additionally, it also has good electrical insulation, the ease of processing by various methods such as extrusion, injection molding, and blow molding makes it a suitable candidate for the manufacturing process [1]. In addition to flexibility, it also offers good UV resistance and low-cost processing, contributing to its widespread use in the device industry. The first OLED device was built by Braun and Heeger when they coated PANI on PET substrate, which paved a way for flexible displays [26]. This development led to opening up the pathway for flexible solar cells, displays, flexible integrated circuits, etc [25, 27–29]. While PET offers numerous advantages for flexible displays and solar cells, it is important to note that its electrical properties, especially its low conductivity, can be a limiting factor.

Therefore, additional layers or coatings might be needed to enhance electrical performance, and in some cases, other materials like transparent conductive films (e.g., indium tin oxide or Ag nanowires (AgNWs)) might be used in conjunction with PET to address these limitations [30] (figure 1.4). Tough PET is a good candidate for numerous applications; wettability, adhesiveness, and printability are the parameters that need to be addressed due to low surface energy of PET and thus call for surface modification of PET [21]. This was achieved using chemical treatment, heat treatment, and, most commonly used, plasma treatment [1]. The PET film was treated with oxygen plasma to generate and activate active spots on the surface for good adhesion [31]. To achieve the smoothness of PET, it was also treated chemically with a solution of sulfuric acid and hydrogen peroxide in an adequate concentration and achieved significant improvement in surface microstructure, enhancing the smoothness of PET for electronic and optoelectronic devices [32].

1.1.2.2 PI substrate

PI is a polymer containing imide groups commonly abbreviated as PI polymer. The heat resistance property of PI makes it a suitable candidate for diverse applications such as high-temperature fuel cells and displays. PI possesses excellent mechanical properties, excellent chemical stability, and good heat resistance, as high as 400 °C

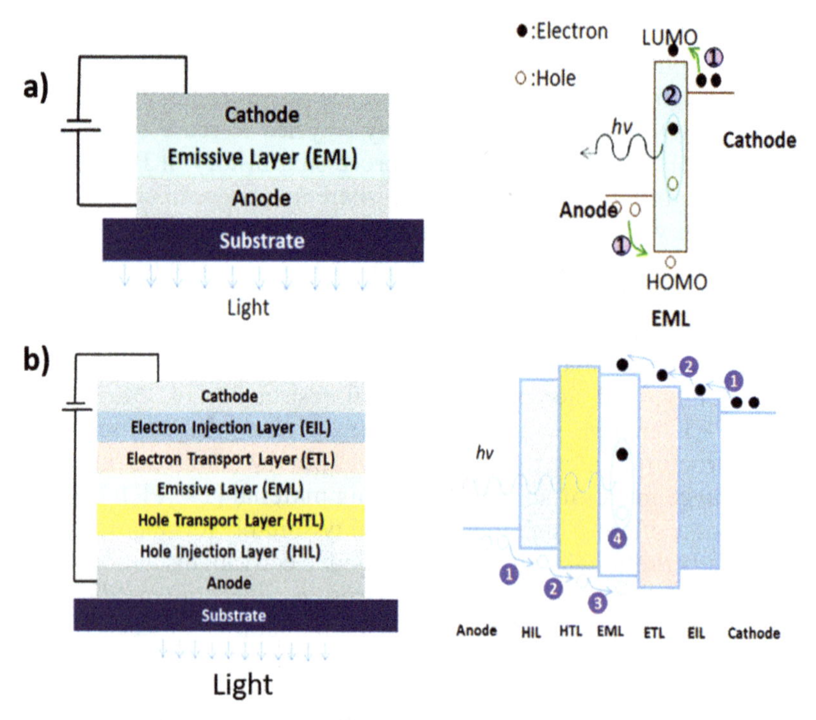

Figure 1.4. Schematic of (a) single-layer and (b) multilayer OLEDs, and corresponding energy level diagram [30]. Reproduced from [30]. CC BY 3.0.

or above with a low thermal expansion coefficient [1]. Bogert and Renshaw made a significant discovery in 1908 when they observed that 4-amino phthalic anhydride does not liquefy upon heating. Instead, it releases water as it transforms into a high molecular weight PI [33]. The yellow color of traditional PI films is mainly due to strong intramolecular and intermolecular charge transfer complexes, which cause strong light absorption in the ultraviolet-visible light range [1]. However, a lot of research has been focused on making PI transparent or rather colorless, also known as colorless PI (CPI) for fully transparent flexible displays and solar cells [34]. To accomplish this, extensive efforts have been undertaken to synthesize CPI (copolyimide) by incorporating unsymmetrical bonds into the primary structure of PI [34]. This approach introduces bends, helical structures, and sizeable or pendant substituents into the polymer chain, along with the utilization of aromatic fluorinated monomers [13, 28, 35]. The researchers made super-thin 1.3 micrometer and bendable, flexible solar cells using a CPI as the base [36]. This thinness made the solar cells very flexible and could be useful for things like electronic clothes and soft robots that power themselves. CPI possess strong adhesion properties with conductive materials like AgNWs and graphene [36]. The researchers figured out that putting conductive stuff into the base material is a good way to make things stick together better [14, 37].

1.1.2.3 PDMS substrates

PDMS substrates are flexible and transparent materials often used for their unique properties, versatility, and ease of processing. It is an excellent elastomer with elongation of up to 160%–180% and has good thermal stability, which can be used for an extended period of time at 150 °C [1, 38]. The soft and elastomeric nature, biocompatibility, and optical transparency of PDMS makes it suitable for applications that require deformable surfaces and structures such as microfluidics [39], soft robotics, and biomedical devices [40]. PDMS substrates offer a convenient platform for creating microstructures, enabling the development of intricate devices and systems. Similar to PET, PDMS also has poor adhesion with its active materials due to low surface energy. The chemical modification and surface treatment is necessary to improve adhesive strength. Qi and colleagues created a bionic elastomer microporous film using a single-step soft photolithography copying technique [19]. They achieved uniformly distributed micropores in the film and modified the PDMS substrate by introducing dopamine and functionalized silane. Similarly, You and coauthors employed a PDMS substrate to create a sandwich structure with AgNWs [41]. They utilized polyurethane urea as the top layer, incorporating 2,2-bis (hydroxymethyl) butyric acid [41]. PDMS is known for its impressive ability to allow oxygen and water vapor to pass through, which makes it a great choice for creating wearable items and devices that can be implanted into the body [17, 42]. However, this very characteristic that makes it breathable also limits its use in certain electronic applications. For instance, PDMS's high permeability makes it unsuitable for use in electronics like OLEDs, light-emitting electrochemical cells, and organic photovoltaic devices [43]. Researchers have explored PDMS-based microdevices for energy harvesting from mechanical vibrations. Its flexibility and ability to deform under stress make it suitable for converting mechanical energy into electrical energy [44].

PDMS is used in the fabrication of microfluidic devices on rotating discs [43]. These systems allow for sample processing, mixing, and analysis within a small, portable platform. PDMS is used as a mold or template in microfabrication processes. Researchers create microstructures on a PDMS surface and then use it to transfer those structures onto other materials, enabling the fabrication of intricate patterns. PDMS optical transparency makes it suitable for creating micro-optical devices, such as lenses and waveguides, for applications in photonics and telecommunications [19, 44, 45]. PDMS-based microgrippers and manipulators have been designed for tasks such as microassembly, microsurgery, and the delicate handling of objects within confined spaces. As a result of these attributes, PDMS has found extensive application in a range of areas. These include micropumps [46], catheter surfaces [7], dressings and bandages [8], microvalves [16], optical systems [47], disease studies conducted *in vitro* [9, 48], implants [49], and microfluidics and photonics [50–52]. Furthermore, the utilization of PDMS in microelectromechanical systems applications and microfluidic components has been greatly propelled by soft-lithography technology [51–53]. Notably, PDMS is the predominant material employed in the production of microfluidic devices, which are pivotal for advancements in fields like drug delivery, DNA sequencing, clinical diagnostics, point-of-care testing, and chemical synthesis [54]. PDMS has also been investigated in the field of medical implants [55–58]. Implants are commonly composed of Ti or its alloys, but they often lack optimal osseointegration. To improve this, PDMS has been explored for creating coatings with microscale features to enhance the bonding between implants and bones [45]. PDMS presents a promising avenue for significant advancements in various applications. To progress in this field, research should focus on exploring methods for producing PDMS-based devices on a larger scale, facilitating their market entry. Moreover, current PDMS hydrophilic treatments require further refinement because their durability is often limited [45]. Therefore, there is a need to innovate new approaches or enhance existing ones to achieve a more enduring and effective hydrophilic characteristic for PDMS.

1.1.3 Conductive and organic semiconducting polymers in flexible electronics

The polymers conduct electrical conductivity due to the presence of conjugated double bonds along their polymer chains, termed as conductive polymers. The polymers are classified into two main categories, inherently conductive polymers, which exhibit electrical conductivity without the need for additional chemical modifications or doping such as PANI, PPy, and polythiophene, and intrinsically conductive polymers, which are typically insulating polymers that have been chemically modified or doped to introduce conductivity [1]. These conductive polymers with unique capabilities such as flexibility, light weight, and low-cost manufacturing find diverse applications in organic photovoltaics, organic LEDs [12], wearable electronics [59], actuators, energy storage devices, printed sensors, and flexible electrodes. Due to the rising demand in electronic devices for communication, national security, and personal health monitoring, conductive polymers serve as an attractive candidate [60]. The technological advancement offered by conducting polymers drives the field of organic electronics by low-cost manufacturing of organic solar cells on a large scale by employing printing technology with its widespread use in consumer devices, be it a window or building structure or digital screen application using OLED.

The viable alternative to inorganic silicon semiconductors is organic semiconducting polymers, mainly due to their adaptability, lightweight nature, and adjustable optoelectronic characteristics [61, 62]. Organic semiconductors are well suited for organic thin-film transistors and for use in biomedical engineering applications [61] (figure 1.5).

The polarity of organic semiconductors (such as p-type, n-type, and ambipolar) can be manipulated by the manner in which they are layered at the molecular level and, hence, their electrical conductivity [61]. The practical utilization of organic semiconductors in biomedical engineering is still limited by the poor electrical performance when compared with their inorganic semiconductors' counterparts [63, 64]. Nonetheless, traditional silicon-based biomedical devices exhibit limited biocompatibility due to their toxicity and limited reliability within *in vivo* systems [65]. However, with the recent advancements in functional materials, varying thin-film formation technologies, and optimization of device structure, the development

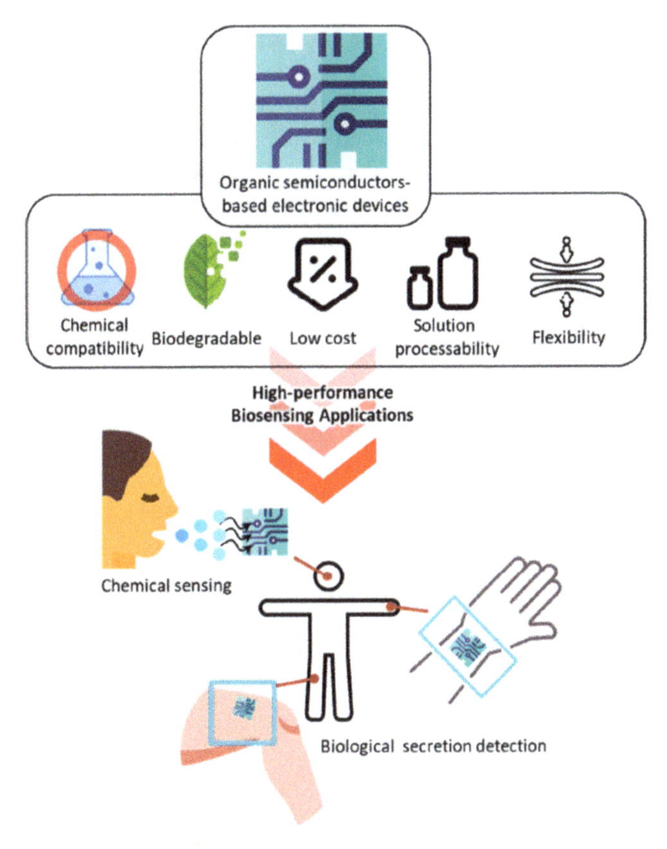

Figure 1.5. Schematic diagram of the use of organic semiconductors in biomedical applications. Reproduced from [61]. CC BY 4.0.

and incorporation of an insulating layer resulted in reported achievements in the improved device performance in biomedical engineering [61].

1.1.4 Polymer conducting composites

Intrinsic conducting polymers are categorized as either semiconductors or conductors based on their level of conductivity [1]. Polymer semiconductors are further classified as p-type or n-type. P-type polymer semiconductors primarily employ holes (positive charges) to facilitate charge transfer between molecules. The examples of p-type polymers include polythiophene (PT), polysilane, and pyrrolopyrrolidone. In contrast, n-type polymer semiconductors employ electrons (negative charges) as carriers for charge transfer, such as imide polymers [1]. The polymeric conductors are those that demonstrate electrical conductivity after doping. The most common and well-known conducting polymers gaining attention worldwide are polyaniline (PANI), polypyrole (PPy), polythiophene (PT), and poly (3,4-ethylenedioxythiophene) (PEDOT), with their distinct p-p conjugate structure [1, 66–68]. The conductive polymeric composites are made using various processing methods such as solution-based, melt blending, *in situ* polymerization, spray deposition, spin coating, and dip coating [1]. The processing techniques harness the functional attributes of conductive fillers and mixed with polymer matrix. PDMS, TPU, and poly (styrene-butadiene-styrene) are frequently used as the polymeric matrix [19, 69, 70], while the conductive fillers mainly consist of carbon-based materials (graphite, graphene, CNTs, etc), metal conductors (AgNWs/nanoparticles, gold nanosheets, and Cu nanowires), and conductive polymers (PANI, PPy, and PEDOT: polystyrene sulfonate (PSS)) [18, 42, 46, 71, 72].

1.1.4.1 Poly (3-hexylthiophene)

Poly (3-hexylthiophene) (P3HT) functions as a light-absorbing material and is particularly used as an active layer material for flexible solar cells [1]. P3HT possesses a wide bandgap (1.9 eV) and an absorption cut-off wavelength at 650 nm, which imposes limitations on the overall efficiency of flexible solar cells. To overcome this, non-fullerene acceptor was specifically designed to improve the performance together, and the efficiency of the device reached 6.4% [73]. Lee *et al* clearly demonstrated the excellent device performance and reliability of the stamping method for implementing electronic circuits by fabricating an organic/inorganic hybrid complementary inverter operating at low voltage (1 V or less) by transferring pressed p-type P3HT and inorganic n-type ZnO electrolytic transistors [74]. Beyond photovoltaics, P3HT has found applications in organic field-effect transistors, LEDs, and sensors [75].

1.1.4.2 PEDOT: PSS

The conductivity of PEDOT: PSS films depends on ratio and the particle size of PEDOT: PSS dispersed in water. The conductivity of PEDOT: PSS is limited by PSS content; once the PEDOT: PSS film is formed, its conductivity will be limited by the content of PSS, which is usually less than 10 S cm^{-1} [76] and can be significantly improved by introduction of additives including acids (sulfuric acid, methanesulfonic acid, etc) [77] or by post-treatment with various solutions such as dimethyl sulfoxide, ethylene glycol, diethylene glycol, etc [24, 78] and surfactants [76]. The PEDOT: PSS film without any additive or solution processing offers low stretchability (6%) [79]. Thus, it is very difficult to maintain a balance between high conductivity and stretchability because the resistance of the film is greatly affected by the applied stain. An attempt was made to improve the conductivity by Lee *et al* by modifying PEDOT: PSS by triblock copolymer followed by post-treatment by sulfuric acid. The triblock copolymer used was poly (ethylene glycol)-B-poly (propylene glycol)-B-poly(ethylene glycol) and prepared a transparent PEDOT: PSS film with conductivity 1700 S cm^{-1} [80].

1.1.4.3 PDMS conductive composites

PDMS-based conductive polymer composites has a multifaceted range of applications, from enhancing human-computer interaction to personal healthcare and motion performance monitoring. PDMS as piezoresistive sensors is attributed to its innate pressure-sensitive characteristics and ease of its integration on curved surfaces

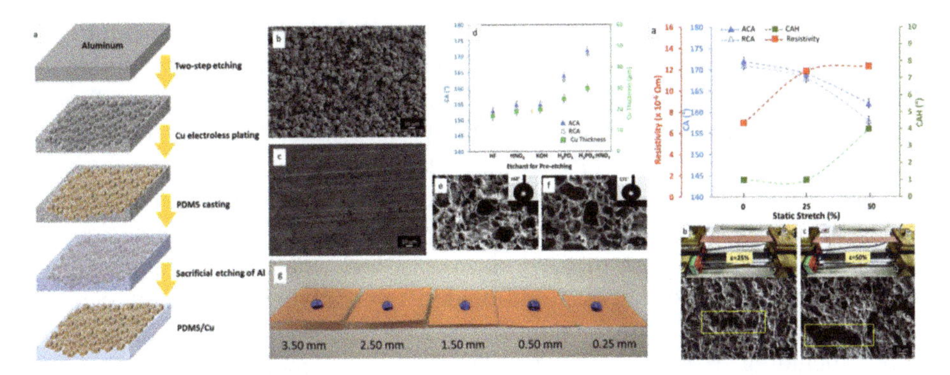

Figure 1.6. (1) Fabrication of PDMS/Cu superhydrophobic composite: (a) Schematic representation of the fabrication process to obtain PDMS/Cu superhydrophobic composite material. The scanning electron microscope (SEM) images represents the different molar concentration of HCl etched Al (b) 3 M HCl and (c) 1 M HCl. (d) represents the effect of etchant types on contact angles and plated Cu thickness. The SEM images show the morphologies of the PDMS/Cu surface obtained by (e) one-step and (f) two-step etching of Al substrate and 1 h Cu plating. (g) Digital photograph of composite. (2) Mechanical and surface properties evaluation: (a) illustrates the resistivity of samples in both unstretched and stretched conditions at 25% and 50% tensile strain, along with contact angles and contact angle hysteresis. Digital photographs depicting the stretching test setup, accompanied by SEM images depicting samples stretched to (b) 25% and (c) 50% of their tensile values [84]. Reproduced from [84]. CC BY 4.0.

such as the human body and surgical tools. Hoshian *et al* developed robust superhydrophobic TiO_2/PDMS utilizing etched aluminum, which has been coated with TiO_2 through atomic layer deposition [81, 82]. The TiO_2 is then transferred to PDMS by sacrificially etching away the aluminum substrate. The TiO2/PDMS was found to be robust against mechanical abrasion and other environmental stressors [81]. For broader applicability, PDMS with conductive composites such as Cu extends fabricated by wet etching, electroless plating, and polymer casting. The fabricated PDMS/Cu composite, as shown in figure 1.6, is found to be superhydrophobic (contact angle > 170°, sliding angle < 7° with 7 μl droplets), electrically conductive, elastic, and wear-resistant with self-healing capabilities [83–85].

Similarly, Fu *et al* [86] demonstrated a strain sensor using silanized cellulose nanocrystals–CNT/PDMS composite. The composite showed a high strain range and good strain sensitivity, with a gauge factor of 37.11 at 501 °C, providing long-term stability and durability [86]. The combined attributes of conductivity and elasticity offer avenues for electrical, electro-thermal, or mechanical dislodgment of biofilms. Leveraging PDMS's self-adhesive nature and adjustable thickness, the material holds the potential to function as conductive superhydrophobic tape, further broadening its practical utility.

1.1.4.4 TPU conductive composites

TPU substrate is a thermoplastic elastomer that is a flexible and durable polymer. Due to its remarkable characteristics, it has found significant usage in the textile, electronics, and automotive industries as well as for medical tools and devices [1, 87]. Its inherent flexible characteristics and outstanding elongation at break makes it a promising candidate for applications such as stretchable clothing, flexible packaging, and robust adhesives [1, 87]. TPU's polymers are also resistant to abrasion and chemicals, making it highly durable and enhancing its performance in footwear and industrial components, where durability is essential [87, 88]. Due to their excellent electrical properties, TPU gained attention and has been extensively used in the electronics industry. The most common usage is cable coating, insulating coating and protective cases. The biocompatibility of TPU offers significant advantages in the realm of medical devices, which includes crafting catheters, wound dressings, and prosthetics. Additionally, TPU polymers are found to enhance safety and functionality with biological systems. The diverse array of processing methods including extrusion, injection molding, blow molding, and 3D printing empowers manufacturers to create very detailed and custom products with careful accuracy and meticulous precision [89].

A lot of work has been done to improve the performances of TPU substrate by adjusting the type and proportion of polyol (ether) and diisocyanate, making TPU an important candidate for the fabrication of wearable pressure sensors and intelligent instruments [17, 90]. TPU polymers can be mixed with conductive materials including multiwall CNTs (MWCNT), Ag nanoparticles, and PANI nanoparticles on the TPU by electrospinning or homogenization techniques. The MWCNT/TPU nanocomposites [87] were prepared using solution-processed mixing along with a high-speed homogenization process as shown in figure 1.7. The strain sensors fabricated using MWCNT/TPU composite can achieve adjustable strength, sensitivity, and strain tolerance.

Due to the rise in wearable devices for health monitoring, a lot of attention has been garnered by wearable electronics. Despite its growing focus, it still poses some limitations and challenges to facilitate the designing of flexible strain sensors with a high and wide range of sensitivity, which is ecofriendly and scalable. Li *et al* demonstrated the method to fabricate flexible strain sensors by coupling AgNWs and reduced graphene oxide (rGO) on an electrospun TPU mat [20] (figure 1.8).

The rGO/AgNWs/TPU strain sensor show good stability and durability for different applied strains and long-term stretching cycles. The results provided a guided platform and potential in healthcare, wearable sensors, and soft robotics.

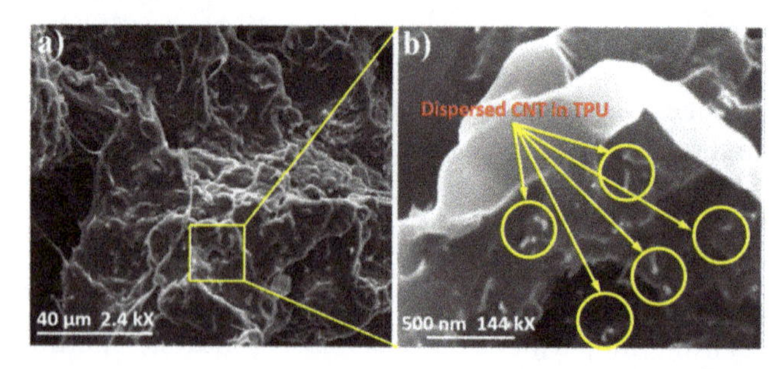

Figure 1.7. Scanning electron microscopy images of TPU/CNT with 0.1% MWCNT loading. [87]. Reproduced with permission from reference [87]. Reprinted from [87] with permission from Elsevier, Copyright (2019).

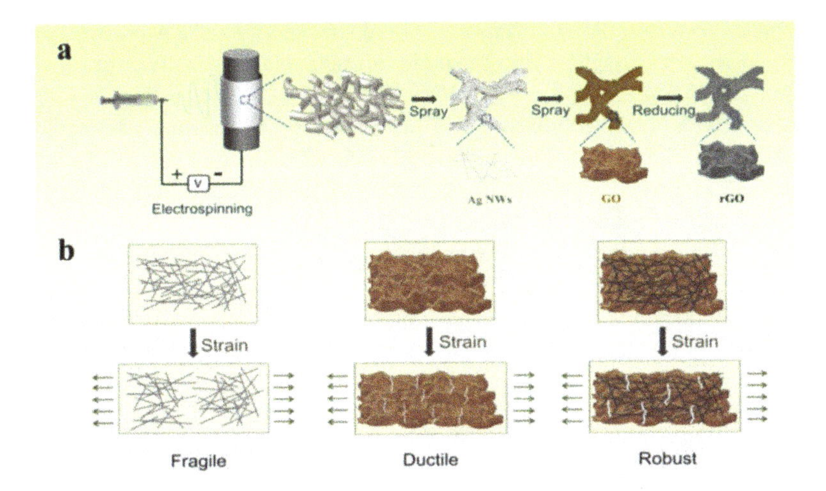

Figure 1.8. Schematic diagrams of (a) the fabrication process for rGO/AgNWs/TPU strain sensors [20]. Reproduced from [20] with permission from the Royal Society of Chemistry.

1.2 Conclusion

In conclusion, polymeric materials have revolutionized the field of flexible and printed electronics, providing a foundation for innovative and adaptable electronic devices. Their unique combination of mechanical properties, electrical conductivity, and optical transparency has enabled the realization of lightweight, flexible, and cost-efficient technology spanning from wearables, solar cells, and flexible displays to biomedical technology. As technology continues to advance, the integration of polymeric materials is expected to further accelerate, driving new opportunities and transforming industries in the years to come. As the realm of electronics continues to evolve, and considering the surge in human machine interface devices, this research sets a promising trajectory for the integration of flexible and transparent electronic components, promising a future where innovation knows no boundaries and is set to explode.

References

[1] Li L, Han L, Hu H and Zhang R 2022 A review on polymers and their composites for flexible electronics *Mater. Adv.* **4** 726–46

[2] Abbasipour M, Kateb P, Cicoira F and Pasini D 2023 Stretchable kirigami-inspired conductive polymers for strain sensors applications *Flex. Print. Electron.* **8** 024003

[3] Balani K, Verma V, Agarwal A and Narayan R 2015 Physical, thermal, and mechanical properties of polymers *Biosurfaces* (New York: Wiley) pp 329–44

[4] Ravindranadh K and Rao M C 2013 Physical properties and applications of conducting polymers: an overview *Int. J. Adv. Pharm. Biol. Chem.* **2** 190–200

[5] Frias J, Pagter E, Nash R and O'Connor I 2018 *Standardised protocol for monitoring microplastics in sediments characterization of molecular and histological effects of MP particles in tissue (sections) of aquatic and terrestrial model organisms view project* Deliverable D4.2

[6] Campanale C, Savino I, Pojar I, Massarelli C and Uricchio V F 2020 A practical overview of methodologies for sampling and analysis of microplastics in riverine environments *Sustainability* **12** 6755

[7] De Rossi F *et al* 2021 Modified P3HT materials as hole transport layers for flexible perovskite solar cells *J. Power Sources* **494** 229735

[8] Sim K, Rao Z, Ershad F and Yu C 2020 Rubbery electronics fully made of stretchable elastomeric electronic materials *Adv. Mater.* **32**

[9] Liu H *et al* 2018 Electrically conductive polymer composites for smart flexible strain sensors: a critical review *J. Mater. Chem.* C **6** 12121–41

[10] Guarino V, Zuppolini S, Borriello A and Ambrosio L 2016 Electro-active polymers (EAPs): a promising route to design bio-organic/bioinspired platforms with on demand functionalities *Polymers* **8** 185

[11] Beygisangchin M, Rashid S A, Shafie S, Sadrolhosseini A R and Lim H N 2021 Preparations, properties, and applications of polyaniline and polyaniline thin films—a review *Polymers* **13** 2003

[12] Chen H W, Lee J H, Lin B Y, Chen S and Wu S T 2017 Liquid crystal display and organic light-emitting diode display: present status and future perspectives *Light: Sci. Appl. 2018* **7** 17168–8

[13] Wu X *et al* 2020 Optically transparent and thermal-stable polyimide films derived from a semi-aliphatic diamine: synthesis and properties *Macromol. Chem. Phys.* **221** 1900506

[14] Khayrudinov V *et al* 2020 Direct growth of light-emitting III–V nanowires on flexible plastic substrates *ACS Nano* **14** 7484–91

[15] Leppäniemi J, Eiroma K, Majumdar H and Alastalo A 2017 Far-UV annealed inkjet-printed In2O3 semiconductor layers for thin-film transistors on a flexible polyethylene naphthalate substrate *ACS Appl. Mater. Interfaces* **9** 8774–82

[16] You I, Kong M and Jeong U 2018 Block copolymer elastomers for stretchable electronics *Acc. Chem. Res.* **52** 63–72

[17] Zhang Z *et al* 2022 Durable and highly sensitive flexible sensors for wearable electronic devices with PDMS-MXene/TPU composite films *Ceram. Int.* **48** 4977–85

[18] Ding Y, Xu W, Wang W, Fong H and Zhu Z 2017 Scalable and facile preparation of highly stretchable electrospun PEDOT:PSS@PU fibrous nonwovens toward wearable conductive textile applications *ACS Appl. Mater. Interfaces* **9** 30014–23

[19] Qi D, Zhang K, Tian G, Jiang B and Huang Y 2021 Stretchable electronics based on PDMS substrates *Adv. Mater.* **33** 2003155

[20] Li Y *et al* 2020 Flexible TPU strain sensors with tunable sensitivity and stretchability by coupling AgNWs with rGO *J. Mater. Chem. C Mater.* **8** 4040–8

[21] Das S K *et al* 2021 Plastic recycling of polyethylene terephthalate (PET) and polyhydroxybutyrate (PHB)—a comprehensive review *Mater. Circ. Econ.* **3** 9

[22] Hiraga K, Taniguchi I, Yoshida S, Kimura Y and Oda K 2019 Biodegradation of waste PET *EMBO Rep.* **20** e49365

[23] Soong Y H V, Sobkowicz M J and Xie D 2022 Recent advances in biological recycling of polyethylene terephthalate (PET) plastic wastes *Bioengineering* **9** 98

[24] Fan X *et al* 2017 Highly conductive stretchable all-plastic electrodes using a novel dipping-embedded transfer method for high-performance wearable sensors and semitransparent organic solar cells *Adv. Electron. Mater.* **3** 1600471

[25] Rogers J A, Bao Z, Dodabalapur A and Makhija A 2000 Organic smart pixels and complementary inverter circuits formed on plastic substrates by casting and rubber stamping *IEEE Electron. Devices Lett.* **21** 100–3
[26] Braun D and Heeger A J 1991 Visible light emission from semiconducting polymer diodes *Appl. Phys. Lett.* **58** 1982–4
[27] Wen P *et al* 2021 A simple and effective method via PH1000 modified Ag-Nanowires electrode enable efficient flexible nonfullerene organic solar cells *Org. Electron.* **94** 106172
[28] Liu Y *et al* 2020 A novel family of optically transparent fluorinated hyperbranched polyimides with long linear backbones and bulky substituents *Eur. Polym. J.* **125** 109526
[29] Song M *et al* 2013 Highly efficient and bendable organic solar cells with solution-processed silver nanowire electrodes *Adv. Funct. Mater.* **23** 4177–84
[30] Amruth C, Pahlevani M and Welch G C 2021 Organic light emitting diodes (OLEDs) with slot-die coated functional layers *Mater. Adv.* **2** 628–45
[31] Chang W-S, Chang T-S, Wang C-M and Liao W-S 2022 Metal-free transparent three-dimensional flexible electronics by selective molecular bridges *ACS Appl. Mater. Interfaces* **14** 22826–37
[32] Zhang Z *et al* 2021 Ultra-smooth and robust graphene-based hybrid anode for high-performance flexible organic light-emitting diodes *J. Mater. Chem. C Mater.* **9** 2106–14
[33] Bogert M T and Renshaw R R 1908 4-amino-0-phthalic acid and some of its derivatives *J. Am. Chem. Soc.* **30** 1135–44
[34] Chen C K *et al* 2021 High performance biomass-based polyimides for flexible electronic applications *ACS Sustain. Chem. Eng.* **9** 3278–88
[35] Tao J *et al* 2020 Highly transparent, highly thermally stable nanocellulose/polymer hybrid substrates for flexible OLED devices *ACS Appl. Mater. Interfaces* **12** 9701–9
[36] Xu X *et al* 2018 Thermally stable, highly efficient, ultraflexible organic photovoltaics *Proc. Natl Acad. Sci. USA* **115** 4589–94
[37] Wang Y *et al* 2022 Ultrathin flexible transparent composite electrode via semi-embedding silver nanowires in a colorless polyimide for high-performance ultraflexible organic solar cells *ACS Appl. Mater. Interfaces* **14** 5699–708
[38] Wang X, Li J, Song H, Huang H and Gou J 2018 Highly stretchable and wearable strain sensor based on printable carbon nanotube layers/polydimethylsiloxane composites with adjustable sensitivity *ACS Appl. Mater. Interfaces* **10** 7371–80
[39] Tony A *et al* 2023 The additive manufacturing approach to polydimethylsiloxane (PDMS) microfluidic devices: review and future directions *Polymers (Basel)* **15** 1926
[40] Miranda I *et al* 2021 Properties and applications of PDMS for biomedical engineering: a review *J. Funct. Biomater.* **13** 2
[41] You B, Han C J, Kim Y, Ju B K and Kim J W 2016 A wearable piezocapacitive pressure sensor with a single layer of silver nanowire-based elastomeric composite electrodes *J. Mater. Chem. A Mater.* **4** 10435–43
[42] Wang Y *et al* 2019 Printable liquid-metal@PDMS stretchable heater with high stretchability and dynamic stability for wearable thermotherapy *Adv. Mater. Technol.* **4** 1800435
[43] Lamberti A, Marasso S L and Cocuzza M 2014 PDMS membranes with tunable gas permeability for microfluidic applications *RSC Adv.* **4** 61415–9
[44] Lin L and Chung C-K 2021 PDMS microfabrication and design for microfluidics and sustainable energy application: review *Micromachines* **12** 1350

[45] Miranda I *et al* 2022 Properties and applications of PDMS for biomedical engineering: a review *J. Funct. Biomater.* **13** 2

[46] Zheng S *et al* 2021 Highly sensitive pressure sensor with broad linearity via constructing a hollow structure in polyaniline/polydimethylsiloxane composite *Compos. Sci. Technol.* **201** 108546

[47] Li S, Cong Y and Fu J 2021 Tissue adhesive hydrogel bioelectronics *J. Mater. Chem.* B **9** 4423–43

[48] Xu X *et al* 2018 Thermally stable, highly efficient, ultraflexible organic photovoltaics *Proc. Natl. Acad. Sci. USA* **115** 4589–94

[49] Fan F R, Tian Z Q and Lin Wang Z 2012 Flexible triboelectric generator *Nano Energy* **1** 328–34

[50] Wang Z L and Song J 2006 Piezoelectric nanogenerators based on zinc oxide nanowire arrays *Source: Sci., New Series* **312** 242–6

[51] Sahoo R *et al* 2020 An approach towards the fabrication of energy harvesting device using Ca-doped ZnO/ PVDF-TrFE composite film *Polymer (Guildf)* **205** 122869

[52] Liang X *et al* 2019 Highly transparent triboelectric nanogenerator utilizing *in situ* chemically welded silver nanowire network as electrode for mechanical energy harvesting and body motion monitoring *Nano Energy* **59** 508–16

[53] Kim H *et al* 2021 Threshold voltage instability and polyimide charging effects of LTPS TFTs for flexible displays *Sci. Rep.* **11** 8387

[54] You I *et al* 2020 Artificial multimodal receptors based on ion relaxation dynamics *Science (1979)* **370** 961–5

[55] Lin Y *et al* 2016 A highly stretchable and sensitive strain sensor based on graphene–elastomer composites with a novel double-interconnected network *J. Mater. Chem. C Mater.* **4** 6345–52

[56] Wang M *et al* 2017 Enhanced electrical conductivity and piezoresistive sensing in multi-wall carbon nanotubes/polydimethylsiloxane nanocomposites via the construction of a self-segregated structure *Nanoscale* **9** 11017–26

[57] Shen J *et al* 2021 A bioinspired porous-designed hydrogel@polyurethane sponge piezoresistive sensor for human–machine interfacing *Nanoscale* **13** 19155–64

[58] Wang F X *et al* 2018 An ultrahighly sensitive and repeatable flexible pressure sensor based on PVDF/PU/MWCNT hierarchical framework-structured aerogels for monitoring human activities *J. Mater. Chem. C Mater.* **6** 12575–83

[59] Oh J Y and Bao Z 2019 Second skin enabled by advanced electronics *Adv. Sci.* **6**

[60] Luscombe C K, Maitra U, Walter M and Wiedmer S K 2021 Theoretical background on semiconducting polymers and their applications to OSCs and OLEDs *Chem. Teach. Int.* **3** 169–83

[61] Kim K *et al* 2022 New opportunities for organic semiconducting polymers in biomedical applications *Polymers* **14** 2960

[62] Hussain S *et al* 2019 An optoelectronic device for rapid monitoring of creatine kinase using cationic conjugated polyelectrolyte *Adv. Mater. Technol.* **4** 1900361

[63] Ohayon D and Inal S 2020 Organic bioelectronics: from functional materials to next-generation devices and power sources *Adv. Mater.* **32** 2001439

[64] Zeglio E, Rutz A L, Winkler T E, Malliaras G G and Herland A 2019 Conjugated polymers for assessing and controlling biological functions *Adv. Mater.* **31** 1806712

[65] Bonaventura G *et al* 2019 Biocompatibility between silicon or silicon carbide surface and neural stem cells *Sci. Rep.* **9** 1–13
[66] Jun Kim H *et al* 2017 Solution-assembled blends of regioregularity-controlled polythiophenes for coexistence of mechanical resilience and electronic performance *ACS Appl. Mater. Interfaces* **9** 14120–8
[67] Rashid I A *et al* 2020 Stretchable strain sensors based on polyaniline/thermoplastic polyurethane blends *Polym. Bull.* **77** 1081–93
[68] Wang M, Baek P, Akbarinejad A, Barker D and Travas-Sejdic J 2019 Conjugated polymers and composites for stretchable organic electronics *J. Mater. Chem.* C **7** 5534–52
[69] Wang X *et al* 2018 Highly sensitive and stretchable piezoresistive strain sensor based on conductive poly(styrene-butadiene-styrene)/few layer graphene composite fiber *Compos. Part A Appl. Sci. Manuf.* **105** 291–9
[70] Amirkhosravi M, Yue L and Manas-Zloczower I 2020 Dusting thermoplastic polyurethane granules with carbon nanotubes toward highly stretchable conductive elastomer composites *ACS Appl. Polym. Mater.* **2** 4037–44
[71] Park M *et al* 2012 Highly stretchable electric circuits from a composite material of silver nanoparticles and elastomeric fibres *Nat. Nanotechnol.* **7** 803–9
[72] Zhu S *et al* 2021 Inherently conductive poly(dimethylsiloxane) elastomers synergistically mediated by nanocellulose/carbon nanotube nanohybrids toward highly sensitive, stretchable, and durable strain sensors *ACS Appl. Mater. Interfaces* **13** 59142–53
[73] Holliday S *et al* 2016 High-efficiency and air-stable P3HT-based polymer solar cells with a new non-fullerene acceptor *Nat. Commun.* **7** 1–11
[74] Lee S, Kim H and Kim Y 2021 Hole injection role of p-type conjugated polymer nanolayers in phosphorescent organic light-emitting devices *Electronics* **10** 2283
[75] Griggs S, Marks A, Bristow H and McCulloch I 2021 n-Type organic semiconducting polymers: stability limitations, design considerations and applications *J. Mater. Chem.* C **9** 8099–128
[76] Vosgueritchian M, Lipomi D J and Bao Z 2012 Highly conductive and transparent PEDOT: PSS films with a fluorosurfactant for stretchable and flexible transparent electrodes *Adv. Funct. Mater.* **22** 421–8
[77] Song W *et al* 2018 All-solution-processed metal-oxide-free flexible organic solar cells with over 10% efficiency *Adv. Mater.* **30** 1800075
[78] Song Y *et al* 2015 Pushing the cycling stability limit of polypyrrole for supercapacitors *Adv. Funct. Mater.* **25** 4626–32
[79] Lang U, Naujoks N and Dual J 2009 Mechanical characterization of PEDOT:PSS thin films *Synth. Met.* **159** 473–9
[80] Ho Lee J *et al* 2018 Highly conductive, stretchable, and transparent pedot:pss electrodes fabricated with triblock copolymer additives and acid treatment *ACS Appl. Mater. Interfaces* **10** 28027–35
[81] Hoshian S, Jokinen V and Franssila S 2016 Robust hybrid elastomer/metal-oxide superhydrophobic surfaces *Soft Matter* **12** 6526–35
[82] Hoshian S, Jokinen V and Franssila S 2016 Novel nanostructure replication process for robust superhydrophobic surfaces *Proc. of the IEEE Int. Conf. on Micro Electro Mechanical Systems (MEMS)* pp 547–9

[83] Jing X *et al* 2021 Synthesis and fabrication of supramolecular polydimethylsiloxane-based nanocomposite elastomer for versatile and intelligent sensing *Indus. Eng. Chem. Res.* **60** 10419–30
[84] Mirmohammadi S M, Hoshian S, Jokinen V P and Franssila S 2021 Fabrication of elastic, conductive, wear-resistant superhydrophobic composite material *Sci. Rep.* **11** 1–10
[85] Wang L *et al* 2019 Fluorine-free superhydrophobic and conductive rubber composite with outstanding deicing performance for highly sensitive and stretchable strain sensors *ACS Appl. Mater. Interfaces* **11** 17774–83
[86] Fu X *et al* 2021 A high-resolution, ultrabroad-range and sensitive capacitive tactile sensor based on a CNT/PDMS composite for robotic hands *Nanoscale* **13** 18780–8
[87] Kumar S, Gupta T K and Varadarajan K M 2019 Strong, stretchable and ultrasensitive MWCNT/TPU nanocomposites for piezoresistive strain sensing *Composites* B **177** 107285
[88] Gupta T K *et al* 2015 Superior nano-mechanical properties of reduced graphene oxide reinforced polyurethane composites *RSC Adv.* **5** 16921–30
[89] Christ J F, Aliheidari N, Ameli A and Pötschke P 2017 3D printed highly elastic strain sensors of multiwalled carbon nanotube/thermoplastic polyurethane nanocomposites *Mater. Des.* **131** 394–401
[90] Gao Q *et al* 2021 Flexible multilayered MXene/thermoplastic polyurethane films with excellent electromagnetic interference shielding, thermal conductivity, and management performances *Adv. Compos. Hybrid Mater.* **4** 274–85

IOP Publishing

Advances in Flexible and Printed Electronics
Materials, fabrication, and applications
Shanmuga Sundar Dhanabalan and Arun Thirumurugan

Chapter 2

New generation polymer nanocomposites for flexible electronics

J Shahitha Parveen and M Thirumurugan

The realm of flexible electronics has experienced an exponential expansion over the last 20 years, with a wide range of applications. The implementation of flexible electronics is progressively escalating across multiple domains that reap the rewards of their diminutive dimensions, lightweight structure, and advantageous dielectric attributes. This chapter comprehensively covers the introduction and fundamentals of flexible electronics and types and manufacture techniques for flexible electronics. This chapter focuses on novel polymeric materials, nanomaterials, polymer hybrids, and polymer-sandwich nanocomposites used in flexible electronics. It also discusses topics including electrospinning, three-dimensional (3D) printing techniques, and roll-to-roll (R2R) processing for flexible electronics. The chapter highlights the role of electrospun nanofibrous composites in energy storage devices and concludes with recent trends, advancement, and applications of flexible electronics.

2.1 Introduction and overview of flexible electronics

Flexible electronics are a cutting-edge technology that enables the production of electronic devices on adaptable substrates including plastic films and paper. This state-of-the-art technology has the capability to completely transform our interaction with electronics by virtue of the fact that it can be utilized to fashion devices that are comparatively lighter, more resilient, and more portable than the conventional electronic devices. For the construction of flexible electronics, several distinct materials are used. Among these, organic semiconductors derived from carbon-based molecules are the most prevalent. The organic semiconductors are comparatively low cost and have facile processability, so they are eminently suited for mass production. Additionally, metal oxides, polymers, and composites represent alternative materials for the production of flexible electronics. The process of manufacturing flexible electronics is akin to that of traditional electronics.

doi:10.1088/978-0-7503-5492-9ch2

Notwithstanding, various challenges must be surmounted to produce high-caliber flexible electronic devices. One of the formidable obstacles is the development of materials that exhibit both flexibility and conductivity. Additionally, the development of manufacturing processes that can be utilized to produce large-area flexible electronic devices presents another challenge [1–4] (table 2.1).

Despite these challenges, the realm of flexible electronics is expanding at an exponential rate. Multiple companies, such as Samsung, LG, and Apple, are engaged in the development of flexible electronic devices for diverse applications, which include wearable devices, displays, and sensors.

Table 2.1. Contributions in the field of flexible electronics.

Year	Description of contributions
1970	Flexible solar cells from thin Si
	Flexible thin film transistors on paper and Mylar
	Developed Si:H
1970–80	Developed conductive polymers
1980–90	Amorphous silicon solar cells on polymer substrates
	Flexible e-skin sensors
	Developed organic electrochemical transistor
	Indium tin oxide–polyethylene terephthalate for optoelectronics
	Cadmium telluride/cadmium sulfide solar cells on flexible substrates
	Development of organic light-emitting diode
	Development of organic field-effect transistor
	Solution processed organic light-emitting diode (OLED)
1990–2000	Development of poly(3,4-ethylenedioxythiophene) polystyrene sulfonate
	Flexible OLED from conductive polymers
	Si:H thin-film transistor on polyimide (PI) substrates
	Solution processed bulk heterojunction organic solar cells
	Poly-Si thin-film transistor devices (TFTs) on plastic substrates
	Large-scale organic complimentary integrated circuits
2000–10	Graphene exfoliation
	Multifunctional skin-like systems
	Flexible supercapacitors
	Graphene transparent electrodes for dye sensitized solar cells
	Perovskite solar cells
	Stretchable OLEDs
	Bioresorbable organic TFTs
	Flexible batteries based on carbon nanotube (CNT)
	Epidermal electronics

2010–20	Self-healable tactile sensor
	Stretchable rechargeable soft batteries
	Biodegradable cellulose-based electronics
	Flexible near field communication
	Self-powered organic electrochemical transistor sensors
	Printed integrated circuits based on organic electrochemical transistors (OECTs)
	3D-printed self-healing thermoelectric generators
	Self-powered sweat sensor patch
	Paper-based glucose sensor from saliva
	Printed nano fullerene organic solar cells
	Electronic skin integrated soft robotic hand
	Integrated soft robotics

The future of flexible electronics appears to be very promising. This technology possesses the potential to transform our interaction with electronics. As the technology progresses to its advanced state, even more creative and captivating applications for flexible electronics are expected.

2.1.1 Types of flexible electronics

There are different types of flexible electronics, including:

- **Flexible printed circuits**: These are circuits that are printed on a flexible substrate such as plastic or paper. Flexible printed circuits are used in a wide range of applications including wearable electronics, medical devices, and displays.
- **Flexible sensors**: These are sensors that are made on a flexible substrate. Flexible sensors are applied in monitoring of environmental/weather conditions, medical diagnostics, and industrial process control.
- **Flexible displays**: These are displays that are made on a flexible substrate. Flexible displays are used in wearable electronics, displays in automotive vehicles, and signage.
- **Flexible batteries**: These are batteries that are made on a flexible substrate. Flexible batteries are used in wearable electronics, medical devices, and portable electronics.

2.1.2 Manufacturing techniques of flexible electronics

Different manufacturing techniques can be used to create flexible electronics. The most common techniques include:

- **Screen printing**: This is a low-cost, high-volume technique that can be used to print conductive patterns on a flexible substrate.
- **Inkjet printing**: This is a high-resolution technique that can be used to print complex patterns on a flexible substrate.

- **Direct writing**: This is a technique that uses a laser or other beam to write conductive patterns on a flexible substrate.
- **Thin-film deposition**: This is a technique that uses vacuum deposition or chemical vapor deposition to deposit conductive materials on a flexible substrate.

The choice of manufacturing technique depends on the specific application and the desired properties of the flexible electronic device.

Two basic methods are developed to produce flexible electronics: (1) transfer and bonding of finished circuits to a flexible substrate and (2) direct fabrication of circuits on the flexible substrate.

In the transfer-and-bond technique, the structure is manufactured using standard techniques on a carrier substrate such as a Si wafer or a glass plate. It is then transferred to or fluidic self-assembled on a flexible substrate. The transfer-and-bond method has been expanded to bond ribbons of Si and GaAs devices to a stretched elastomer, which forms a 'wavy' semiconductor upon relaxation that is capable of being stretched and relaxed reversibly. The transfer methods have the advantage of producing high-performance devices on flexible substrates. However, they suffer from small surface area coverage and high cost.

Direct fabrication on flexible substrates is a hub of process research. Novel process techniques include the printing of etch masks, the additive printing of active device materials, and the introduction of electronic functions by local chemical reaction. Nanocrystalline silicon and printable polymers for organic light-emitting diodes (OLEDs) are also materials of intense research.

Flexibility can encompass various properties for both manufacturers and users, and as a mechanical characteristic, it conveniently falls under three categories: (1) bendable or rollable, (2) permanently shaped, and (3) elastically stretchable. Microfabrication tools have been developed primarily for flat substrates, resulting in all current manufacturing being done on a flat workpiece that is shaped as late as possible in the process. This approach benefits from the extensive technology base established by the planar integrated circuit and display industries.

2.1.3 Applications of flexible electronics

Flexible electronics have a wide range of potential applications, including:

- **Wearable electronics**: Flexible electronics can be used to create wearable electronics, such as smart watches, fitness trackers, and smart clothing. Wearable electronics can track our physical activity, monitor our health, and provide us with information and entertainment.
- **Medical devices**: Flexible electronics can be used to create medical devices such as flexible sensors and implants. Flexible sensors can be used to monitor our health, while flexible implants can be used to treat medical conditions.
- **Displays**: Flexible electronics can be used to create flexible displays such as rollable and bendable TVs. Flexible displays can be used in a variety of settings including homes, businesses, and transportation.

- **Energy harvesting**: Flexible electronics can be used to create energy harvesting devices such as solar cells and batteries. Energy harvesting devices can collect energy from the environment, such as sunlight or vibrations, and use this energy to power electronic devices.
- **Security and surveillance**: Flexible electronics can be used to create security and surveillance devices such as flexible antennas and sensors. Flexible antennas can be used to detect and track radio signals, while flexible sensors can be used to detect motion and other events.

2.1.4 Future of flexible electronics

The field of flexible electronics is still in its early stages, but it has the potential to revolutionize many industries. As the technology continues to develop, we can expect to see even more innovative and groundbreaking applications for flexible electronics. Some of the potential future applications of flexible electronics include:

- **Foldable smartphones**: Foldable smartphones are already on the market, and they are likely to become more popular in the future. Flexible electronics will make it possible to create even more flexible and durable foldable smartphones.
- **Rollable televisions**: Rollable TVs are another potential application of flexible electronics. Rollable TVs can be rolled up and stored when not in use, and they can be unrolled to create a large-screen TV.
- **Wearable medical devices**: Flexible electronics can be used to create wearable medical devices, such as smart patches that can monitor our health.

As digital technology continues to rapidly advance, the role of electronics in our daily lives is becoming increasingly prominent, leading to the expansion of the electronics industry. The development of smaller, lighter, and more portable devices requires innovative thinking as well as modern materials and manufacturing techniques. Sensors, for instance, are now being produced at lower costs, in smaller sizes, and with greater accessibility through the use of flexible electronics, a technology that is widely implemented in the health care, automotive, and consumer electronics industries. In fact, flexible and printed electronics are now being utilized in a variety of applications including smart packaging, smart clothing, electrochemical testing strips, wearables and skin patches, and next-generation vehicles. Additionally, the Internet of things would not be possible without the utilization of flexible and printed electronics. One of the key benefits of this technology is its flexibility, which is largely dependent on the type and thickness of the material used.

2.2 Novel polymer materials for flexible electronics

Novel polymer materials are being developed for flexible electronics. These materials offer a number of advantages over traditional inorganic materials, such as:

- **Flexibility**: Polymers can be easily bent and folded without breaking, making them ideal for wearable and flexible electronics.

- **Transparency**: Many polymers are transparent, which allows light to pass through them, making them suitable for displays and other optical devices.
- **Lightweight material**: Polymers are much lighter than inorganic materials, making them ideal for portable electronics.
- **Cost-effectiveness**: Polymers are often less expensive than inorganic materials, making them a more cost-effective option for many applications.

Essentially, flexible electronic circuits, also referred as flex circuits, involve mounting the electronics circuit on a polymer flexible substrate, which typically consists of a base material, bonding adhesive, and metal foil. The base material, which is typically a flexible polymer film, serves as the foundation for the laminate, with most films ranging in thickness from 12–125 μm. Thinner films are generally more flexible, whereas thicker films tend to be stiffer. Adhesives are used to create a laminate, and like the base materials, bonding adhesives come in varying thicknesses. Finally, the most common conductive material used in flexible circuits is metal foil, which is typically etched to create circuit paths. Flexible circuits can be categorized into several types based on their construction, including single-sided flex circuits, double-sided flex circuits, multilayer flex circuits, and rigid flex circuits.

Candidate polymers suitable for flexible substrates consist of thermoplastic semicrystalline polymers such as polyethylene terephthalate (PET) and polyethylene naphthalate (PEN), thermoplastic noncrystalline polymers including polycarbonate (PC) and polyethersulfone (PES), and high-Tg materials such as polyarylates (PAR), polycyclic olefin (PCO), and polyimide (PI). PC, PES, PAR, and PCO have relatively high Tg and are optically clear compared to PET and PEN. However, the higher (50 ppm C^{-1}) coefficient of thermal expansion or higher and poor chemical resistance are considered to be drawbacks.

There are several materials used as base films, including PET, PI, PEN, and polyetherimide, as well as various fluropolymers and copolymers.

The base material commonly used in rigid printed circuit boards is woven fiberglass impregnated with epoxy resin. Although these boards are referred to as 'rigid,' a single laminate layer still possesses a reasonable amount of elasticity. The board's rigidity is primarily due to the cured epoxy. As a result of the use of epoxy resins, these boards are often referred to as organic rigid printed circuit boards. However, this material is not flexible enough for many applications. For simple assemblies where there is no constant movement, it may be suitable. The most popular material of choice used as a flex printed circuit board (PCB) substrate is PI. This material is highly flexible, tough, and exhibits exceptional heat resistance.

PET is a flex-circuit material that is commonly used. However, it is not tolerant enough of high temperatures to survive soldering. In certain low-cost electronics, PET is used for the flexible part with printed conductors, where the material cannot handle the heat of lamination. In such cases, contact is made by crude pressure with an isotropic conductive elastomer.

When using the flexible part of the circuit to reduce manufacturing time and costs by removing cabling and connectors, the usual laminated copper foil (electro-deposited) for rigid board use is sufficient. This type of foil may also be used when heavier copper weights are desired to keep high current–carrying conductors to the minimum viable width, such as in planar inductors.

In bonding copper foil to PI or other films, adhesives are traditionally required. This is because unlike a typical FR -4 (standard for flame retardant glass - reinforced epoxy resin laminate) rigid board, there is less 'tooth' in the annealed copper, and heat and pressure alone are not enough to form a reliable bond. Manufacturers offer pre-laminated single- and double-sided copper-clad films for flexible circuit etching using acrylic or epoxy-based adhesives with typical thicknesses of ½ and 1 ml. Adhesives are specially developed for flexibility. 'Adhesiveless' laminates are becoming more prevalent due to newer processes involving copper plating or deposition directly onto the PI film. These films are chosen when finer pitches and smaller vias are needed, as in high - density interconnect (HDI) circuits. Silicones, hot-melt glues, and epoxy resins are also used when protective beads are added to the flex-to-rigid joins or interfaces [5–8].

Some of the most promising novel polymer materials for flexible electronics include:

- **Polythiophene**: Polythiophene is a conducting polymer that is known for its high electrical conductivity and transparency. It has been used to create a variety of flexible electronic devices including OLED displays, solar cells, and transistors.
- **Polyaniline**: Polyaniline is another conducting polymer that is known for its high electrical conductivity and flexibility. It has been used to create a variety of flexible electronic devices including batteries, sensors, and actuators.
- **Polypyrrole**: Polypyrrole is a conducting polymer that is known for its high electrical conductivity and chemical stability. It has been used to create a variety of flexible electronic devices including batteries, sensors, and actuators.

As research in this area continues, it is likely that even more innovative and promising materials will be developed in the future. The development of novel polymer materials for flexible electronics is a rapidly growing field with the potential to revolutionize the way we interact with technology. As these materials continue to evolve, we can expect to see even more innovative and exciting applications in the years to come. Novel polymer materials for flexible electronics are also being developed with the following characteristics in mind:

- **Processability**: The material should be easy to process into the desired shape and size.
- **Durability**: The material should be able to withstand the stresses and strains of everyday use.
- **Environmental stability**: The material should be stable in a variety of environmental conditions including high temperatures and humidity.
- **Flexibility**: Polymers can be easily bent and stretched, making them ideal for applications such as wearable electronics and flexible displays.
- **Light weight**: Polymers are much lighter than inorganic materials, making them ideal for applications where weight is a concern, such as in wearable electronics and medical devices.
- **Low cost**: Polymers are much less expensive than inorganic materials, making them ideal for mass-produced electronics.

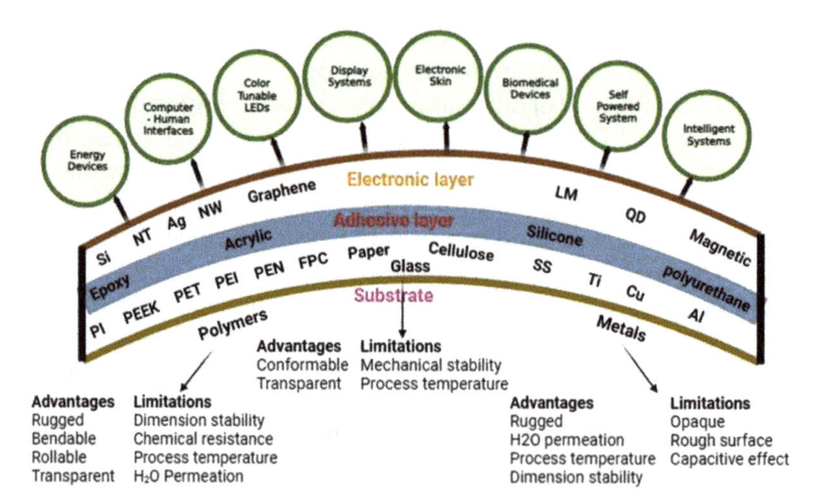

Figure 2.1. Description of materials used in different parts of the flexible electronic system. (Abbreviations used: Top row: Si- Silicon , NT- NanoTube, Ag- Silver, NW- NanoWires, LM -Liguid Metal, QD – Quantum Dot Bottom Row: PI- Polyimide, PEEK – Polyether ether ketone, PET- Polyethylene terephthalate, PEI – Polyethylenimine, PEN - Polyethylene naphthalate, FPC- Flexible Printed Circuit, SS- Stainless Steel, Ti-Titanium, Cu- Copper , Al- Aluminium)

In addition to these, novel polymer materials can also be tailored to meet the specific needs of different applications (figure 2.1).

2.3 Nanomaterials consideration for flexible electronics

Nanomaterials are materials with at least one dimension in the nanoscale (1–100 nm). They have unique properties that make them ideal for use in flexible electronics. For example, they are often very strong and lightweight, and they can be easily integrated into flexible substrates.

There are many different types of nanomaterials that can be used in flexible electronics. Some of the most common include:

- **Graphene**: Graphene is a two-dimensional material made up of a single layer of carbon atoms. It is incredibly strong and lightweight, and it is also an excellent conductor of electricity.
- **Carbon nanotubes**: Carbon nanotubes are one-dimensional materials made up of rolled-up sheets of carbon atoms. They are very strong and lightweight, and they can also conduct electricity.
- **Metal nanowires**: Metal nanowires are long, thin wires made up of metal atoms. They are very conductive, and they can be used to create flexible electrodes.
- **Conducting polymers**: Conducting polymers are polymers that can conduct electricity. They are often used to create flexible sensors and actuators.

Nanomaterials offer a number of advantages over traditional materials for use in flexible electronics. They are often stronger, lighter, and more flexible than traditional materials. They can also be easily integrated into flexible substrates. This makes them ideal for a wide range of applications such as wearable electronics, medical devices, and smart textiles.

However, there are also some challenges associated with using nanomaterials in flexible electronics. One challenge is that nanomaterials can be expensive to produce. Another challenge is that nanomaterials can be difficult to process and integrate into devices.

Despite the challenges, nanomaterials offer a promising future for flexible electronics. As the cost of nanomaterials decreases and the technology for processing and integrating them into devices improves, nanomaterials will become increasingly common in flexible electronics.

Here are some examples of how nanomaterials are being used in flexible electronics:

- **Graphene-based touch screens**: Graphene is being used to create touch screens that are more flexible and durable than traditional touch screens.
- **Carbon nanotube–based batteries**: Carbon nanotubes are being used to create batteries that are more lightweight and powerful than traditional batteries.
- **Metal nanowire–based solar cells**: Metal nanowires are being used to create solar cells that are more flexible and efficient than traditional solar cells.
- **Conducting polymer–based sensors**: Conducting polymers are being used to create sensors that are more sensitive and flexible than traditional sensors.

These are just a few examples of the many ways that nanomaterials are being used in flexible electronics. As the technology continues to develop, we can expect to see even more innovative applications for nanomaterials in flexible electronics.

2.4 Electrospinning, 3D printing techniques, and R2R processing for flexible electronics

2.4.1 Electrospinning

An innovative, effective, and flexible technique for producing polymer nanofibers with diameters ranging from a few micrometers to tens of nanometers is electrospinning. It is an electrohydrodynamic process during which a liquid droplet is electrified to generate a jet, followed by stretching and elongation to generate fibers. As illustrated in figure 2.2, the electrospinning setup involves three components: (a) a high-voltage power source, (b) a capillary tube equipped with a thin needle, and (c) a metallic collector. An electrically charged jet of polymer melt or solution is released during electrospinning. The liquid body becomes charged when a high-potential electrical charge is applied, and electrostatic repulsion works to counteract surface tension and stretch the droplet. At the point of criticality, a liquid stream is discharged from the surface, and the location of discharge, termed the Taylor cone, is presented in figure 2.3. Following solvent evaporation, the polymer fibers are deposited onto the revolving drum. In the case of a molten state, solidification of the expelled jet occurs during its traversal through the air before ultimately accumulating on the electrically grounded metal collector.

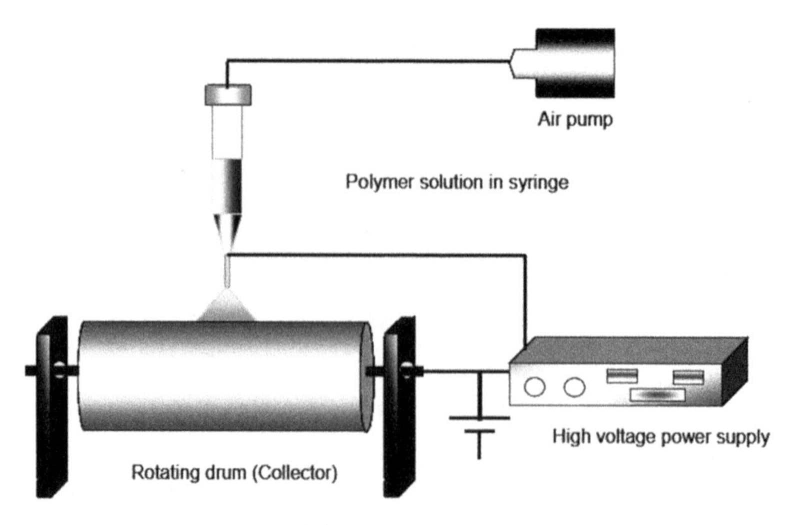

Figure 2.2. Schematic diagram of an electrospinning setup.

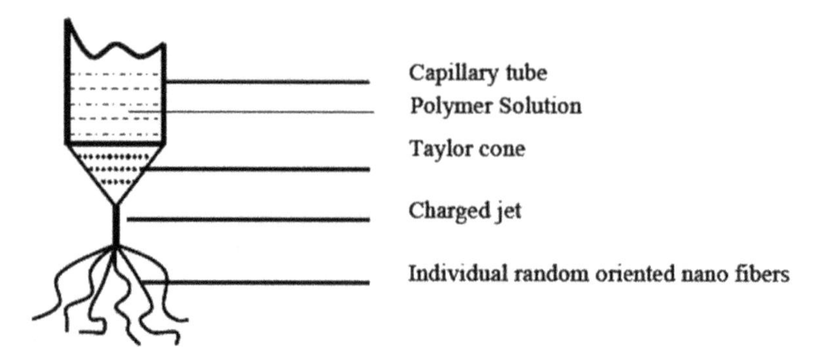

Figure 2.3. Schematic diagram of the formation of the Taylor cone.

The transformation of polymer solutions into nanofibers within the electrospinning process entails numerous parameters and processing variables. The system parameters consist of factors such as molecular weight, viscosity, molecular weight distribution, elasticity, polymer architecture, conductivity, surface tension, dielectric constant, and the charge borne by the spinning jet. The parameters utilized during the process encompass the voltage applied at the capillary tip, as well as the hydrostatic pressure present within the capillary tube. Additional factors include the distance between the tip and the collector, the rotational velocity of the collector, the temperature of the solution, and measurements of both air humidity and velocity within the electrospinning chamber [9].

Through optimization of electrospinning parameters, it is possible to control the diameter of fibers in a manner that ensures consistency and is free from defects. The electrospun nanofibers are used in sensors, supercapacitors, protective clothing, biomedical applications, solar cells, and micro- and nano-optoelectronics such as light-emitting diodes and photocells, etc [10].

2.4.2 Effects of parameters on electrospinning technique

The electrospinning technique can be categorized into two distinct classes, namely, solution parameters and process parameters.

2.4.2.1 Solution parameters

2.4.2.1.1 Concentration

At low concentrations of polymer solution, electrospraying, rather than electrospinning, occurs due to the solution's low viscosity and high surface tensions. With an increase in the concentration of the solution, a combination of beads and fibers is obtained. An appropriate concentration of the solution results in smooth fibers, while a very high concentration produces helix-shaped microribbons [11]. The diameter of the fiber is proportional to the concentration of the solution. The concentration of poly(vinylidene-co-hexafloropropylene) (PVdF-HFP) at 15 wt% under the applied voltage of 16 kV is illustrated in figure 2.4. Analysis of the scanning electron microscope (SEM) micrograph revealed that the fibrous membrane possessed a microporous architecture comprising of interconnected multi-fibrous layers and a 3D network structure. The fiber diameter was estimated to be approximately 100–300 nm.

2.4.2.1.2 Molecular weight

The influence of the polymer's molecular weight on the electrospun fiber morphologies is significant. The solution viscosity, which is related to the entanglement of polymer chains in fluids, is reflected by the polymer's molecular weight. The transformation of bead formation in fibers to smooth fibers can be achieved by means of reducing the surface tension of the solution. Conversely, increasing the molecular weight will result in the production of smooth fibers. Furthermore, microribbons will form as the molecular weight is further increased.

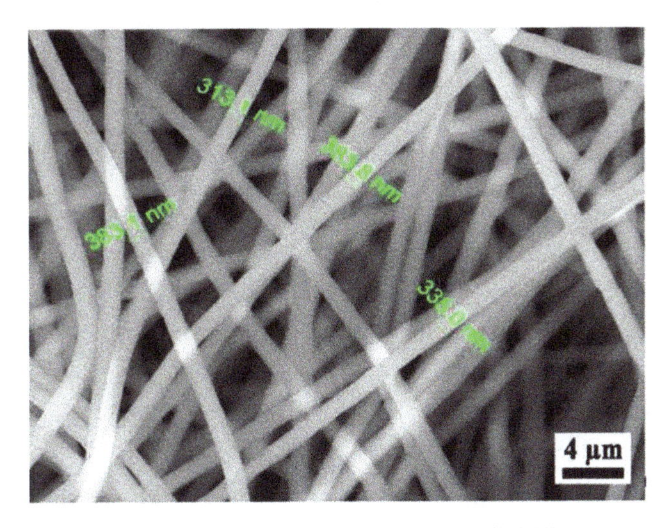

Figure 2.4. SEM image of electrospun PVdF-HFP.

2.4.2.1.3 *Viscosity and surface tension*

The solution viscosity is a crucial determinant of fiber morphology. Extensive research has demonstrated that excessively low viscosity fails to produce fibers that are continuous and smooth. Conversely, excessive viscosity leads to the hard ejection of jets from the solution. In general, changing the solution's polymer concentration can be used to alter the viscosity. The bead formation of fibers could be converted to smooth fibers by reducing the surface tension of the solution. The surface tension and solution viscosity can be optimized by altering the mass ratio of solvent and solute and fiber morphology. The surface tension determines the upper and lower boundaries of the electrospinning window.

2.4.2.2 *Process parameters*

2.4.2.2.1 *Voltage*

During the process of electrospinning, a major factor of significance is the voltage applied to the solution. The expulsion of the solution as charged jets from Taylor cones is contingent upon the voltage applied exceeding the threshold voltage ($\sim$1 kV cm^{-1}, which is dependent upon the polymer solution). The voltage applied has the stream and ultimately culminates in a reduction of fiber diameter. It has been found that changing the applied voltage will alter the initial drop's shape, changing the structure and morphology of the fibers. The potency of the interactions between the charged polymer solution and the voltage, as well as the quantity of charges conveyed by the electrically charged jet and the magnitude of electrostatic repulsion between the charges, are directly influenced by the applied voltage. Increasing voltage decreases the mean fiber diameter. This phenomenon could potentially be clarified by the amplified charge conveyed by the charged solution at higher external electric fields, which would subsequently increase electrostatic and Coulomb repulsive forces, thereby decreasing the diameter of the fibers.

2.4.2.2.2 *Flow rate*

The flow rate is regarded as the paramount factor in controlling the fiber diameter and its distribution, the shape of the initiation droplet, and the jet trajectory and in maintaining the Taylor cone and deposition area. Larger droplets occur at high flow rates, increasing average fiber diameters and bead sizes. Lower flow rates are preferable because they provide the solvent enough time to evaporate. High flow rates, however, result in the needle tip drawing out more solution, which takes longer to dry. The remaining solvent in this case may cause the fibers to converge and form webs rather than fibers.

2.4.2.2.3 *Collector*

For electrospinning, a variety of shapes of collector have been employed. These consist of a grid, a revolving disc, parallel plates, a static plate, and a rotating drum or mandrel. As per the findings of Sahay *et al* [12], electrospinning setups intended to generate aligned nanofibers frequently employ a rotating mechanism as the collector. In order to physically stretch the fibers and aid in their alignment, the rotating device serves as a collector.

2.4.2.2.4 Distance between needle tip to collector

During the process of electrospinning, the solvent undergoes evaporation as the droplets travel towards the collector subsequent to their ejection from the needle's apex. The optimization of the distance between the needle tip and the collector serves as a means of controlling the diameter and morphology of the nanofiber. It is essential to ensure that the electrospinning process incorporates a sufficiently short distance to facilitate satisfactory solvent evaporation prior to the arrival of the fibers at the collector. In the process of electrospinning, it is necessary to identify the optimal distance that allows for adequate solvent evaporation to occur prior to fiber deposition onto the collector. Before the polymer is collected by the collector, it reduces or freezes, transforming into fibers. The diameter of the nanofiber and morphology can also be controlled by optimizing the gap between the needle tip and the collector. In electrospinning, the shortest distance between the needle tip and the collector allows adequate solvent evaporation before the fibers reach the collector.

Thinner fibers have been successfully generated over an extensive distance, albeit with the formation of beads occurring when the spacing was either too large or too small. This occurrence may be attributed to the hindered drying capacity of the polymer jet stream over abbreviated distances, which is expected to influence the production of wet and/or dense nanofibers. The impediment of jet stream formation at the needle tip is a consequence of increased distances, as they reduce the electric field intensity between the nozzle and the collector. Overcoming this issue will necessitate increased power application. To facilitate the solidification and elongation of the polymer jet, which are crucial for generating fine and desiccated strands, it is recommended that the spinneret–collector distance be tailored to suit the specific polymer solution.

Short lengths of the polymer jet stream may impede its ability to dry, which could potentially affect wet and/or thick nanofibers. Longer distances, on the other hand, can lead to a decrease in the electric field intensity between the nozzle and collector, thus hindering the formation of a jet stream at the needle point. As a solution to this issue, an increase in the applied voltage may be required. Generally speaking, the distance between the spinneret and collector should be adjusted for a specific polymer solution in order to facilitate the solidification and stretching of the polymer jet, which are crucial for the production of thin and dry fibers. The electrospinning technique exhibits the capacity to achieve substantial surface-to-volume ratios and desirable physical and chemical properties, rendering it a suitable candidate for a diverse range of applications. These applications encompass sensors [13], antibacterial surfaces, scaffolds [14], photocatalysts, nanofilters, anticounterfeiting applications, and waterproof fabric [15], as well as solar energy applications.

2.4.3 3D printing of flexible electronic devices

Flexible electronic devices produced via 3D printing exhibit intricate geometries with meticulously designed microarchitectures and exceptional mechanical properties, surpassing those of traditional, simplistic planar or tubular flexible devices generated through spin coating, casting, or extrusion. This allows them to satisfy a wide range of individual needs. The complete realization of 3D-printed flexible electrical devices is still impeded by the hindrance presented by the flexibility and conductivity of the materials employed in 3D printing. Consequently, the amalgamation of the flexibility of 3D-printed flexible devices with the functionality of conductive materials or hard silicon-based electronic devices is being researched to create 3D-printed flexible electronic devices. These adaptable electrical devices fabricated through 3D-printing technology possess the capacity to be extensively utilized in various sectors such as personal wearable technology, prosthetic organs for individuals with disabilities, and interfaces that connect humans with computers. Future trends in 3D-printed flexible electronics present a huge possibility for the development of electrical devices across a variety of industries from soft robotics to biomimetic gadgets.

2.4.4 R2R processing for flexible electronics

The fabrication technique employed in the creation of flexible electronics involves a series of processes including embedding, coating, printing, and laminating diverse applications onto a malleable rolled substrate material. As a flexible substrate is transported through a roller-based processing line, it is continuously processed in R2R materials manufacturing [16]. The product is rolls of completed material that are created in an economical and effective way.

The R2R process flow is driven by the enabling technologies of flexible substrate materials, state-of-the-art patterning methods, and sturdy barrier layer packaging. The deposition, patterning, and packing are the three essential phases in R2R production. The bottom contact thin film transistor device (TFT) shown in the figure 2.5 has the flexible substrate covered with a transparent conducting oxide layer that serves as the TFT's gate electrode, adhering to the bottom contact of the TFT device. Indium tin oxide is presently favored in the industry due to its superior environmental durability and low electrical resistivity (ranging from 1 to 3×10^4 cm), as well as its exceptional transparency to visible light (at a film thickness of 100 nm, exceeding 90%). Before printing the organic semiconductor layer, a thin insulating dielectric film of SiO_2 and the metallic source and drain electrodes are produced in the following patterning processes. Instead of traditional photolithography methods, soft lithography methods, laser ablation, and inkjet printing methods appear to be

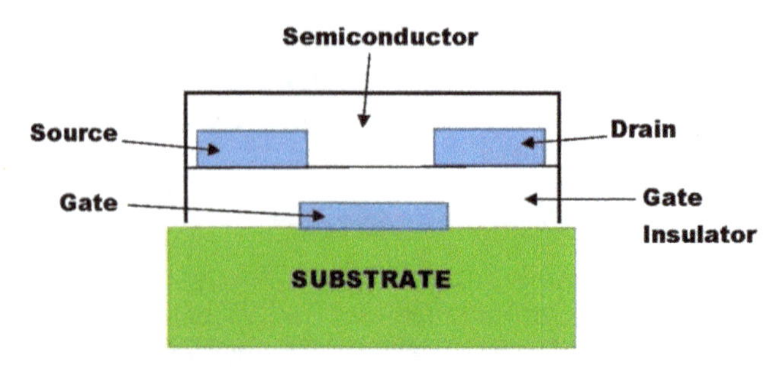

Figure 2.5. Layers of a bottom contact TFT.

promising technologies suitable for large-area flexible displays. Ultimately, the devices must be enclosed in a barrier layer to prevent oxygen and moisture contamination of the organic semiconductor layers. Flexible material is necessary for many applications in order to complete a final product. R2R manufacturing is a widely used technique for generating a diverse range of items including electronic devices, solar panels, thin-film batteries, fuel cell membranes, fabric coatings with agents, and medicinal or scientific instruments.

2.5 Polymer-sandwich nanocomposite for flexible electronics

Electric power systems and microelectronics require energy storage capacitors with elevated breakdown strengths and dielectric constants, indicating a great need for such components in these domains. A study has been done on all-organic film capacitors because of their high breakdown strength and minimal dielectric loss [17]. In this study, sandwich-structured composites were made using poly(vinylidene fluoride) (PVDF) possessing exceptional breakdown strength, which was employed as the outer layers, and poly(vinylidenefluoride-ter-trifluoroethylene-ter-chlorotrifluoroethylene) (P(VDF-TrFE-CTFE)) possessing an elevated dielectric constant, which was utilized as the interlayer. These composites were meticulously developed utilizing a layer-by-layer solution–cast approach with varying terpolymer compositions of PVDF/P(VDF-TrFE-CTFE)/PVDF. When the terpolymer concentration is 45 vol%, the sandwich-structured composites' maximum dielectric constant at 1 kHz can reach up to 18.61. With increasing terpolymer content in the same electric field, the discharged energy density rises. The composite containing 25 vol% terpolymer produces the highest discharged energy density at 660 kV mm^{-1}, which is 20.86 J cm^{-3}. Recently, to produce high-k (high dielectric constant) polymer nanocomposite materials, there has been a growing trend of fusing technology with cogent conventional methods. However, enhancing the dielectric responses of such materials while maintaining their physicochemical and electromechanical stability under conditioned stimulus is a challenging task. The stoichiometric aspect ratio and surface properties of nanoparticles in the dielectric matrix govern the final properties of these materials for specialized applications. Therefore, developing functional organic moieties to alter the inorganic nano-architectures is imperative for obtaining optimized high-k nanoparticles. In order to ensure the reliability of organic electronics applications, a substantial level of synthetic proficiency is essential. As a result, the advancement of higher-k materials frequently becomes imperative, calling for an elevated level of reliability, thus yielding higher-k materials that exhibit a greater degree of dependability for applications in organic electronics.

Recent advancements in the field of soft actuators and 3D-printable artificial muscles that possess electromechanical responsiveness necessitate a certain level of technical expertise and proficiency to effectively support the development of comprehensive dynamic structures. Dielectric elastomeric materials have garnered significant attention in the scientific community for their ability to construct sensitive smart materials that respond to external stimuli, owing to their exceptional capacity for recoverable deformation. When functionalized high-k dielectric nanoparticles were included, the characteristics of the dielectric elastomers were studied less. Flexible electronics, often known as flex circuits, is a technique that allows for the assembly of electrical circuits on flexible or elastic surfaces. This provides fresh requests for proposals for designing stretchable and flexible displays, flexible photovoltaic cell array panels, electronic circuits on fabrics, flexible wearable battery devices, etc [18–21]. It could be contended that the optimal methodology for producing innovative functional materials involves the incorporation of high-k dielectric nanoparticles into a variety of malleable polymeric polymers. Accomplishing successful fabrication is also imperative for enhancing the dielectric features while considering the principal obstacle of the inferior dielectric constant tendencies of polymer dielectric substances. This has been a pivotal realm of investigation in the development of superior high-k polymer dielectric materials.

2.6 Electrospun nanofibrous composites in energy storage devices and flexible electronics

Critical attention is given to the improvements in electrospun nanofiber applications for several important flexible electrochemical energy storage devices such as super-capacitors, metal-ion batteries, and metal-air batteries. Although electrospun fibers are frequently employed as the cathode, anode, and separator, there are no fully flexible batteries that use electrospun nanofibers for all of their component elements. The majority of the electrochemical energy storage prototypes developed lack suitable packaging, which reduces the overall performance, dependability, and even basic security of the device. The electrospun nanofibers are developed from polyacrylonitrile (PAN), polyvinylidene fluoride, polyvinyl alcohol, and polyvinyl chloride (PVC). Wei Liu *et al* [22] studied the effect of inorganic fillers such as Li^+-conductive nanowires in various polymer composites. The significant increase in conductivity is attributed to an efficient ion-conducting channel that runs along the surfaces of the aligned nanowires without crossing any junctions. Additionally, the introduction of nanowires enhances structural stability over the long term.

Chin-Shen Lim [23] explored the preparation based on poly(vinyl alcohol) incorporated with lithium perchlorate and antimony trioxide (Sb_2O_3) by solution casting methods. The investigation of the dielectric behavior displays the polymer electrolytes' non-Debye characteristics. The frequency-dependent conductivity measurement reveals that the low-frequency dispersion is a result of interfacial resistance. Incorporating Sb_2O_3 into the polymer matrix has been demonstrated to expand the maximum operational potential region, as proven by linear sweep voltammetry. The cyclic voltammetry test indicates rectangular shapes with slight distortion, and there is no evidence of any redox currents on either anodic or cathodic sweeps, which suggests the typical behavior of electric double layer capacitance (EDCL). Both EDCL cells display good cyclability over 200 cycles, with specific capacitance retention exceeding 90%.

Deka *et al* (2011) synthesized varied concentrations of polyelectrolytes through the introduction of intercalating poly(vinylidene fluoride) into organically modified montmorillonite (MMT) clay. Their findings indicate that higher clay loading leads to an augmentation of ionic conductivity. Additionally, an enhancement of both electrochemical and interfacial properties was observed [24]. Chen *et al* (2011) conducted a study on the impact of silica aerogel powder (SAP) on the electrochemical properties of nanocomposite polymer electrolytes based on PAN. The findings of the study revealed that the use of SAP resulted in a noteworthy enhancement in both the ionic conductivity and the electrochemical stability and cyclability [25].

Zheng Zhong *et al* (2012) produced a mixture of poly(vinyl difluoride)-PVC nanofibrous membranes, which were subsequently immersed in an electrolyte solution to create polyelectrolytes. The study demonstrated that the incorporation of PVC led to an increase in both the electrolyte absorption and the ionic conductivity of the resultant composite polyelectrolytes [26]. A Karmakar *et al* (2012) conducted an investigation into the performance of PEO-$LiClO_4$ electrolytes, examining the impact of varying lithium-ion concentration. The study revealed that higher lithium salt concentration resulted in greater conductivity, as evidenced by the dielectric permittivity and conductivity relaxation results. The modulus data was fitted using a nonexponential Kohlrausch–Williams–Watts function, which led to the conclusion that PEO-Li salt-based electrolytes possess charge carriers that are independent of temperature and salt concentration [27].

Kazem Jeddi *et al* (2013) successfully synthesized poly(methyl methacrylate) (PMMA) by incorporating inorganic trimethoxy silane. The resulting mixture was subsequently blended with PVdF-HFP to develop a polymer electrolyte for use in lithium-sulfur batteries. Upon investigation, the cyclability of the battery cell was observed to be stable [28–30].

Bing Sun *et al* (2014) successfully synthesized poly(trimethylene carbonate) (PTMC) polyelectrolyte through the process of ring-opening polymerization. Their study focused on the investigation of the impact of various lithium salts with different ratios on PTMC. The outcomes revealed favorable electrochemical stability and notable ionic conductivity at 1:13 and 1:8 ratios of Li+ to PTMC [31].

Jonas Mindemark *et al* (2015) conducted a study involving the preparation of polyelectrolytes from poly(ε-caprolactone). To reduce the crystallinity, carbonate repeating units were incorporated into the poly(ε-caprolactone). As a result, it was observed that the ionic conductivity increased over a broader range of temperatures. Additionally, the transport properties were deemed favorable, with a high cation transference number [32]. Yogitha P Mahant *et al* (2015) conducted an investigation on the synthesis of PVdF/PMMA blends with varying ratios and the blends were electrospun into fibers for battery applications. The porous nature of the nanofibers led to an increased electrolyte uptake, resulting in a significant enhancement in ionic conductivity [33]. Lizhen Long *et al* (2016) conducted a comprehensive analysis of solid polymer electrolytes and gel polymer electrolytes. They evaluated the ion transfer mechanisms of various modified polyelectrolytes while also exploring techniques to improve the ionic conductivity of polymer electrolytes. Additionally, the authors compared the performance of various types of lithium salts and plasticizers incorporated into the polymer electrolytes [34].

Daniel T Hallinan Jr *et al* [35] conducted a synthesis of polystyrene-b-poly (ethylene oxide), a block copolymer, via living anionic polymerization.

Incorporation of lithium bis-trifluoromethanesulfonimide into the block copolymer resulted in the preparation of an electrolyte. Subsequently, a lithium battery cell was fabricated and subjected to electrochemical performance analysis. Linear sweep voltammetry was utilized to investigate copolymer electrolytes across varying potentials, which were analyzed using the Butler–Volmer model. The study revealed that copolymer electrolytes exhibit satisfactory electrochemical oxidative stability.

Zhen Li *et al* [36] carried out the modification of poly(ether ether ketone) by chloromethylation process and electrospun it into nanofibrous mats. The dibenzyl ether group employed to cross-link the polymer matrix provides the membrane high durability and good antishrinkage properties. The nonwoven membrane's interwoven structure efficiently contributes to the high electrolyte uptake of 215.8%. Compared to the electrolyte-swelled Celgard membrane, the electrospinning-fabricated membrane has a 51% greater ionic conductivity. The lithium-ion battery with this nonwoven membrane therefore displays improved rate performance (up to 42.5% higher than the lithium-ion battery with a polypropylene (PP) separator) and acceptable cycling performance. Research is also focused on incorporation of various ionic liquids (ILs) such as *N*-methyl-*N*-propyl piperidinium bis(trifluoromethane-sulfonyl imide), silane-based ILs, poly(ionic liquid) hydrogels etc into polymers to develop electrolytes [37]. ILs, which are comprised of voluminous organic cations and adaptable anions, have been increasingly employed in environmentally sustainable energy storage applications. Supercapacitors, regarded as competitive devices with high-power capabilities, have attained significant attention owing to their capacity for high-rate energy harvesting and long-term durability [38]. The electric energy stored in supercapacitors is achieved through the ion dynamics and physicochemical interactions at the electrolyte–electrode interface. To meet the high-energy demands required for the development of superior supercapacitors, ILs have emerged as a standout option owing to their characteristic negligible vapor pressure and molecular designability, as well as several intriguing features such as a highly ionized environment, good thermal/chemical stability, and universal solubility/affinity.

2.7 Recent development and advanced applications of flexible electronics

Flexible electronics have a diverse array of applications including photodetectors, biosensors, batteries, supercapacitors, triboelectric nanogenerators, flexible paper-based electronic devices, and health care etc. By the end of 2029, it is anticipated that the commercial flexible electronics industry will have amassed a total revenue of $77.3 billion, exhibiting a noteworthy compound annual growth rate of 8.5% between 2018 and 2029. Dan-Liang Wen *et al* [39, 40] developed silk fibroin (SF) with superior properties such as excellent biodegradability, biocompatibility, optical transmittance, mechanical robustness, and water solubility. The versatility and ease of processing make it extremely suitable for the construction of next-generation biocompatible flexible electronic devices. Various processing techniques, such as electrospinning, 3D printing, spin coating, soft lithography, freeze drying, particulate leaching, and ultrasonic induction, have been devised to process SF materials into diverse forms, including silk fibers, films, sponges, and hydrogels for numerous application fields. SF, due to its required biological characteristics, is frequently employed as substrates, encapsulating materials, and scaffolds for flexible wearable and implantable electronic devices such as electronic skins, bioabsorbable electronics, and therapeutic electronics.

SF has been discovered to possess piezoelectric characteristics and an exceptional ability to expel electrons. As a consequence, it has been utilized as a dielectric material for constructing both triboelectric and piezoelectric energy harvesters. The flexible sensors that incorporate SF functional components include humidity sensors, temperature sensors, pressure sensors, airflow sensors, and electrochemical sensors. This is attributed to the distinctive features of SF that render it responsive to environmental variables. Despite its superiority in terms of flexibility and biocompatibility over conventional silicon-based electronics, greater endeavors are required to enhance the performance of SF-based flexible electronic devices to such an extent that they can replace rigid silicon-based electronics. It is observed that SF undergoes a reduction in mechanical strength under conditions of high water content and a decrease in flexibility under conditions of low water content. Consequently, an optimum equilibrium state must be explored in order to preserve the exceptional flexibility and high mechanical strength of SF, thereby necessitating the development of novel techniques.

Flexible substrates utilized in the production of flexible antennas, particularly in the field of biomedical applications, ought to possess certain characteristics such as biodegradability, inertness, biocompatibility, and nontoxicity [41]. The polydimethylsiloxane (PDMS) based flexible substrate is widely acknowledged as an optimal substrate for flexible wearable devices. Its exceptional features set it apart from the existing flexible substrates and have the potential to transform the entire field of flexible, stretchable, and wearable technology.

Flexible substrate made of high-temperature PDMS (HT-PDMS) is created and its surface engineering for flexible electronic applications is expounded upon. The HT-PDMS possesses exceptional qualities such as high thermal stability up to 400°C, remarkable mechanical stability, and great transparency. The HT-PDMS is inherently hydrophobic, but to alter it to hydrophilic, various surface modification treatments are conducted. The combined treatment of oxygen-plasma and sodium dodecyl surfactant has the capability of transforming the HT-PDMS surface to hydrophilic in perpetuity. The HT-PDMS substrate reported herein supports a high-temperature fabrication process and has the potential to be utilized in flexible optoelectronic devices [42].

Flexible OLEDs are widely regarded as the most appropriate choice for lighting devices of the future. A novel flexible and transparent substrate has been synthesized through the use of organic materials such as tetraphenylphosphonium modified lithium saponite, synthetic saponite, and *N*, *N*-dimethylformamide. This substrate boasts an average transmittance of 60%–70% in the visible region (300–700 nm), rendering it highly suitable for use in flexible displays. Moreover, its refractive index is found to be similar to that of glass, further enhancing its potential for this application. Additional parameters, such as extinction coefficient (k) and absorption, are also meticulously computed in this research [43].

Recent advancements in the realm of flexible electronics have prompted researchers to pursue the development of a novel flexible substrate that can replace the current rigid glass and inflexible plastics. The utilization of flexible substrates presents considerable advantages in terms of the ability to fabricate sturdy, slimmer, adaptable, light, and storable electronic devices. This study presents the synthesis of a new flexible and transparent substrate using organic materials such as PDMS and tetra ethyl ortho silicate. The substrate synthesized demonstrates a transmittance of approximately 90%–95% in the visible region (400–700 nm). Additionally, the substrate exhibits superior thermal characteristics, withstanding temperatures of up to 200°C without significant degradation [44].

With the progression of contemporary society, the development of pliant flexible electronic devices is flourishing incessantly in the guise of portable detectors, flexible and portable medical supervision mechanisms [45], adaptable energy repositories and transmuters [46], semiconductor nanowires [47], and food monitoring sensors [48]. In this exposition, the focus is placed on four distinct and representative applications of flexible electronics as they pertain to cardiovascular diseases. These applications include blood pressure monitoring, electrocardiogram monitoring, echocardiogram monitoring, and direct epicardium monitoring [49].

The conductive hydrogels, fabricated by Ruxue Yang *et al* [50] through employment of 3D printing technology, hold significant potential as prospective materials for flexible electronic devices due to their biocompatible nature, physicomechanical properties, commendable electrical conductivity, and a multitude of stimulus response characteristics. Recent progress has also been made in the development of organic photodetectors (OPDs) for use in flexible electronics. The implementation of OPD technology has the potential to yield wearable devices, compact information sensors, and self-monitoring health devices, as well as flexible optical communication systems and large-area image sensors, all at commercially viable costs. In comparison to wafer-based PD techniques, solution-processable flexible OPD technologies offer a considerable cost advantage. The solution-processable OPDs' lifetime and long-wavelength photodetection, however, are still insufficient for commercial viability. It is crucial to create low-bandgap organic photoactive materials, comprehend charge transport characteristics under various operating situations, find ways to reduce noise current, and improve operation stability in order to maximize the performance and longevity of OPDs. Further development of OPDs focuses on the development of infrared OPDs, device integration through solution-fabricated processes, and stability of solution-processable OPDs. Thus, thin, lightweight, and flexible electronics are currently considered a significant milestone in the progression of cutting-edge technological innovations.

2.8 Conclusions

This chapter provides an exposition on the latest generation of polymers employed in the realm of flexible electronics. The topics covered herein span from the rudiments of flexible electronics to the most recent advancements in this field. The introduction section delves into the various materials and device/components that have been developed since 1960, as well as the different types and manufacturing techniques that have been employed in flexible electronics. The section on novel polymer materials expatiates on the imperative characteristics of flexible materials and different materials used in the various layers of flexible electronics parts. Furthermore, the discourse on the role of different nanomaterials in flexible electronics is presented, and the novel manufacturing methods of electrospinning, 3D printing, and R2R processing are expounded upon along with their corresponding processing parameters. The use of polymer sandwich composites is briefly discussed, with emphasis on the utilization of electrospun nanofibrous composites in energy storage applications. Lastly, the final section of this chapter provides references to the recent material and manufacturing processes that have been employed in the development of flexible electronics.

References

[1] Sircar A and Kumar H 2021 An introduction to flexible electronics: Manufacturing techniques, types and future *J. Phys. Conf. Ser.* **1913** 012047

[2] Metin Uz K, Jackson M S, Donta J, Jung M T, Lentner J A, Hondred J C, Claussen and Mallapragada S K 2019 *Sci. Rep.* **9** 10595

[3] Corzo D, Tostado-Blázquez G and Baran D 2020 *Flex. Electron.* **1** 594003

[4] Zhou Z, Zhang H, Liu J and Huang W 2021 *Giant* **6** 100051

[5] Li L, Han L, Hu H and Zhang R 2023 A review on polymers and their composites for flexible electronics *Mater. Adv.* **4**(3) 726–746

[6] Hussain A M 2022 *Introduction to Flexible Electronics* 1st edn (Boca Raton, FL: CRC Press)

[7] Li R-W and Liu G 2020 *Flexible and Stretchable Electronics Materials, Designs, and Devices* (Singapore: Jenny Stanford Publishing Pte. Ltd)

[8] Khanna V K 2019 *Flexible Electronics, Mechanical Background, Materials and Manufacturing* **vol 1** (Bristol: IOP Publishing)

[9] Doshi J and Reneker D H 1995 Electrospinning process and applications of electrospun fibers *J. Electrostat.* **35** 151–60

[10] Li X, Cheruvalli G, Kim J-K, Choi J-W, Ahn J-H, Kim K-W and Ahn H-J 2007 Polymer electrolytes based on an electrospun poly (vinylidene fluoride-co-hexafluoropropylene) membrane for lithium batteries *J. Power Sources* **167** 491–8

[11] Wang X, Drew C, Lee S-H, Senecal K J, Kumar J and Samuel-son L A 2002 Electrospun nanofibrous membranes for highly sensitive optical sensors *Nano Lett.* **2** 1273–5

[12] Sahay R, Thavasi V and Ramakrishna S 2011 Design modifications in electrospinning setup for advanced applications *J. Nanometer.* **2011** 317673

[13] Im J S, Kang S C, Lee S H and Lee Y S 2010 Improved gas sensing of electrospun carbon fibers based on pore structure, conductivity and surface modification *Carbon* **48** 2573–81

[14] Chew S Y, Mi R, Hoke A and Leong K W 2008 The effect of the alignment of electrospun fibrous scaffolds on Schwann cell maturation *Biomater* **29** 653–61

[15] Gratzel M 2005 Dye-sensitized solid-state heterojunction solar cells *MRS Bull.* **30** 23–7

[16] Palavesam N, Martin S, Landesberg C and Hemmetzberg D 2018 Roll-to-roll processing of film substrates for hybrid integrated flexible electronics *Flex. Prin. Electron.* **3** 014002

[17] Wang L, Luo H, Zhou X, Yuan X, Zhou K and Zhang D 2019 Sandwich-structured all-organic composites with high breakdown strength and high dielectric constant for film capacitor *Composites* A **117** 369–7
[18] Gu Y, Zhang T, Chen H, Wang F, Pu Y, Gao C and Li S 2019 Mini review on flexible and wearable electronics for monitoring human health information *Nanoscale Res. Lett.* **14** 263
[19] Zou M, Ma Y, Yuan X, Hu Y, Liu J and Jin Z 2018 Flexible devices: from materials, architectures to applications *J. Semicond.* **39** 011010
[20] Wu W 2019 Stretchable electronics: functional materials, fabrication strategies and applications *Sci. Technol. Adv. Mater.* **20** 187–221
[21] Huang S, Liu Y, Zhao Y, Ren Z and Guo C F 2018 Flexible electronics : stretchable electrodes and their future *Adv. Funct. Mater.* **29** 1805924
[22] Liu W, Lee S W, Lin D, Shi F, Wang S, Sendek A D and Cui Y 2017 Enhancing ionic conductivity in composite polymer electrolytes with well-aligned ceramic nanowires *Nat. Energy* **2** 17035
[23] Lim C-S, Teoh K H, Liew C-W and Ramesh S 2014 Electric double layer capacitor based on activated carbon electrode and biodegradable composite polymer electrolyte *Ionics* **20** 251–8
[24] Deka M and Kumar A 2011 Electrical and electrochemical studies of poly (vinylidene fluoride)-clay nanocomposite gel polymer electrolytes for Li-ion batteries *J. Power Sources* **196** 1358–364
[25] Chen-Yang Y W, Wang Y L, Chen Y T, Li Y K, Chen H C and Chiu H Y 2008 Influence of silica aerogel on the properties of polyethylene oxide-based nanocomposite polymer electrolytes for lithium battery *J. Power Sources* **182** 340–8
[26] Zhong Z, Cao Q, Jing B, Wang X, Li X and Deng H 2012 Electrospun PVdF-PVC nanofibrous polymer electrolytes for polymer lithium-ion batteries *Mater. Sci. Eng.* B **177** 86–91
[27] Karmakar A and Ghosh A 2012 Dielectric permittivity and electric modulus of polyethylene oxide (PEO)-$LiClO_4$ composite electrolytes *Curr. Appl. Phys.* **12** 539–43
[28] Jeddi K, Ghaznavi M and Chen P 2013 A novel polymer electrolyte to improve the cycle life of high performance lithium–sulphur batteries *J. Mater. Chem.* A **1** 2769–72
[29] Yang Y, Zheng G and Cui Y 2013 Nanostructured sulfur cathodes *Chem. Soc. Rev.* **42** 3018–32
[30] Libo L, Jiesi L, Shuo Y, Shaowen G and Peixia Y 2014 Gel polymer electrolytes containing ionic liquids prepared by radical polymerization *Colloids Surf.* A **495** 136–41
[31] Sun B, Mindemark J, Edstrom K and Brandell D 2014 Polycarbonate-based solid polymer electrolytes for Li-ion batteries *Solid State Ionics* **262** 738–42
[32] Mindemark J, Sun B, Torma E and Brandell D 2015 High-performance solid polymer electrolytes for lithium batteries operational at ambient temperature *J. Power Sources* **298** 166–70
[33] Mahant Y P, Kondawar S B, Bhute M and Nandanwar D V 2015 Electrospun poly (vinylidene fluoride)/poly(methyl methacrylate) composite nanofibers polymer electrolyte or batteries *Procedia Mater. Sci.* **10** 595–602
[34] Long L, Wang S, Xiao M and Meng Y 2016 Polymer electrolytes for lithium polymer batteries *J. Mater. Chem.* A **4** 10038–69
[35] Hallinan D T, Rausch A and McGill B 2016 An electrochemical approach to measuring oxidative stability of solid polymer electrolytes for lithium batteries *Chem. Eng. Sci.* **154** 34–41

[36] Li Z, Wang W, Han Y, Zhang L, Li S, Tang B, Xu S and Xu Z 2018 Ether modified poly (ether ether ketone) nonwoven membrane with excellent wettability and stability as a lithium ion battery separator *J. Power Sources* **378** 176–83
[37] Miao L, Song Z, Zhu D, Li L, Gan L and Liu M 2021 Ionic liquids for supercapacitive energy storage : a mini-review *Energy Fuels* **35** 8443–55
[38] Park H, Lee S, Jeong S H, Jung U H, Park K, Lee M G, Kim S and Lee J 2018 Enhanced moisture-reactive hydrophilic-PTFE-based flexible humidity sensor for real-time monitoring *Sensors* **18** 921
[39] Wen D-L *et al* 2021 Recent progress in silk fibrion-based flexible electronics *Microsyst. Nanoeng.* **7** 35
[40] Sudheshwar A, Malinverno N, Hischier R, Nowack B and Som C 2023 The need for design-for-recycling of paper-based electronics—a prospective comparison with printed circuit boards *Resour. Conserv. Recl.* **189** 106757
[41] Dhanabalan S S, Sitharthan R, Madurakavi K, Thirumurugan A, Rajesh M, Avaninathan S R and Carrasco M F 2022 Flexible compact system for wearable health monitoring applications *Comput. Electr. Eng.* **102** 108130
[42] Dhanabalan S S, Thirumurugan A, Periyasamy G, Dineshbabu N, Chidhambaram N, Avaninathan S R and Carrasco M F 2022 Surface engineering of high-temperature PDMS substrate for flexible optoelectronic applications *Chem. Phys. Lett.* **800** 139692
[43] Shanmuga Sundar D, Sivanantha Raja A, Sanjeeviraja C and Jeyakumar D 2016 Synthesis and characterization of transparent and flexible polymer clay substrate for OLEDs *Mater. Today Proc.* **3** 2409–12
[44] Shanmuga Sundar D, Raja A S, Sanjeeviraja C and Jeyakumar D 2017 High temperature processable flexible polymer films *Int. J. Nanosci.* **16** 1650038
[45] Sasmal A and Arockiarajan A 2023 Recent progress in flexible magnetoelectric composites and devices for next generation wearable electronics *Nano Energy* **115** 108733
[46] Gao Q, Agarwal S, Greiner A and Zhang T 2023 Electrospun fiber-based flexible electronics: fiber fabrication, device platform, functionality integration and applications *Prog. Mater. Sci.* **137** 101139
[47] Chen K, Pan J, Yin W, Ma C and Wang L 2023 Flexible electronics based on one-dimensional inorganic semiconductor nanowires and two-dimensional transition metal dichalcogenides *Chi. Chem. Lett.* **34**(11) 108226
[48] Yue C, Wang J, Wang Z, Kong B and Wang G 2023 Flexible printed electronics and their applications in food quality monitoring and intelligence food packaging: recent *Adv. Food Control* **154** 109983
[49] Zhang T, Liu N, Xu J, Liu Z, Zhou Y, Yang Y, Li S, Huang Y and Jiang S 2023 Flexible electronics for cardiovascular healthcare monitoring *Innovation* **4** 100485
[50] Yang R, Chen X, Zheng Y, Chen K, Zeng W and Wu X 2022 Recent advances in the 3D printing of electrically conductive hydrogels for flexible electronics *J. Mater. Chem.* B **10**(14) 5380–99

IOP Publishing

Advances in Flexible and Printed Electronics
Materials, fabrication, and applications
Shanmuga Sundar Dhanabalan and Arun Thirumurugan

Chapter 3

Flexible thin films for electronics

M Alagappan, T Sridarshini and P S Sowbarnika

Flexible thin films provide a way to develop various electronic circuits and devices on flexible substrate, which can outsmart the complexity of traditional electronics in terms of human–machine interface, affordability, flexibility, customizability, innovation, and portability. In addition, flexible electronics are ideal for integration without the requirement for extrinsic packages and provides sustainable circuit board assemblies with the ability to reduce e-waste about 95%. This emerging technique enables design and developments that are impossible with conventional semiconductor materials and traditional substrates. Hence this chapter deals with the need for flexible thin films and devices, advantages of flexible thin film-based devices, and their material characteristics. Furthermore, the recent advancements and technical developments in the areas of health care, photovoltaics, sensors, and displays are elaborated as a breakthrough in the product development for societal needs.

3.1 Glimpse of thin films and flexible electronic devices

Intelligentization has played a vital role in the advancement of science and technology during the past two decades. Electronics have increased the current application possibilities tremendously. To date, nonflexible electronic devices, particularly sensors, have been utilised in a multitude of applications, although they have exhibited drawbacks such as rigidness, processing and production challenges, and high operating expenses [1]. In this way, flexible electronics have opened promising opportunities for the development of exciting devices.

Generally, materials of thin layers with thicknesses that vary from a few nanometres to several micrometres are referred as thin films. Malleable substrates coated with extremely thin layers of functional materials enable individuals to create devices in all the sectors that were not previously possible to achieve. These devices are popularly called flexible electronic devices or flexible thin film devices and offer significant advantages over conventional silicon systems.

doi:10.1088/978-0-7503-5492-9ch3

Flexible electronics possess various potential attributes, including but not limited to being thin, lightweight, stretchable, disposable, easily processed with low cost, and capable of miniaturisation [2]. The noteworthy advantage of flexible electronics lies in their robustness, which merits attention [3]. The aforementioned has noteworthy implications for different implementations, encompassing notebooks, wearables, and various other electronic goods that traditionally integrated with sensors made from flexible materials.

The materials utilised in the production of these devices are capable of being bent within specific ranges without compromising their properties. In addition, it should be noted that these devices are designed to be worn, providing individuals with a comfortable and protective experience. Furthermore, their low elastic modulus enables them to maintain their mechanical properties even when subjected to malleable objects.

As a consequence, there is a significant market demand for flexible electronic devices. It has been projected that the market valuation of printed and flexible electronics will exceed 75 billion USD by the year 2025 and that the market value of wearable flexible sensors (WFS) will exceed 40 million USD by the year 2025. However, the development process requires a significant amount of time and unwavering dedication [1].

3.2 Flexible thin film devices: materials and their characteristics

Flexible electronics are often known as printed electronics (PE) that describe functional electronics made by developing conductive lines with one of numerous printing techniques. The utilisation of this prominent and cost-effective technique has resulted in a noteworthy reduction in production costs, leading to the development of various devices including sensors, such as pressure sensors, temperature and humidity sensors, gas sensors, and accelerometers etc [2]. The following essential elements should typically be included in PE: a flexible (circuit) board to serve as the substrate, the utilisation of inks (semiconductors), and a continuous printing processing technology such as the slot-die coating process [4].

The role of substrate materials is significant in the field of flexible electronics technology. Materials such as plastic foils, cloths, glass, paper, and metals are serving this purpose. Polymeric substrates are favouring the industries for device development due to their ease of processing, minimal infrastructure requirements, lower production costs, and ability to achieve small bending radii of curvature [2]. Polyethylene terephthalate (PET), polyethylene naphthalate (PEN), polycarbonate (PC), and polyimide (PI) are frequently utilised polymeric substrates [1, 2]. PET and PEN are preferred over other organic polymers due to their transparency and relatively lower cost.

In contrast to conventional rigid substrates, flexible polymer substrates exhibit a significantly reduced glass transition temperature (Tg). Therefore, it is of the greatest importance to take into account the fabrication temperature of the printing technology. The Tg of PI (value of up to 360°C) is notably higher than that of various other polymer substrates, including PET (120°C), PEN (150°C), PC (155°C), polyethersulfone (230°C), and polyether ether ketone (250°C). Due to the limited ability of flexible polymer substrates to withstand high-temperature processing or treatment, alternative low-temperature post-processing techniques have been developed, including photonic sintering and microwave sintering etc [5, 6].

According to the research conducted by Sichao Tong and colleagues, the flexible transistors that were placed on a polydimethylsiloxane (PDMS) substrate with a thickness of approximately 100 μm exhibited the ability to bend with a minimal curvature ranging from 12–3 mm. Furthermore, the performance parameters of the transistors remained relatively stable even after undergoing 450 deformation cycles. Additionally, a study showcased the fabrication of copper phthalocyanine (CuPc) nanowire transistors on a PDMS substrate through transfer printing. The transistors, which were 260 μm thick, maintained their device performance even after undergoing 700 cycles of bending with a small curvature of 5 mm [5]. Researchers have reported on the utilisation of inkjet-printed single-walled carbon nanotubes (CNTs) as semiconductors and reduced graphene oxide (rGO) patterns as source/drain/gate electrodes on a PET substrate for the construction of thin-film transistors (TFTs). The resulting TFTs exhibited exceptional performance, with an I_{on}/I_{off} ratio of $\sim 1 \times 10^4$ and a mobility of $\sim 8\ cm^2\ V^{-1}\ s^{-1}$ [5, 7].

Organic materials have been utilised to produce transistors that substitute the conventional silicon and hard ceramic materials with flexible, carbon-based materials. Organic TFTs (OTFTs) possess inherent flexibility and exhibit superior performance in comparison to amorphous silicon. The high-performance OTFTs exhibit the ability to undergo processing at temperatures below 100°C. This feature facilitates the utilisation of thin-film substrates that are less expensive, thereby enabling the development of flexible devices.

In the instance of thermal sintering, the operating temperature (usually between 100 and 400°C) has to be lower than the softening temperature of the substrates. A few nanometers' thickness of an organic film between the conducting particles is sufficient to prevent electrons moving from one particle to another by lowering the conductivity. In this case, the organic covering must be removed at high temperatures. Hence, the sintering temperature of nanoparticle-based inks plays a vital role in the development of plastic electronic devices when PET and PC are utilised as substrate materials. The nanoparticle-based inks are conductive due to the presence of conductive particles in their composition. The popular choices are metallic particles (i.e. gold, silver, and copper), carbon, and conducting polymers. These nanoparticles possess excellent electrical conductivity and sometimes do not fit into the low-cost printable electronics applications [6].

Notably, the organic and large area electronics (OLAE) process is a contemporary method employed for fabricating electronic devices, which are printed in thin layers through the utilisation of functional inks. OLAE devices are composed of carbon-based materials that are abundantly accessible, comparatively inexpensive, and less hazardous than conventional silicon-based electronics [1]. Parallel to this, metal nanoparticles, such as aluminium, silver, gold, and copper, have been identified as promising candidates for electrical wiring. Also, core materials, such as nanoparticles, nanowires, nanotubes, and thin films, can be chosen for their suitability in certain processing environments. Furthermore, it is recommended that these materials undergo processing procedures at low temperatures not exceeding 150°C [4].

Flexible batteries have been developed in response to the ongoing demand for power supply in monitoring devices. Hybrids and nanocomposites incorporating materials, such as graphene, CNTs, carbon cloth, and cellulose, have been utilised in the development of flexible lithium-ion batteries [1]. CNTs are extensively utilised in the fields of energy storage and microelectronics due to their outstanding electrical and mechanical properties.

Zinc oxide (ZnO) possesses a wide band gap of 3.37 eV [8] and exhibits unique optical, catalytic, and electrical properties. As a result, it is extensively employed in various applications such as energy storage, sensing, and photo detection with dielectric properties [9].

Subsequently, devices that utilise paper as a substrate have been created for implementation in health care and environmental contexts. The utilisation of radio frequency identification (RFID) technology on paper has also been documented in academic literature. Various techniques have been developed to achieve an optimal substrate, including the conversion of hydrophilic paper to hydrophobic paper through the use of photopaper, the modification of papers using organosilanes, and the reduction of surface roughness. The utilisation of paper substrate presents several benefits such as being cost-effective, lightweight, biodegradable, and flexible. These characteristics hold great potential for the development of innovative platforms featuring affordable, portable, and uncomplicated film-based devices [2, 5, 10].

Flexible printed circuit boards (FPCBs) were also developed for the purpose of conducting in real-time perspiration analysis [1]. The nature of needs for modern electronic devices is sometimes conflicting. For example, the items must be both light and able to tolerate heat and vibration [11]. Furthermore, in order to manufacture the product at a cheap cost, it needs to be machine assembled. Reliability and durability are required to compete. Electronic gadgets' physical dimensions have shrunk over the last several decades [12]. That implies that the printed circuit board, or PCB, inside must be smaller, as well. Although it may not be feasible to reduce the size of a rigid board, shifting the circuit's layout to an FPCB provides the benefit of allowing for greater use of the device's limited area. In truth, FPCBs provide the designer a lot of flexibility, enabling them to come up with durable products as well as lightweight circuit boards [13].

3.2.1 A perspective on device fabrication techniques

Basically, the following two methods are adopted to develop flexible electronic devices: (1) Designed circuits are transferred and bonded to flexible substrate, and (2) Circuits are fabricated directly on the flexible substrate [14].

Furthermore, the flexible electronics i.e. the PE techniques are classified as contact or noncontact methods. In contact methods, the surface of the printing plate comes into close association with the substrate, whereas in the noncontact methods, only the deposited substance comes into interaction with the substrate.

Soft lithography, flexography, gravure printing, and screen printing are examples of contact method. Inkjet printing, laser direct writing, and aerosol printing techniques are coming under the category of noncontact methods [6]. Some of the parameters, such as sensor design, processing, and operation, become more important than others depending on the type of device to be printed, the materials and substrate used, the size of the device, and the volume of production. Research and development efforts are still being made to address them.

3.3 Flexible electronics in the health sector

Living creatures, particularly humans, are involved in a complex cognitive process in which any disease may signify the release of important physiological signals. As a result, developing flexible electronics for the health sector focuses on accurately grasping physiological signals upon the human body. Before comprehending the flexible electronics manufacturing processes and the challenges of detecting physiological signals, we must first understand the various forms of physiological signals. The physiological signs and their classifications are shown in table 3.1.

3.3.1 Bioelectrical signal monitoring

Bioelectrical signals, such as electroencephalogram (EEG)/electrocorticogram (ECoG), electrocardiogram (ECG), electromyogram (EMG), and electrooculogram (EOG), are important signals because they possess sets of systematic cues about various diseases and human physiological activities. EEG and ECoG (which is also known as intracranial electroencephalography) have really been important methods for understanding brain functioning in cognitive neuroscience. They are also used to study neurological disorders in clinical settings such as Parkinson's disease, epilepsy, and Alzheimer's disease, among others. EEG signals are classified as α (8–13 Hz), β (13–30 Hz), γ (30–80 Hz), δ (0.1–4 Hz), and θ (4–8 Hz) according to their frequency ranges [15] and are traditionally measured by infusing surface contacting electrode pads containing conductive gel on the scalp, while ECoG is assessed via probing microneedles. In spite of the tremendous excellence in EEG as well as ECoG-related studies, advancements in these techniques are always needed in the following areas:

(1) Enhancing high resolution
(2) Examining an area by employing high-density electrode arrays
(3) Improving signal strength with intrinsically conductive interaction

Table 3.1. Categories of commonly detectable physiological signals.

S. No.	Physiological signal	Types
1	Bioelectrical signals	EEG, ECoG, EOG, ECG, EMG
2	Biophysical signals	Body temperature, skin strain/pressure, blood flow, skin modulus, skin hydration
3	Biochemical signals	Blood glucose, sweat composition

(4) Lower surface inductances
(5) Establishing protracted as well as consistent EEG and ECoG records of signal with energy-efficient, biocompatible wireless devices.

Recent breakthroughs in flexible electronics have demonstrated the possibilities in addressing the above issues. As a result, lowering the thickness of the electrode serves as the initial step for close contact, lower sensitivity, and ultimately superior signals [16]. Larger area and flexible electronic devices utilizing polymer substrates, particularly PI, are more appealing than standard electrodes because of improved connection and the ability to capture the information or signal in greater regions. Flexible electronics can be utilised to identify and cure heart problems including heart failure and cardiac arrhythmias [17]. A great deal of effort has gone into the development of flexible devices for electrotherapy and cardiac electrophysiology mapping. This includes intracorporeal devices like multipurpose power sources and balloon catheters, which can also be implanted into the heart, as well as extracorporeal devices. In terms of epidermal electronic devices for electrocardiography, gel-type multifunctional sensors are deemed the most advanced approach when contrasted to those that necessitate contact gel, owing to benefits such as adaptability, portability, mobility, adhesion, and flexibility [18].

EMG is the measurement of electromagnetic impulses produced by muscle tissue, which is useful in research and clinical settings. Flexible and stretchable EMG detectors that are incorporated with both the skeletal muscles and epidermal systems are preferred because they provide significantly lower and much more consistent contact impedance, particularly with vibrant measuring capability. Surface EMG (sEMG) and intramuscular EMG (iEMG) are the two different types of EMG. For the measurement of sEMG, dermal methods, such as tattoo-like patches that are worn on the human neck and face, are developed [19]. Guvanasen *et al* designed methods for monitoring iEMG signals. The researchers have developed a stretchy microneedle electro network for generating and tracking intramuscular electromyographic activity [20].

3.3.2 Biophysical signal monitoring

Attaching elastic and foldable biomedical devices to the skin allows the biophysical signals, such as temperature, pressure, and strain, to be measured. Temperature is the essential characteristic that doctors are most worried about. Constant and long-term measurement of baseline ambient temperature, localized temperature surrounding injuries, and temperature measurement routing around dermal bloodstream provide essential data for diagnosis of the disease. Furthermore, pressure and strain on the patient's mental health can be utilised to represent many physiologic processes that produce cellular displacement, such as physical movement, breathing, pulses, and vocalisations. Numerous epidermal devices have been designed to assess the temperature, pressure, and strain of the human body via skin integration.

The key obstacles with these epidermal systems are (1) the biomaterial issue for protracted incorporation into skin surface, (2) separating the distortion created by physical movement from the delicate desired signal, and (3) integrating advanced functionality in a single device, as well as many others. Innovative epidermal systems are being used in a wide range of situations including injury evaluation and treatment, blood circulation monitoring, and cutaneous condition assessment, notably. Therefore, it is essential to discuss the body temperature and strain by embedding elastic and flexible peripherals into the skin, which focus upon the aforementioned issues and uses.

3.3.2.1 Temperature

In view of the significance and strong link with many human body pathologies, body temperature is a basic parameter in physiology. Long-term genuine temperature monitoring is critical in biomedical research, notably for assessing the health status of anaesthetic patients or infants. Previously, body temperature was characterized as that of the supplemental temperatures recorded by several thermometers. Nevertheless, this procedure is indeed not appropriately convenient for newborns. Noncontact infrared thermometers are now being employed to detect body temperature promptly and noninvasively. As a matter of fact, such instruments are expensive and cannot effectively discern the body temperature. Temperature equipment that is wearable and flexible is a viable and practical option.

By reducing external impacts, epidermal devices may be likely bonded to the skin, thus precisely recording human body temperature. Various wearable temperature sensors have been developed based on the epidermal electronics concept [21]. For example, adaptable multifunctional detector sheets are designed to be fitted on skin; such sensors are expected to be the major function in detecting disease. This device operates by incorporating a flexible temperature sensor with an ion selective field effect transistor [22]. Temperature sensors like this can be added to the flexible electronic gadgets that monitor and manage the human body. Moreover, device-level stretchy and flexible circuits have indeed been recommended and manufactured as innovative solutions to public health concerns.

3.3.2.2 Blood flow

Skin-like devices can play a vital role in wound monitoring and blood flow applications aside from body temperature. For the purpose of comparison, flow of blood is considered; standard diagnostic techniques confront a typical issue that impedes precision. To be specific, in order to ensure correctness, a device must be forced against the human body to achieve proper contact; nevertheless, this force will modify the flow of blood regionally, and noncontact assessments are indeed extremely susceptible to deformation and body movements. Epidermal devices can overcome this problem because they are able to function just on skin as a physically undetectable and accurate manner regardless of whether the body is moving or deformed. Several scientists have proposed epidermal devices that utilize the thermal approaches based on this notion in order to map circulation of blood in an accurate, noninvasive, and uninterrupted manner [23].

3.3.2.3 Wound monitoring

Wound healing is a constantly changing phenomenon that can span anywhere between weeks to months [24]. The injury recovery process is classified in three different stages:

(i) The inflammatory stage (which lasts for approximately three to five days)
(ii) The proliferation stage (that usually takes about one week to ten days) and
(iii) The maturation stage (further lasts several months to years)

The continuous improvements of long-term injury measurement techniques will be aided by the rapidly expanding need for wound healing. Primarily, the injury surface remains moist, fluid filled, and pliable. As a result, the wet adhesive characteristics and chemical stability of substances for IoT (Internet of things) bandages would also be taken into account.

Furthermore, the timeline for tissue repair varies depending on the type of wound. For instance, infection occurs within 48 h to one week following a burn wound, but injury develops at any point during the healing process for a patient with diabetes. As a result, particular time frames must be fulfilled for injury monitoring devices. Current injury tracking devices have a life span from only several days to over a month and are suitable for cutaneous inflammation. They are, however, insufficient for chronic wounds like diabetic foot wounds, which might linger for months or even years.

3.3.2.4 Strain and pressure

The fundamental variable that must be monitored on the skin is passive or active distortion. It could be specifically pressure or strain induced by certain physiological activities from which we can identify pulses, body movements, breathing, articulation, and skin mechanical properties among other things. Capacitive, triboelectric, and piezoelectric sensors are used to measure pressure and strain [25]. Various materials demand the device's predictability, elastic property, sensitivity, detection limit, and range differently when any of these sensors are used for the analysis. To fulfil such requirements, the material type, design, and pattern and the device mechanical characteristics might be examined. As the deformation is rather substantial, monitoring joint mobility often necessitates a device that is dynamically responsive, stretchy, and reliable, with a wide range of detection. As a result, nonmetallic substances are preferred.

3.3.2.5 Acoustic signal

Stretchable acoustic systems, which use acoustic energy in humans as detecting items, provide an important element towards the advancement of device applications. There are two main forms of elastic acoustical components. One kind, known as microphones or flexible acoustic sensors, detects vibration upon the skin surface triggered by human organs including the throat and the heart. For clinical connections with gastrointestinal functioning, gut sounds are also significant. Yetisen *et al* created a foldable wearable wireless device for real-time observation of intestinal noises in 2018, which delivers the audio signals to a smartphone via

Bluetooth. Also, they demonstrated the use of foldable devices for the collection and categorization of bowel noises generated independently during the digestion process. Empirical trials on a healthy subject with artificial intestinal blockage have proven that the system can record the features of small bowel obstructions. This shows that the devices may be used as a benchmark for future systems in the supplemental diagnostics of intestinal issues [26].

Ultrasonography technologies are the foundation for other categories of elastic acoustical devices. Ultrasonography is indeed a noninvasive, widely accepted secure way to monitor the well-being of individuals. When comparing with physical and biochemical signals, ultrasonography may reach a far deeper level within the body. As a result, an elastic and malleable ultrasonic array for constant centralized measurement of blood pressure was developed. Also, it demonstrated the measurements of blood pressure in several arteries on the neck, arms, wrist, and foot, which is consistent with the results from a commercialized tonometer. This study blends ultrasonic technology with flexible electronics to provide excellent tissue penetration ability for monitoring human tissues deeply.

3.3.3 Biochemical signal monitoring

Chemical constituents inside the human system are significant factors for assessing the health of the body. Many disorders require the collection and examination of biological fluids or blood samples, which could involve greater infection risk and trauma. A common example is 'finger-pricking' in diabetes glucose testing. To track the fluctuations of blood glucose at various points in time, it must be done frequently throughout the day, causing both mental and physical suffering to the individual.

Even though the technology is advanced, there is a need for virtual health care and long-term illness monitoring devices at home. A few more easily available biological secretions, such as saliva, interstitial fluid (ISF), ocular fluid, sweat, or urine, can provide lot of relevant input for noninvasive periodic health care applications. Many of these factors contribute to the advancement of elastic and foldable electronic devices for less invasive or noninvasive real-time consistent monitoring systems.

3.3.3.1 Saliva

Several conformal biochemical-based sensors have been created to noninvasively detect indicators in human bodily fluids in order to get deeper meaningful data on human physiological parameters. A readily accessible fluid that may be studied for wellness and case detection is saliva. It is difficult to identify biomarkers in saliva in real time or continuously because of its complexity behind the tooth structure. Additionally, the wet environment in the mouth and the substantial movements of the mouth make it challenging to integrate devices into mouth when munching and speaking. As a result, the saliva-based monitors used today are typically attached to mouth guards or the teeth [27]. Therefore, the diagnostics value of saliva in recognizing illnesses has been demonstrated through sufficient evidence. Yet constant access to bodily fluids is necessary for ongoing real-time illness monitoring and control.

Several scientists have successfully developed biomaterials based on mouth guards that facilitate the measurement of uric acid levels and salivary lactate. Even though some randomized trials show a strong link between blood glucose and salivary glucose, the structure of saliva is rapidly changed by contaminating foods as well as beverages. Therefore, there are limits to its application when treating patients with diabetes.

3.3.3.2 Tears (ocular fluid)

Tears are desirable body samples for prospective applications in noninvasive and continuous glucose monitoring. But they are difficult to access due to the high correlations between tears and blood concentrations. As they enable continual contact with tear fluids, contact lenses provide a distinctive IoT environment for circuit integration, thereby further minimizing irritation to the eyes and wearer discomfort. The subject's field of vision may be impacted by the installation of highly integrated hardware and software components into such a small region of the soft contact lens framework [28].

3.3.3.3 Sweat

Sweat, which includes tens to hundreds of chemicals that are dispersed across the body and may be accessible continually. It plays a significant role in the measurement of particular electrolyte concentrations. Its biochemical constituents provide a wealth of essential research data about a human's health condition. It is being used to assess ailments such as diabetes, cystic fibrosis [29], immune disorders, and diseases that result from an elevated heavy metal concentration in the blood. The following three things together constitute the overall construction of stretchy and elastic electrochemical devices:

1. The sweat-uptake layer,
2. Synthetically produced sensory electrodes, and
3. Encapsulation

The sweat-uptake framework makes contact with human skin, which absorbs and retains sweat via capillarity. Regarding dermatological perspiration or sweat analysis, a variety of microporous materials, like regenerated polyurethane sponge, cellulose sponge, cellulose paper, polyvinyl alcohol sponge, and sponge silicone, have been investigated as sweat-uptake layers in epidermal sweat sensors. Sensing electrodes that have been chemically changed translate biological signals to electrical impulses, and epidermal surface nanocrystallisation may boost sensitivity. Thirdly, encapsulation provides gentle support and protection while wearing the device.

Tattoo-based electrochemical devices are manufactured using the screen printing technique, which provides continuous monitoring via close interaction with the skin. Based on this approach, to monitor the sweat lactate concentration during exercises, an electrochemical sensing element had been developed. The detector exhibited greater sensitivity for sweat lactate concentration and strong resistance to deformation caused by physical exercise. Subsequently, this manufacturing process was expanded to measure the level of pH, sodium level, alcohol level and glucose level in sweat [30]. Recent development has demonstrated that dermatological sweat-based galactose monitoring system and epidermal medication management can be incorporated on a wearable patch. To increase the electrochemical properties of detectors, highly permeable golden electrodes are created via electro-deposition, enhancing the surface area and reinforcing enzymatic entrapment. A hybrid structure-based electrode is designed to increase the surface area whilst keeping the device's elasticity and high conductivity.

A variety of deeply integrated smart technologies for perspiration analysis have now been fostered to monitor numerous chemical compositions in sweat at once. Such detectors use chemically synthesized electrodes for preferentially tracking magnesium, potassium, sodium, toxic metals, lactates, and hyperglycemia in sweat upon the flexible PET substrate. For important signal processing, treatment, and wireless connectivity, the detecting element is attached to a foldable PCB. Furthermore, to eliminate the impact of pH or temperature change, the incorporated model would include pH and temperature sensors to compensate for the device measurements. As a result, such completely interconnected systems offer wearable, concurrent, and noninvasive perspiration monitoring, with the possibility of improvement on the traditional examination of biochemical constituents in the human body, which is reliant on large and costly devices.

3.3.3.4 Glucose

Due to the prevalence of diabetes and the requirement for long-term and precise glycemic control, noninvasive glucose detection techniques are required. Consequently, several schemes, including electrochemical as well as optical ones, have indeed been explored. Wearable electrochemical glucose sensors are indeed the group that is plagued by the greatest safety concerns. However, some bodily fluids, including tears, sweat, saliva, skin ISF, and others that contain galactose, can be used as the measurement targets.

3.3.4 Challenges and strategies to overcome

3.3.4.1 Accuracy

For reliable and constant observation of physiological parameters in the human body, inherent and external problems with its techniques should be addressed. Low signal-to-noise ratio (SNR) and poor relevancy are inherent obstacles for reliable monitoring. Stretchable interaction can significantly improve the noise ratio of a signal. Moreover, a highly permeable electrode can be employed to improve the SNR throughout the perspiration elemental analysis [31].

The continuous distortion of the human body, external variables like temperature, light, and internal interfaces, as well as human-to-device contact, are critical for accurate long-term signal collection. The inspired answers to these challenges are microstructure improved devices and tattoo-like electrodes that allow closer connections.

3.3.4.2 Compatibility

Another critical aspect of medical technology is compatibility. While collecting signals, it should be physiologically, mechanically, and thermodynamically favourable to the human body. Conformal contact and stretchability are the foundations for mechanically invisible devices to the human. A device that mimics the stress–strain behaviour of the skin surface must be secured and invited to use for long periods of time. To be physiologically compatible, products that function on the skin must be permeable using a microporous substrate. Stretchable resorbable biomaterials and degradable substrates provide implanted electronics with the bioactivity required within the human. Eliminating negative response from internal and external devices is accomplished by encapsulating them in biomaterials such as silicone, PDMS, PI, silk protein, and hydrogel, among others. Furthermore, for unique working environments, such as wounds, the systems need to be bacteria resistant. Silver nanoparticle substrates have been shown to be useful in such circumstances [32].

Stretchy devices, unlike traditional electronics, are frequently enclosed by high thermoconductive materials. Regarding the susceptibility of body cells to global temperature, systems temperature rise is an unavoidable challenge as integration complexity grows, especially for devices incorporating major elements. To address this issue, heat sinks and thermal diffusivity substrates have been developed [33]. Hence, flexibility in electronic devices is not only helping the human health sector but also alters our overall approach to bioscience research.

3.4 Flexible thin film devices in photovoltaic applications

Solar cells will be a suitable candidate for an efficient replacement of nonrenewable energy sources to have a cost-effective renewable and green energy resource. Among the different and various components and devices used for trapping solar energy, flexible thin film–based photovoltaic cell and devices will be of great use because of its low weight, flexibility, stretch ability, twistability, ease of transportation and storage, integration feasibility in curved surfaces, and ease of process ability [34–37]. Flexible photovoltaic cells can be classified into the following groups as perovskite solar cells, dye-sensitized solar cells, organic solar cells, and fiber solar cells. All the categories will be made of an anode, cathode, and a photoactive layer.

3.4.1 Perovskite solar cells

These structures are based on the perovskite layer as the photoactive region. The schematic of a thin-film perovskite solar cell is shown in figure 3.1. When solar energy is incident on the photoactive layer of the cell, charges are generated, and it can be then extracted from the perovskite region.

The perovskite cell absorption layer is very thin and has a high efficiency. The efficiency of these solar cells was reported to be 25.2% [38]. The structure of the photovoltaic cell has its active layer sandwiched between the n and p regions. Chemical formula XYZ_3 is used for the photoactive layer of the structure, where X, Y, and Z may be a metal cation, halide anion, or monovalent cation. A huge number of perovskite layer structures have been developed for different bandgap ranges

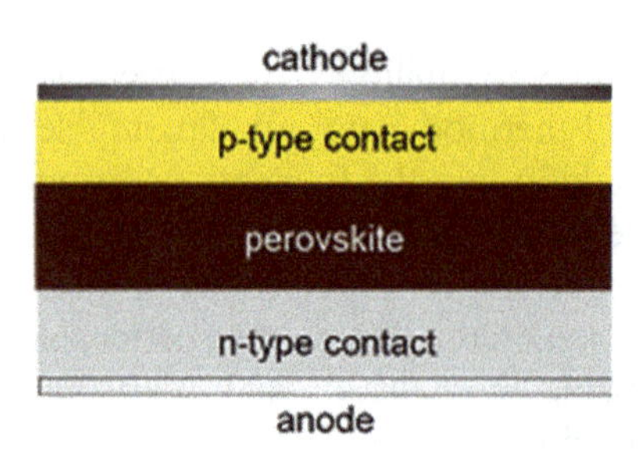

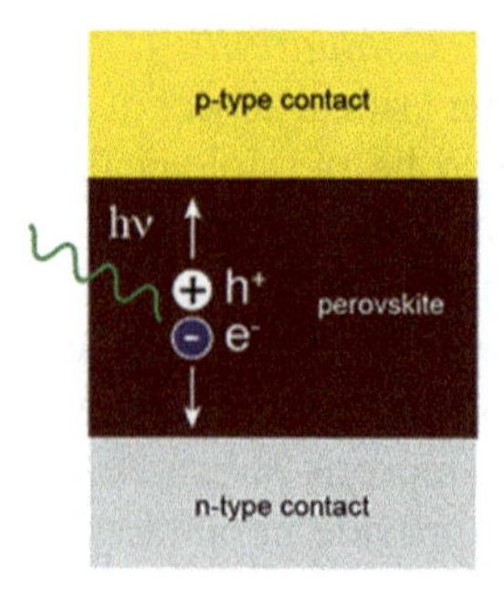

Figure 3.1. (a) Schematic of a thin-film perovskite solar cell. (b) Charge generation and extraction in the perovskite thin film. This perovskite solar cell architecture image has been obtained by the author(s) from the Wikipedia website where it was made available under a CC BY-SA 4.0 licence. It is included within this article on that basis. It is attributed to Sevhab.

Table 3.2. Research and works involved with perovskite photovoltaic cells.

Researchers	Description of the research/work
Feng *et al* [40]	Dimethyl sulfide as the additive in flexible perovskite solar cells (PSCs), with the power conversion efficiency (PCE) of 18.4 % from 16.9 % with flexibility and 1.72 times more resistance against humidity
Huang *et al* [41]	Improved the device to overcome the flaws such as rigidity, which greatly affects the flexibility
Bag *et al* [42]	Semitransparent perovskite photovoltaic (PV) cells with PCE of 12.6%
Dagar *et al* [43]	Cost-effective solar cells with high PCE of 26.9%
Gao *et al* [44]	Improved performance of PSC by imprinting pillar-shaped nanostructures
Ma *et al* [45]	Developed stretchable, bendable, twistable, and compressible PSCs, which are highly resistant against humidity and fully functional even at humidity rate of 78%
Li *et al* [46]	Developed a flexible PV, which is a photorechargeable device
Ryn *et al* [47]	Structured a nanocrystalline metal organic framework, which resulted in the development of PSC with PCE of 18.94% and with excellent mechanical robustness
Wei *et al* [48]	Lead halide PSC with high performance and long-term stability

(1.33–1.55 eV). Flexible perovskite plays a major role in self-powered energy sources, wearable electronics, and integrated photovoltaics in buildings [39]. The solid nature and low-temperature production procedure makes this a suitable candidate for flexible devices that are portable and wearable. One of the main issues associated with this is the defect-free production of thin-film perovskite structure in low processing temperature. The various research in perovskite photovoltaic cells is tabulated in table 3.2.

Enormous research work has been done in perovskite photovoltaic cells and devices so as to have the improved power conversion efficiency and to have good mechanical and other physical properties of the cells. Design and development of photorechargeable devices will mark the revolution in electronics industries for the innovation of self-powered devices that can be used in various other fields.

3.4.2 Dye-sensitized solar cells

O'regan and Grätzel in 1991 [49] reported flexible dye-sensitized solar cells. These are highly economical, have a very simple production technique, and are based on the generation of excitation electron hole pairs. They are highly flexible, thereby making them a suitable candidate for wearable solar cells. The schematic of a dye-sensitized solar cell is shown in figure 3.2. It consists of transparent substrates with conductive oxides. The cathode, electrolyte, and the photoanode or photoelectrode form the three main parts of the solar cell. For the photoanode structure, wide bandgap nanocrystalline metal oxide semiconductors, such as tin (IV) oxide, zinc oxide, and titanium dioxide, are used to coat the transparent conductive glass substrate. This photoanode is very sensitized against the dye molecules. For the absorption of the sunlight and photoelectron generation, Ru-based chemical compounds were used. Generally, platinised conductive glass is used for the cathode to have the catalytic reaction that takes place between liquid electrolyte mediators (such as iodide redox/triiodide) and to generate charges. The disadvantage of using the liquid is that it affects the flexibility of the structure and the stability of the device [50]. The process involved in dye-sensitized solar cells is the absorption of the light and the excitation of the dye molecules generating charges, transportation and the collection of the generated charges, and regeneration of the oxidized dye molecules by the catalytic interaction between transparent conductive oxide substrate and the liquid electrolyte [51].

Some of the notable research in this field includes the work by Xu *et al* [53], where they developed a large-area highly flexible CuS-based flexible dye-sensitized photovoltaic cell. This structure can be produced at low temperature and can be adapted to any substrate. It is highly tolerant to temperature and pH variations. Lee *et al* [54]

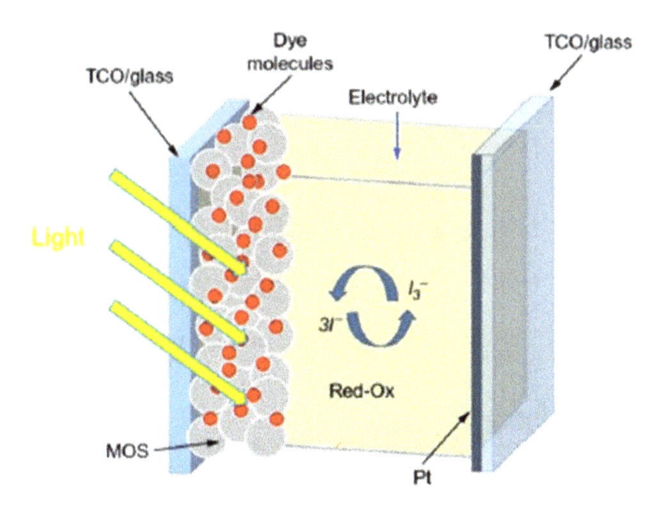

Figure 3.2. Schematic of a dye-sensitized solar cell. Reproduced from [52] with permission from the Royal Society of Chemistry, Copyright (2023).

designed and developed a cost-effective technique for fabrication of solar cells, which improved the PCE to 1.92% from 1.05% along with the improvement in the fill factor. Tao *et al* [55] developed wire-shaped cells, which can improve the overall performance of the structure.

3.4.3 Organic solar cells

Organic solar cells consist of a polymeric molecule as electron donor, and the electron acceptors are the nanomaterials derived from fullerene [56]. It is a heterojunction in structure. The electron donor and electron acceptor are sandwiched between the anode and cathode. The structure has an electron transport layer (ETL) and hole transport layer (HTL), which are played between the electrode and the active layer. The electron hole pairs generated are moved through the ETL and HTL to their respective electrodes. The main disadvantage of organic photovoltaic cells is that they have much lower PCE than any other type of solar cell [57]. A way to improve their PCE has been mentioned: replacing the fullerene-derived nanomaterials by a processed solution of non-fullerene electron acceptor. This showed the improvement of PCE by 12% along with the device stability and its features [58, 59]. Figure 3.3 shows the structure of organic solar cells. Table 3.3 describes the research and works involved with organic flexible solar cells.

3.4.4 Fiber solar cells

Along with the need for flexible electronics, an efficient, reliable flexible electrical source is also required to energise these wearable devices. One of the promising technologies that can pave a way to this is the fiber-based solar cells, which can end up in self-powered structures and devices. A fiber solar cell seems to be a suitable candidate to address the concern about electrical power supply. Researchers have devised one-dimensional solar cells using fiber by the light harvesting capability of the fibers along with high stability and flexibility characteristics [65–67]. These structures can have various applications in woven fabrics, clothes etc [68]. Further advantages of these types of photovoltaic cells are that they

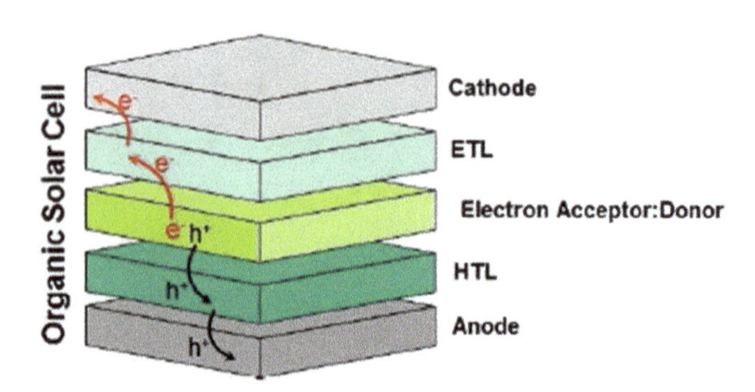

Figure 3.3. Functioning mechanism of organic solar cells. Reproduced from reference [52] with permission from the Royal Society of Chemistry, Copyright (2023).

Table 3.3. Research and works involved with organic flexible solar cells.

Researchers	Description of the research/work
Wu [60]	Description of 'all-polymer solar cells', which have polymer donors and acceptors and have features like greater flexibility, robustness, and stability
Ding *et al* [61]	Fully polymeric organic solar cells with PCE of 6%–7%.
Yang *et al* [62]	Tunable PCE was achieved by varying the organic cathode interlayer. This work emphasised the importance of interlayer interaction in PCE of the cells.
Shin *et al* [63]	Highly flexible and transparent solar cell using transparent conductive graphene electrode with grapheme quantum dot
Hsieh *et al* [64]	Devised skin-attachable organic solar cells based on buckle-on-elastomer. The structure used ultrathin PEN layer with 3M elastomer, which showed a PCE of up to 5.61%.

show excellent flexibility, chemical performance, and mechanical stability and obtain a PCE of about 22.95% [69].

Highly flexible solar cells coupled with a luminescent solar concentrator has been reported by Peng *et al* [70]. This has a titanium wire around which a platinum layer is twisted. It showed a PCE of about 4.03%–22.95%. Dye-sensitized mechanisms play a major role for the wire-shaped solar cells. In order to overcome the cons of corrosion of the metal wire, this mechanism was used. Numerous work has been done by researchers and scientists to develop three-dimensional wire-shaped energy harvesting devices or storage devices [71]. Thus, fiber flexible solar cells have so many applications including the energy sector, smart energy textiles, and self-powered wearable devices, thereby to revolutionizing people's lifestyles.

3.5 Flexible thin film sensors

Flexible sensing devices are divided into four categories based on their historical development: (1) wearable gadgets, (2) e-skins, (3) devices that can be implanted, and (4) sophisticated sensors with enhanced features like transparency, self-healing, and self-power etc. Several manufacturing processes using polymeric materials are presently available to achieve great flexibility in sensing devices, such as solution processing, low-temperature deposition, transfer printing, and micromolding or nanomolding [72].

The area where significant work has been done recently is the use of battery-operated WFS. Sensors made of flexible materials were connected to a person, together with a system with embedded components, to track a parameter and send crucial information back to the monitoring unit for further analysis. Wearable sensors have played an essential part in monitoring a person's physiological characteristics in order to prevent any malfunction in the body [73]. The process of developing wearable health and medical devices is utilised due to its ability to provide multifunctional, ultra-lightweight, ultra-thin, and flexible systems. The challenge for the corporations is to build the systems in such a way that the total fabrication cost of the systems is lessened. One approach to do this is to use low-cost, safe, and biocompatible materials in the design [74].

The use of tattoos on the skin to track biological fluids like perspiration and saliva is another significant feature of the use of WFS. Nanosensors are attached to the mouth and wrist, respectively. By integrating the wireless sensor with a contact lens, these sensors were also utilised to electrochemically monitor glucose from a person's tears. Numerous applications, such as potentiometric and amperometric sensor-based systems, have made extensive use of tattoo-based sensors [1]. A report described electrochemical biosensors that resembled tattoos and showed chemical selectivity in response to lactate concentrations during exercise. One of the biomarkers for tissue oxygenation is lactate, which may be measured during exercise to evaluate physical performance. To measure the forearm's real-time lactate concentration profile, a tattoo lactate sensor was placed there. Lactate oxidase's enzymatic response on lactate allowed for the detection of changes in lactate levels [75].

Similar to this, electrochemical sensing principles are commonly employed in flexible electrodes based on metallic nanoparticles. Gold, silver, and copper nanoparticles are commonly employed in research and development endeavours. In contemporary times, CNTs have played a significant role in the advancement of electrodes for flexible sensors, particularly in the field of biosensing applications. This is due to their curved sidewalls and hydrophobic properties, which facilitate a robust interaction through π-bonding. Another important aspect of the CNT is its radical conductivity change even for the little amount of chemical species adsorption. Due to its high surface area, stability, and greater conducting property, CNTs are successfully used for the purpose of detecting various analytes such as cholesterol, glucose, lactate etc [76].

Recent studies show that organic electrochemical transistors (OECTs) have the potential to serve as a promising platform for biosensor applications. The scope of its application encompasses biomolecules, enzymes, bacteria, viruses, cells, and nucleotide detectors. OECT sensors have been employed for the purpose of evaluating saliva through the conversion of biochemical signals into electrical signals [77]. Flexible sensors may also be used to examine circadian cycles and track changes in brain temperature during mental activity. Then, various piezoresistive and piezoelectric sensor types have been created to this point to track a variety of physiological parameters by wearing them as gloves, bandages etc.

Flexible wearable sensors are also developed for clothing to monitor dangerous gases. For safety reasons, carbon monoxide (CO) and carbon dioxide (CO_2) gas sensors were installed in the clothing or boots of persons like firemen. Similar to this, oxygen (O_2) detecting devices were created and installed on a person's wrist to monitor the constant change in oxygen level occurring in haemoglobin during breathing. At the same time, the sensor patches would need to be larger and more flexible in order to monitor the physiological characteristics of a person, such as limb movements, motions like walking, running etc, and gait analysis. But the sensors would need to be delicate and sensitive in order to detect parameters like respiration, heart rate, and cardio respiratory signals.

The WFS still has certain problems, despite the fact that a lot of work has been done on it. A patient should not experience any pain when wearing the WFS, in the authors' opinion. Regarding the purposes of the monitoring, there should not be any invasion of the patient's privacy. The extensive volume of data produced by the wearable sensor system presents a challenge in terms of management and storage. The vast quantity of monitored data necessitates the implementation of a robust security system to prevent the mishandling of and unauthorised use of the acquired data. The matter of power consumption associated with WFS is a significant concern that requires attention. Ensuring a consistent and uninterrupted power supply to future systems remains a significant challenge that requires attention.

The term 'e-skin' was coined because of its resemblance to human skin and its readiness for usage in a variety of sectors including multitouch sensing in humanoid robots, flexible displays, and prosthetics. Along with advancements in e-skins, skin contact thin electronic gadgets have been developed by integration of many components such as sensors, signal transmitters, light emitting diodes, and electric power producers. Likewise, an implanted device that may be directly linked to an inside organ is being investigated as an alternative clinical method for monitoring electric signals from the interior body, such as epicardial electrogram and electrocorticographic impulses, since a low-invasive and nonpenetrating measuring system with large area protection, excellent performance, and acceptable spatial resolution is needed for in vivo diagnostics and medical devices.

Several unique ideas have been coupled with existing sensing platforms to broaden the domain of flexible electronics. To attain these objectives, the integration of an extra function with different qualities and processes is necessary [72]. To prove this, researchers produced a skin-like and transparent pressure sensor using PDMS elastic sheets of CNTs. CNT solution is coated over an ultraviolet/ozone-treated PDMS sheet using a stencil mask and customized the shape of the CNT bundles [72, 78]. Similarly, the RFID antenna coil, which is in a printed form, is coated with a material that has the ability to absorb moisture. The resonant frequency of the coil is altered due to the variation in capacitance caused by the absorption of moisture.

Capacitive touch sensors and piezoresistive sensors are two types of sensors commonly utilised in modern electronic devices. Capacitive touch sensors are typically located on the top surface of smartphone and tablet displays, while piezoresistive sensors are designed to detect variations in pressure. The integration of these two functionalities can be achieved through the utilisation of a multilayer printed film, resulting in the development of hybrid sensors. These sensors are capable of detecting both proximity and pressure.

3.6 Flexible thin-film displays

The field of PE has attracted growing interest in various applications such as organic light-emitting diodes (OLEDs), neuromorphic devices, capacitive energy storage devices, TFT liquid crystal display (LCD), photodetectors, and others [5]. OLEDs possess the ability to be produced in different geometries and exhibit thin and flexible configurations in contrast to traditional light-emitting diodes. They are extending its applications in visible light communication, lighting, and displays. The characteristics of OLEDs, such as its low weight, extreme thinness, and protective nature of the environment, justify its role in next-generation lighting and display systems. Therefore, polymers are highly suitable material for OLED applications and organic photovoltaic systems rather than glass and plastics [79]. Similarly, white polymer light-emitting devices (PLEDs) are extending its applications in interior lighting sources and flat panel displays. Internal quantum efficiency of high-efficiency PLEDs is nearing 100%, and hence the light extraction out of the device is also possessing maximum efficiency [80].

The attributes of thickness and weight hold significant importance in the context of applications that necessitate expansive area displays, including but not limited to automotive and digital signage. The utilisation of an adaptable substrate in conjunction with organic transistors facilitates the attainment of a reduced overall thickness of the stack, as opposed to the use of glass. An illustration of a low-cost bio-based substrate is triacetyl cellulose, which has a thickness of less than 1/10 of a millimetre. Consequently, the complete thickness of the organic liquid crystal display (OLCD) measures merely 0.3 mm, excluding the backlight [81]. Furthermore, the slender profile and pliability of OLCDs facilitate the creation of borderless screens with curved contours and creased perimeters. This confers significant benefits in use cases such as notebooks and tablets, where a borderless design results in a larger display area without increasing the overall size of the device.

3.7 Conclusion

Even though the flexible electronic devices have been available in the market for a couple of decades, there is tremendous growth in the recent past due to the enormous development of engineered materials, especially polymers. Achievement of bio-compatibility and biodegradability through polymers triggered the new progress in the bio-based systems. The advent of nanomaterials, especially the metals and carbon-based materials, and novel innovations in printing techniques added flavour in developing flexible electronic devices.

In this chapter, current scenarios, advantages, challenges, and opportunities of flexible thin-film electronics technology, which are holding future prospects, are discussed in accordance with the materials, fabrication techniques, and specific field-oriented applications. Performance of some of the devices and problems related to some of the applications are also presented. Since technology is growing every day, expectations of society and needs for industry are increased in all aspects, especially for the applications which require cost reduction. Furthermore, other challenges related to the pilot scale to commercial model are to be addressed.

References

[1] Nag A, Mukhopadhyay S C and Kosel J 2017 Wearable flexible sensors: a review *IEEE Sens. J.* **17** 3949–60

[2] Briand D, Molina-Lopez F, Quintero A V, Ataman C, Courbat J and de Rooij N F 2011 Why going towards plastic and flexible sensors? *Procedia Eng.* **25** 8–15

[3] Wang Y, Sun L, Wang C, Yang F, Ren X, Zhang X, Dong H and Hu W 2019 Organic crystalline materials in flexible electronics *Chem. Soc. Rev.* **48** 1492–530
[4] Rim Y S, Bae S H, Chen H, De Marco N and Yang Y 2016 Recent progress in materials and devices toward printable and flexible sensors *Adv. Mater.* **28** 4415–40
[5] Tong S, Sun J and Yang J 2018 Printed thin-film transistors: research from China *ACS Appl. Mater. Interfaces* **10** 25902–24
[6] Cruz S M F, Rocha L A and Viana J C 2018 Printing technologies on flexible substrates for printed electronics *Flexible Electronics* (London: Intech Open)
[7] Su Y, Du J, Sun D, Liu C and Cheng H 2013 Reduced graphene oxide with a highly restored π-conjugated structure for inkjet printing and its use in all-carbon transistors *Nano Res.* **6** 842–52
[8] Anyaegbunam F N C and Augustine C 2018 A study of optical band gap and associated urbach energy tail of chemically deposited metal oxides binary thin films *Dig. J. Nanomater. Biostructures* **13** 847–56
[9] Sasmal A, Medda S K, Devi P S and Sen S 2020 Nano-ZnO decorated $ZnSnO_3$ as efficient fillers in PVDF matrixes: toward simultaneous enhancement of energy storage density and efficiency and improved energy harvesting activity *Nanoscale* **12** 20908–21
[10] https://idtechex.com/en/research-article/top-five-innovative-printed-flexible-sensor-technologies/23101Dyson M 2021
[11] Zeng W, Shu L, Li Q, Chen S, Wang F and Tao X M 2014 Fiber-based wearable electronics: a review of materials, fabrication, devices, and applications *Adv. Mater.* **26** 5310–36
[12] Guan Z, Hu H, Shen X, Xiang P, Zhong N, Chu J and Duan C 2020 Recent progress in two-dimensional ferroelectric materials *Adv. Electron. Mater.* **6** 1900818
[13] Zhao H, Frijia E M, Vidal Rosas E, Collins-Jones L, Smith G, Nixon-Hill R, Powell S, Everdell N L and Cooper R J 2021 Design and validation of a mechanically flexible and ultra-lightweight high-density diffuse optical tomography system for functional neuroimaging of newborns *Neurophotonics* **8** 15011–1
[14] Cheng I C and Wagner S 2009 Overview of Flexible Electronics Technology ed W S Wong and A Salleo *Flexible Electronics. Electronic Materials: Science & Technology* **vol 11** (Boston, MA: Springer)
[15] Singh P, Joshi S D, Patney R K and Saha K 2016 Fourier-based feature extraction for classification of EEG signals using EEG rhythms *Circuits Syst. Signal Process.* **35** 3700–15
[16] Kuzum D *et al* 2014 Transparent and flexible low noise graphene electrodes for simultaneous electrophysiology and neuroimaging *Nat. Commun.* **5** 5259
[17] Bin Heyat M B, Akhtar F, Abbas S J, Al-Sarem M, Alqarafi A, Stalin A, Abbasi R, Muaad A Y, Lai D and Wu K 2022 Wearable flexible electronics based cardiac electrode for researcher mental stress detection system using machinelearning models on single lead electrocardiogram signal
[18] Chun K Y, Seo S and Han C S 2022 A wearable all-gel multimodal cutaneous sensor enabling simultaneous single-site monitoring of cardiac-related biophysical signals *Adv. Mater.* **34** 2110082
[19] Liu H, Dong W, Li Y, Li F, Geng J, Zhu M, Chen T, Zhang H, Sun L and Lee C 2020 An epidermal sEMG tattoo-like patch as a new human–machine interface for patients with loss of voice *Microsyst. Nanoeng.* **6** 16
[20] Guvanasen G S, Guo L, Aguilar R J, Cheek A L, Shafor C S, Rajaraman S, Nichols T R and DeWeerth S P 2016 A stretchable microneedle electrode array for stimulating and

measuring intramuscular electromyographic activity *IEEE Trans. Neural Syst. Rehabil. Eng.* **25** 1440–52
[21] Gao W, Ota H, Kiriya D, Takei K and Javey A 2019 Flexible electronics toward wearable sensing *Acc. Chem. Res.* **52** 523–33
[22] Nakata S, Arie T, Akita S and Takei K 2017 Wearable, flexible, and multifunctional healthcare device with an ISFETchemical sensor for simultaneous sweat pH and skin temperature monitoring *ACS Sens.* **2** 443–8
[23] Yang G, Pang G, Pang Z, Gu Y, Mäntysalo M and Yang H 2018 Non-invasive flexible and stretchable wearable sensors with nano-based enhancement for chronic disease care *IEEE Rev. Biomed. Eng.* **12** 34–71
[24] Witte M B and Barbul A 1997 General principles of wound healing *Surg. Clin. N. Am.* **77** 509–28
[25] Pierre Claver U and Zhao G 2021 Recent progress in flexible pressure sensors based electronic skin *Adv. Eng. Mater.* **23** 2001187
[26] Yetisen A K, Martinez-Hurtado J L, Ünal B, Khademhosseini A and Butt H 2018 Wearables in medicine *Adv. Mater.* **30** 1706910
[27] Moonla C, Lee D H, Rokaya D, Rasitanon N, Kathayat G, Lee W Y, Kim J and Jeerapan I 2022 Lab-in-a-mouth and advanced point-of-care sensing systems: detecting bioinformation from the oral cavity and saliva *ECS Sens. Plus* **1** 021603
[28] Kim J *et al* 2017 Wearable smart sensor systems integrated on soft contact lenses for wireless ocular diagnostics *Nat. Commun.* **8** 14997.
[29] Emaminejad S *et al* 2017 Autonomous sweat extraction and analysis applied to cystic fibrosis and glucose monitoring using a fully integrated wearable platform *Proc. Natl Acad. Sci.* **114** 4625–30
[30] Kim J, Campbell A S and Wang J 2018 Wearable non-invasive epidermal glucose sensors: a review *Talanta* **177** 163–70
[31] Lee H, Song C, Hong Y S, Kim M, Cho H R, Kang T, Shin K, Choi S H, Hyeon T and Kim D H 2017 Wearable/disposable sweat-based glucose monitoring device with multistage transdermal drug delivery module *Sci. Adv.* **3** e1601314
[32] Chen Y, Lan W, Wang J, Zhu R, Yang Z, Ding D, Tang G, Wang K, Su Q and Xie E 2016 Highly flexible, transparent, conductive and antibacterial films made of spin-coated silver nanowires and a protective ZnO layer *Physica* E **76** 88–94
[33] Zhang Y F, Ren Y J, Guo H C and Bai S L 2019 Enhanced thermal properties of PDMS composites containing vertically aligned graphene tubes *Appl. Therm. Eng.* **150** 840–8
[34] Li Y, Meng L, Yang Y M, Xu G, Hong Z, Chen Q, You J, Li G, Yang Y and Li Y 2016 *Nat. Commun.* **7** 10214
[35] Chen J, Huang Y, Zhang N, Zou H, Liu R, Tao C, Fan X and Wang Z L 2016 *Nat. Energy* **1** 16138
[36] Zhang K, Gao K, Xia R, Wu Z, Sun C, Cao J, Qian L, Li W, Liu S and Huang F 2016 *Adv. Mater.* **28** 4817–23
[37] Jung S, Lee J, Seo J, Kim U, Choi Y and Park H 2018 *Nano Lett.* **18** 1337–43
[38] Fagiolari L and Bella F 2019 *Energy Environ. Sci.* **12** 3427–72
[39] Yoon S, Tak S, Kim J, Jun Y, Kang K and Park J 2011 *Build. Environ.* **46** 1899–904
[40] Feng X, Zhu Z, Yang X, Zhang J, Niu Z, Wang S, Zuo S, Priya S, Liu and Yang D 2018 *Adv. Mater.* **30** 1801418

[41] Huang K, Peng Y, Gao Y, Shi J, Li H, Mo X, Huang H, Gao Y, Ding L and Yang J 2019 *Adv. Energy Mater.* **9** 1901419
[42] Bag S and Durstock M F 2016 *Nano Energy* **30** 542–8
[43] Dagar J, Castro-Hermosa S, Lucarelli G, Cacialli F and Brown T M 2018 *Nano Energy* **49** 290–9
[44] Gao M, Han X, Zhan X, Liu P, Shan Y, Chen Y, Li J, Zhang R, Wang S and Zhang Q 2019 *Mater. Lett.*
[45] Ma M, Tang Q, Chen H, He B and Yang P 2017 *Sol. Energy Mater. Sol. Cells* **160** 67–76
[46] Li C, Cong S, Tian Z, Song Y, Yu L, Lu C, Shao Y, Li J, Zou G and Rümmeli M H 2019 *Nano Energy*
[47] Ryu U, Jee S, Park J-S, Han I K, Lee J H, Park M and Choi K M 2018 *ACS Nano* **12** 4968–75
[48] Wei J, Li H, Zhao Y, Zhou W, Fu R, Leprince-Wang Y, Yu D and Zhao Q 2016 *Nano Energy* **26** 139–47
[49] O'regan B and Grätzel M 1991 *Nature* **353** 737
[50] Roldán-Carmona C, Malinkiewicz O, Soriano A, Espallargas G M, Garcia A, Reinecke P, Kroyer T, Dar M I, Nazeeruddin M K and Bolink H J 2014 *Energy Environ. Sci.* **7** 994–7
[51] Calisir M, Stojanovska E and Kilic A 2018 *Polymer-based Nanocomposites for Energy and Environmental Applications* (Amsterdam: Elsevier) pp 361–96
[52] Hashemi S A, Ramakrishna S and Aberle A G 2020 Recent progress in flexible–wearable solar cells for self-powered electronic devices *Energy Environ. Sci.* **13** 685–743
[53] Xu Z, Li T, Liu Q, Zhang F, Hong X, Xie S, Lin C, Liu X and Guo W 2018 *Sol. Energy Mater. Sol. Cells* **179** 297–304
[54] Lee J-W, Choi J O, Jeong J-E, Yang S, Ahn S H, Kwon K-W and Lee C S 2013 *Electrochim. Acta* **103** 252–8
[55] Tao P, Guo W, Du J, Tao C, Qing S and Fan X 2016 *J. Colloid Interface Sci.* **478** 172–80
[56] Mishra A and Bäuerle P 2012 *Angew. Chem. Int. Ed.* **51** 2020–67
[57] Li Y, Xu G, Cui C and Li Y 2018 *Adv. Energy Mater.* **8** 1701791
[58] Mateker W R, Douglas J D, Cabanetos C, Sachs-Quintana I, Bartelt J A, Hoke E T, El Labban A, Beaujuge P M, Fréchet J M and McGehee M D 2013 *Energy Environ. Sci.* **6** 2529–37
[59] Kong J, Song S, Yoo M, Lee G Y, Kwon O, Park J K, Back H, Kim G, Lee S H and Suh H 2014 *Nat. Commun.* **5** 5688
[60] Wu S 1987 *Polymer* **28** 1144–8
[61] Ding G, Yuan J, Jin F, Zhang Y, Han L, Ling X, Zhao H and Ma W 2017 *Nano Energy* **36** 356–65
[62] Yang Y, Ou J, Lv X, Meng C and Mai Y 2019 *Sol. Energy* **180** 57–62
[63] Shin D H, Seo S W, Kim J M, Lee H S and Choi S-H 2018 *J. Alloys Compd.* **744** 1–6
[64] Hsieh Y-T, Chen J-Y, Shih C-C, Chueh C-C and Chen W-C 2018 *Org. Electron.* **53** 339–45
[65] Song W, Wang H, Liu G, Peng M and Zou D 2016 *Nano Energy* **19** 1–7
[66] Peng M, Yan K, Hu H, Shen D, Song W and Zou D 2015 *J. Mater. Chem.* C*3 2157–65*
[67] Liu B, Liu B, Wang X, Chen D, Fan Z and Shen G 2014 *Nano Energy* **10** 99–107
[68] Peng M, Dong B and Zou D 2018 *J. Energy Chem.* **27** 611–21
[69] Fan X, Zhang X, Zhang N, Cheng L, Du J and Tao C 2015 *Electrochim. Acta* **161** 358–63
[70] Peng M, Yu X, Cai X, Yang Q, Hu H, Yan K, Wang H, Dong B, Zhu F and Zou D 2014 *Nano Energy* **10** 117–24

[71] Lee M R, Eckert R D, Forberich K, Dennler G, Brabec C J and Gaudiana R A 2009 *Science* **324** 232–5

[72] Pang C, Lee C and Suh K Y 2013 Recent advances in flexible sensors for wearable and implantable devices *J. Appl. Polym. Sci.* **130** 1429–41

[73] Liu Y, Wang H, Zhao W, Zhang M, Qin H and Xie Y 2018 Flexible, stretchable sensors for wearable health monitoring: sensing mechanisms, materials, fabrication strategies and features *Sensors* **18** 645

[74] Heng W, Solomon S and Gao W 2022 Flexible electronics and devices as human–machine interfaces for medical robotics *Adv. Mater.* **34** 2107902

[75] Jia W, Bandodkar A J, Valdés-Ramírez G, Windmiller J R, Yang Z, Ramírez J, Chan G and Wang J 2013 Electrochemical tattoo biosensors for real-time noninvasive lactate monitoring in human perspiration *Anal. Chem.* **85** 6553–60

[76] Alagappan M, Immanuel S, Sivasubramanian R and Kandaswamy A 2020 Development of cholesterol biosensor using Au nanoparticles decorated f-MWCNT covered with polypyrrole network *Arab. J. Chem.* **13** 2001–10

[77] Sophocleous M, Contat-Rodrigo L, García-Breijo E and Georgiou J 2020 Organic electrochemical transistors as an emerging platform for bio-sensing applications: a review *IEEE Sens. J.* **21** 3977–4006

[78] Lipomi D J, Vosgueritchian M, Tee B C, Hellstrom S L, Lee J A, Fox C H and Bao Z 2011 Skin-like pressure and strain sensors based on transparent elastic films of carbon nanotubes *Nat. Nanotechnol.* **6** 788–92

[79] Sundar D S, Sivanantharaja A, Sanjeeviraja C and Jeyakumar D 2016 Synthesis and characterization of transparent and flexible polymer clay substrate for OLEDs *Mater. Today Proc.* **3** 2409–12

[80] Shanmuga Sundar D and Sivanantharaja A 2013 High efficient plastic substrate polymer white light emitting diode *Opt. Quantum Electron.* **45** 79–85

[81] Gong X, Lyu G Q, Wang Z and Feng Q B 2023 Investigating the effect of gravity on the optical performance of large-sized liquid crystal displays *Displays* **77** 102406

Chapter 4

A comparative study of carbon-based nanoribbons and MoS_2-based nanoribbons for spintronics-based devices

Sandeep Kumar and Surender Pratap

The two most studied quasi-one-dimensional materials are graphene and MoS_2 nanoribbons. In this chapter, a comparative study of the electronic properties of carbon-based and MoS_2-based nanoribbons is provided. Depending on the edge geometry, the presence of spin–orbit interaction predicts spin polarization in the nanoribbons of both types of materials. The study of these spin-polarized states is important for their performance in spintronics-based devices. Therefore, we investigate the effect of spin–orbit coupling on carbon-based and MoS_2-based nanoribbons. Moreover, we have also incorporated the effect of strain on the electronic properties of both types of nanoribbons. We discuss only the zigzag and armchair nanoribbons for both carbon and MoS_2.

4.1 Introduction

The discovery of a two-dimensional (2D) carbon allotrope known as graphene has opened up new avenues in the fundamental science and technology of 2D materials due to their facile fabrication, remarkable charge transport, and optical, thermal, and mechanical properties. 2D materials have been the subject of substantial research for use in next-generation nanodevices. The majority of the study focused on the electronic properties of graphene, despite the fact that it offers a whole new range of peculiar features to explore. Graphene's dimensionality and extremely unique electronic dispersion relation, in which electrons mimic relativistic particles, are the source of many of its exceptional properties. Because of this, electrons in graphene are commonly referred to as massless Dirac fermions, which can be thought of as electrons having zero rest mass (although being fundamental particles, electrons have characteristic mass). Therefore, graphene is an ideal material to explore relativistic effects in condensed matter physics because of the unusual

doi:10.1088/978-0-7503-5492-9ch4

behavior of electrons [1]. The quantum Hall effect, anomalous quantum Hall effect, Klein paradox, ballistic electron propagation, metal-free magnetism, breakdown of the adiabatic Born-Oppenheimer approximation, possibility of high T_c superconductivity, and observation of relativistic phenomena such as zitterbewegung, or jittery motion of a wave function under the influence of confining potentials, are the most remarkable consequences of the unusual behavior of electrons in graphene [1, 2]. Furthermore, graphene has few electronic states close to the Fermi level because the valence and conduction bands touch at two non-equivalent points (Dirac points). Therefore, graphene is usually known as a zero-gap semiconductor or a semimetal. The band structure of graphene is highly sensitive to any change, such as external electric fields, doping, mechanical deformations, and edge functionalization, which is indispensable for application in sensors [2, 3].

In graphene, each carbon atom has one unhybridized, half-filled p-orbital perpendicular to the graphene plane and forms a sp^2 hybridized network with its three nearest neighbors, each at a distance of 1.42 Å. Graphene is a bipartite lattice because the rhombus unit cell is made up of two atoms from two distinct sublattices. The electronic spins in a bipartite lattice prefer to align antiferromagnetically within the two sublattice points, according to Lieb's theorem [4, 5]. It has been observed that the lower limit of the spin coherence length of electrons injected from ferromagnetic (FM) electrodes into carbon nanotubes is 130 nm, which provides insight into the spintronics applications of carbon-based materials. At room temperature, graphene also has a long spin relaxation time and length for spin-injected electrons. The spin lifetime may be longer than the microsecond regime, according to theoretical predictions based on the weak hyperfine interaction and low spin–orbit coupling (SOC), which theoretically makes it a promising candidate for graphene-based room-temperature spintronic devices. Unfortunately, this has not been seen in experiments. Non-local spin valve measurements at room temperature and low temperature have been used to experimentally obtain upper limits of 1–2 ns [5]. The lack of reproducibility of experimental findings has made carbon-based magnetism a long-standing and contentious field of study. Graphene edge states, however, are a crucial signpost that allows a thorough understanding of carbon-based magnetism. A number of findings suggest that spin polarization occurs near the edges of FM materials [5]. It has been observed that the ground state of graphene is close to an interaction-driven phase transition into an insulating antiferromagnetic (AFM) state. The AFM correlations prohibit the magnetic moments from ordering ferromagnetically, even if they do form, and rule out the possibility of finding hysteresis in conventional magnetic measurements. In addition to this, the Mermin-Wagner theorem states that perfect ordering is impossible in 2D magnetic systems (like the magnetic graphene edges) at any finite temperature [6]. Recently, the experimental realization of tunable edge states in twisted bilayer graphene nanoribbon junctions was by Wang *et al* [7].

Most recent theoretical ideas for graphene-based spintronics applications are rooted in the magnetic properties of nanoribbons. Two different edge geometries, namely zigzag and armchair, are produced when graphene is finitely terminated, resulting in quasi-one-dimensional ribbon-like structures. The edges of nanoribbons intrinsically have dangling bonds. These determine the ribbon properties by producing eigenstates close to Fermi energy. The unique properties of graphene nanoribbons depend on edge structure; therefore, engineering the edge will have an impact on the electronic and magnetic properties of the nanoribbons. It has been observed that close to Fermi energy, pristine zigzag graphene nanoribbons (ZGNRs) form edge states, whereas armchair graphene nanoribbons (AGNRs) do not form edge states. In narrower ZGNRs (<7 nm), the spin polarization at two edges results in band gap opening (0.2–0.3 eV) and switching ZGNRs from metal to semiconductor. The electronic edge states are ferromagnetically ordered but anti-ferromagnetically coupled to each other. Moreover, under the effect of an external electric field, theoretical investigations show that the spin polarization of the edge states makes ZGNR half metallic and stable at room temperature. These findings open the application of ZGNR for room-temperature spintronic devices [8–10].

Graphene exhibits lots of exceptional properties, but it lacks the electronic band gap, which limits its application in field-effect transistors (FETs) due to the low on/off switching ratio. Band gap engineering of graphene is done using a variety of techniques. Nevertheless, these techniques increase complexity and limit the mobility of the charge carrier. The scientific community has therefore concentrated on other single- or few-layered 2D materials [54]. Transition metal dichalcogenides (TMDs) are unique in this regard because of exhibiting band gaps both in bulk and monolayer. Moreover, 2D TMDs are stable in a variety of circumstances and are easily exfoliated. It has been observed that several TMDs possess sizable electronic band gaps in the range of 1–2 eV; therefore, these are suitable for electronic, optoelectronic, and spintronic applications [11, 51]. TMDs are layered materials with the formula MX_2, where M is a transition metal element from group IV (Ti, Zr, Hf, and so on), group V (for example, V, Nb, or Ta), or group VI (Mo, W, and so on), and X is a chalcogen (S, Se, or Te). The crystal structure of these materials consists of hexagonally packed, weakly linked sandwich layers made up of two X layers around an M-atom layer. In layered TMDs, van der Waals interaction, which differs in stacking orders and metal atom coordination, weakly holds adjacent layers together to create the bulk crystal in a range of polytypes (hexagonal (2H), tetragonal (1T), and rhombohedral (3R)). Among various TMDs, MoS_2 is gaining interest in device applications due to its adequate band gap, outstanding flexibility, thermal stability, strong carrier mobility, and compatibility with silicon complementary metal oxide semiconductors [12]. The bulk MoS_2 exhibits an indirect band gap of 1.3 eV. When bulk MoS_2 is thinned to a monolayer, an indirect-to-direct band gap transition takes place. MoS_2 monolayer is a direct band gap semiconductor that has a band gap of 1.8 eV and can be easily exfoliated using cellophane tape or lithium-based intercalation. The high current on/off ratio, carrier mobility up to 200 $cm^2\ V^{-1}\ s^{-1}$, and good sub-threshold swing make this material suitable for fabricating more complex electronic devices with low power consumption [13]. Hence, researchers have been motivated to investigate MoS_2's unexplored features by the fascinating prospect of possible nanoelectronic applications that may benefit from the novel physical properties [12].

Recently, quasi-one-dimensional MoS_2 nanoribbons were experimentally synthesized using an electrochemical approach. The encapsulation of MoS_2 nanoribbons inside carbon nanotubes was reported as a breakthrough in the fabrication of ultranarrow MoS_2-based nanoribbons in 2010 [14]. Theoretical interest in the physical and chemical properties of 2D MoS_2 nanoribbons was sparked by the experimental successes in order to understand the sources of the observed electrical, optical, mechanical, and magnetic properties and help guide the design of MoS_2-based nanodevices. Like graphene nanoribbons, MoS_2-based zigzag nanoribbons (MZNRs) are also FM and metallic, i.e., they have edge states near Fermi energy. In contrast to MZNRs, most of the studies show that MoS_2-based armchair nanoribbons (MANRs) are nonmagnetic (NM) and semiconducting, i.e., they do not have edge states. However, an exception to this reports the metallic MANRs to have hydrogen passivation–dependent magnetic moments [15]. In MZGRs, the metallic edges may have current leakage, which consequently severely impairs normal device function. However, most of the electronic and optical properties of MZNRs are highly sensitive to the modification of these edge states. Moreover, by adjusting the edge structures terminating with hydrogen (H) atoms, their conductivity can be either half metallic or semiconducting (n- or p-type) [14]. H-termination increases the magnetic states of MZNRs, which are substantially stronger than those of ZGNRs. Therefore, based on the theoretical and experimental investigations, MoS_2-based nanoribbons with versatile functions may have applications in nanodevices, energy harvesting, and spintronics [14, 16, 17].

In the literature, several types of nanoribbons are reported depending on their chirality and edge structures [3]. The edge stability and edge-dependent properties of the nanoribbons are still in infancy. In this chapter, starting from the need for nanoribbons and their synthesis techniques, we provide a comparative study on the electronic properties of carbon- and MoS_2-based nanoribbons, including the effect of intrinsic and Rashba spin–orbit interactions. It has been realized that the presence of SOC gives rise to spin-polarized states, which are important for future spintronics-based devices. In addition to this, the effect of strain on the electronic properties of both types of nanoribbons is also explained.

4.2 Why do we need nanoribbons?

The astounding advancement of high-performance computers has been aided over the past 50 years by the incorporation of microelectronic components [18]. By following Moore's law, the transistors in integrated circuits have reduced the size of features to nanometers [18]. As a result, short-channel effects and the limitations of lithography pose obstacles to further device miniaturization [19]. Numerous low-dimensional materials have been put out in recent years in the hopes of increasing computer performance or degree of freedom. The electronic characteristics of low-dimensional materials are primarily determined by size, shape, electronic correlations, and applied boundary conditions [19]. In this regard, nanoribbons hold great promise for future applications in nanodevices due to various structural and electronic properties. Moreover, nanoribbons offer changes in properties by fine-tuning the precursor design, which offers tunability across a wide range [20].

Nanoribbons are quasi-one-dimensional strips with lengths exceeding the width by at least one order of magnitude. Nanoribbons must have a width of less than 100 nm in order to use the prefix 'nano.' Nanoribbons made up of graphene generally have widths of less than 50 nm [21, 22]. As discussed in the introduction, there is no band gap in the case of graphene, which is very much needed as far as standard electronic appliances are concerned [21, 23]. Graphene nanoribbons show sizeable band gaps, high charge carrier mobility, high current-carrying capacity, and versatile electrical properties. These properties are highly influenced by the orientations and edge structures of the nanoribbons. Due to these unique features, graphene nanoribbons are attractive candidates for future electronics applications, such as nanoscale FETs, spintronics, and quantum information processing. Graphene nanoribbons are commonly categorized into three categories depending on their edge structures: (1) zigzag, (2) armchair, and (3) chiral graphene nanoribbons (see figure 4.1). The physical properties of graphene nanoribbons strongly depend on the edge boundaries. It has been observed that ZGNRs (without spin polarization) remain metallic, whereas AGNRs are metallic in addition to depending upon the width of the nanoribbons [24–26].

Like graphene nanoribbons, MoS_2 nanoribbons also show unusual electronic and magnetic properties and are studied theoretically as well as experimentally [14, 27]. MoS_2 nanoribbons also exhibit edge structure–dependent electronic and magnetic properties. MZNRs are metallic and more stable than MANRs. Moreover, unlike graphene nanoribbons, the electronic properties of MANRs weakly depend on width. It has been reported that beyond a certain width of MANRs, the band gap of MANRs saturates to a value of 0.56 eV [28]. The application of an external electric field and strain also affect the physical properties of MoS_2-based nanoribbons.

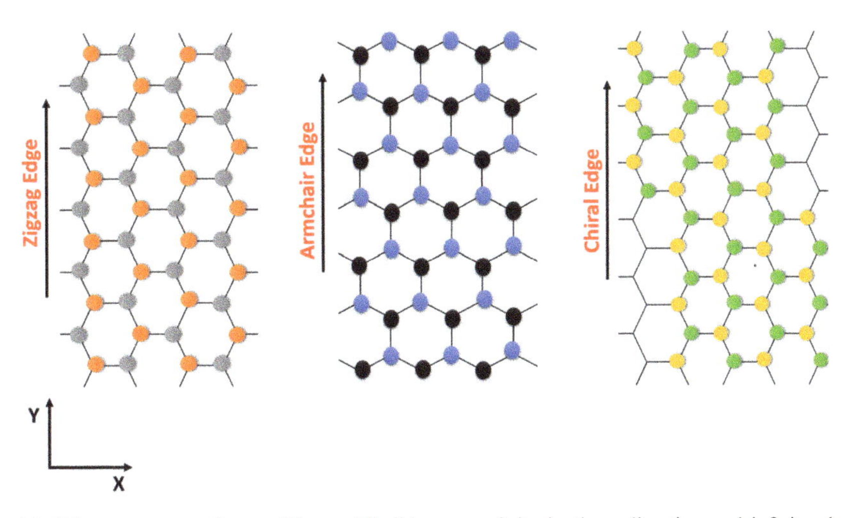

Figure 4.1. Edge structures of nanoribbons. All ribbons are finite in the x-direction and infinite along the y-direction.

The band gap of MANR is discovered to be substantially reduced by an applied external transverse electric field, and beyond a specific critical field, a metal-insulator transition occurs [29]. Theories and experiments show that depending upon the edge structures and widths, both types of nanoribbons (carbon- and MoS_2-based) can have magnetic metallic or NM semiconducting ground states. Recent experimental research has shown that it is possible to successfully synthesize nanoribbons with tunable structural and electronic properties [29]. Therefore, the synthesis and improved understanding of nanoribbons have been the focus of recent efforts.

4.3 Recent synthesis techniques for nanoribbons

The advancement of experimental methods has enabled the synthesis of nanoribbons at the nanoscale in various ways. The way a nanoribbon is synthesized has a significant impact on both the atomic structure and the associated characteristic features of the nanoribbon. Top-down and bottom-up are two basic experimental approaches for the synthesis of nanoribbons. A top-down approach is an extension of the approach used to produce micron-sized particles, in which the bulk material is broken down into nanoscale structures or particles. On the other hand, the large desired entity is created using the bottom-up strategy by assembling the little construction blocks. In other words, top-down synthesis uses physical methods, whereas chemical methods are involved in the bottom-up fabrication of nanomaterials [3].

In top-down synthesis, high-quality graphene nanoribbons can be made in various ways, including lithographic patterning, unzipping carbon nanotubes, and cutting with catalysts. In the lithographic patterning method, graphene nanoribbons are fabricated from epitaxial graphene grown on silicon carbide or graphene deposited chemically on a metal surface. This technique is not adapted for large-scale production of graphene nanoribbons because it is quite difficult to control the edge structure with this approach. In polymer films, the plasma etching of nanotubes produces nanoribbons from multi-walled carbon nanotubes. In this approach, graphene nanoribbons with regulated widths and high-quality edge structures may be produced by carefully arranging and aligning the carbon nanotubes. It has been found that carbon nanotubes produce sub-10 nm graphene nanoribbons with band gaps suitable for room-temperature transistor applications. Moreover, for ribbons 10–20 nm in width, the obtained graphene nanoribbons have the highest electrical conductance and mobility ever recorded (up to $5e^2\ h^{-1}$ and 1.500 $cm^2\ Vs^{-1}$, respectively). Another method for making graphene nanoribbons is to use a catalytic reaction to break apart C–C bonds. Graphene nanoribbons are cut from graphene in this procedure by depositing a particle of the catalytic metal. Using single-layer graphene, Campos *et al* achieved the production of sub-10 nm–wide nanoribbons using thermally activated nickel (Ni) particles [30]. A multi-step organic synthesis based on the cyclization of previously synthesized polymer chains is required for the bottom-up fabrication of graphene nanoribbons. With the help of this technique, incredibly narrow ribbons with atomically exact edge arrangements can be produced. A schematic of the synthesis of semiconducting graphene nanoribbons via the

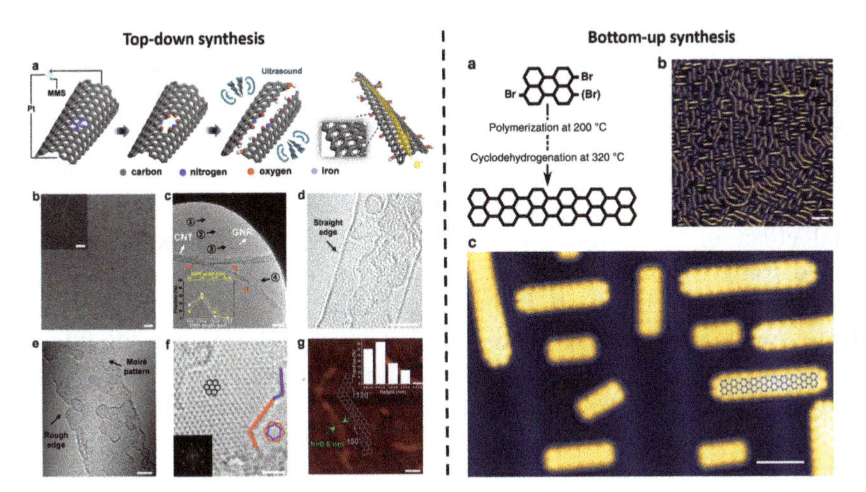

Figure 4.2. Top-down synthesis: (a) A two-step unzipping method for the synthesis of graphene nanoribbons. (b) Unzipped graphene nanoribbons with tilted SEM pictures. (c)–(f) Graphene nanoribbon transmission electron microscopy (TEM) pictures with aberration correction. (c) Monolayer graphene nanoribbon strands with a long, fully unzipped length and residual n-doped carbon nanotubes. (d) Crystalline graphene nanoribbons with a smooth edge. (e) Low edge roughness crystal lattice, less than 2 nm. (f) Fast Fourier analysis of the graphene nanoribbon hexagonal lattice using a magnified TEM. (g) Atomic force microscopy measurement of a monolayer graphene nanoribbon with a 0.6 nm height. Adapted with permission from [31]. Copyright (2019) American Chemical Society. Bottom-up synthesis: (a) Schematic showing the polymerization of the dibromoperylene $C_{20}H_{10}Br_2$ (DBP) precursor to atomically well-defined $N = 5$ AGNRs. (b) After cyclodehydrogenation at 320 °C, an overall STM image of ribbons. (c) Zoomed-in STM topography of various ribbon lengths. Reprinted with permission from [33]. Copyright (2015) Springer Nature Limited.

two-step unzipping method is shown in figure 4.2 (left panel) [31]. Using the bottom-up method, there are two basic methods for fabricating graphene nanoribbons: on-surface synthesis and solution synthesis. There are several benefits to producing graphene nanoribbons using solution synthesis, including efficient fabrication at the gram scale and longer graphene nanoribbons with excellent dispersibility in organic solvents. However, this technique can be efficient in producing very narrow (1–12 nm) and functionalizing graphene nanoribbons, but the deposition of these pure solution-synthesized nanoribbons on the substrate is very difficult [32]. On the other hand, ultra-thin graphene nanoribbon films can be produced by on-surface synthesis, and these films can then be examined using scanning tunneling microscopy (STM) and atomic force microscopy. The on-surface synthesis of graphene nanoribbons can be easily transferred to the substrate for device integration. By using the bottom-up approach, Kimouche *et al* reported the fabrication of ultrathin AGNR with five atoms ($N = 5$) in its width (see figure 4.2 (right panel)). Recently, nanoribbons showing topological quantum states were synthesized using this technique [3].

MOS_2-based nanoribbons are also synthesized experimentally. Commonly, electron irradiation and carbon nanotube–based encapsulation techniques are used for the synthesis of MoS_2-based nanoribbons. A two-step electrochemical/chemical process was used to achieve the first successful synthesis of narrow MoS_2-based nanoribbons [34]. In this process, 50–800 nm wide and 3–100 nm thick

MoS_2-based nanoribbons are produced at temperatures above 800 °C (see figure 4.3(A)). Using optical absorption measurements, it has been realized that these nanoribbons are direct band gap semiconductors with a band gap opening of 1.95 eV. MoS_2-based nanoribbons having widths between 50 nm and 800 nm correspond to the same electronic band gap. This shows that the band gap of MoS_2-based nanoribbons is saturated up to a certain value, and there will be a very minimal effect of the widths on the band gap. The MoS_2-based nanoribbons fabricated using electron irradiation, which uses high-energy electrons, exhibit a uniform width of 0.35 nm. It is found that fabricated nanoribbons are semiconductors with an electronic band gap of 0.77 eV and a Young modulus of 300 GPa and can tolerate 9% tensile strain without fracture. Even though it is very challenging to tune the geometries and properties of nanoribbons below 10 nm width, it is noted that the utilization of electron irradiation in this study is crucial for such goals [28].

Since controlling the dimension of nanomaterials is a challenging issue, mixed and stable low-dimensional MoS_2-based nanoribbons encapsulated in carbon nanotubes are synthesized to control nanoribbon dimensions. The carbon nanotubes–encapsulated MoS_2-based nanoribbons are uniform, with widths lying between 1 and 4 nm, and are known as nanoburritos [14]. Recently, Li *et al* experimentally prepared the one-dimensional MoS_2-based nanoribbons using chemical vapor deposition techniques to study photodetection performance [35]. The smooth ribbon-like structures are observed using scanning electron microscopy (SEM) and atomic force microscopy (see figures 4.3(B)(a) and (b)). The obtained MoS_2-based nanoribbon shows a width of about 600 nm and a height of about 180 nm, which is confirmed by the cross-sectional

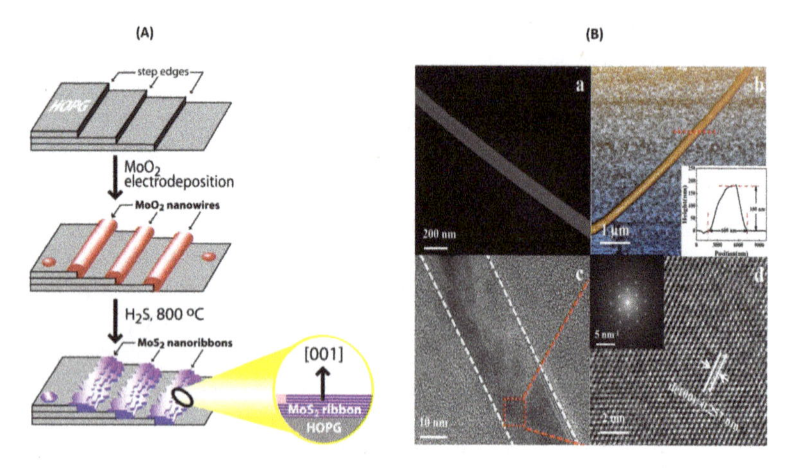

Figure 4.3. (A) The schematic for the electrochemical/chemical process for synthesizing polycrystalline MoS_2 nanowires and nanoribbons on graphite surfaces. Adapted with permission from [34]. Copyright (2004) American Chemical Society. (B) Images of MoS_2-based nanoribbons synthesized by the chemical vapor deposition method. (a) SEM image of the as-grown MoS_2 nanoribbon. (b) MoS_2 nanoribbon image in enhanced color using atomic force microscopy. (c) Low- and (d) high-resolution transmission electron microscopy images of MoS_2 nanoribbons (fast Fourier transform image is shown in the inset). Adapted with permission from [35]. Copyright (2021) American Chemical Society.

atomic force microscopy height profile as shown in the inset of figure 4.3(B)(b). Moreover, the low- and high-magnification transmission electron microscopy images show that MoS_2-based nanoribbons have a lattice space of 0.257 nm (see figures 4.3(B)(c) and (d)), and the high-quality single crystal MoS_2-based is confirmed using a fast Fourier transform (see inset figure 4.3(B)(d)) [35]. There are various experimental investigations that provide information on the physical behavior of nanoribbons. Despite this, there is a large number of theoretical investigations that forecast the physical properties of these nanoribbons, which are crucial for nanodevice applications.

4.4 Carbon-based nanoribbons and MoS_2-based nanoribbons

As aforementioned, the electronic properties of one-dimensional materials are highly tunable with edge termination. Theoretically, analytical approaches such as the tight-binding approach (TBA) and computational approaches such as density functional theory (DFT) are employed to study the edge-dependent physical properties of nanoribbons. In this section, we shall focus on the theoretically calculated electronic properties of carbon- and MoS_2-based nanoribbons, which are necessary for future spintronics-based nanodevices.

4.4.1 Carbon-based nanoribbons

It is very well known that graphene nanoribbons show energy band gaps due to the confinement of charge carriers in one dimension. Using TBA, it has been investigated that ZGNRs always show metallic behavior, whereas the semiconducting or metallic behavior of AGNRs depends on their widths. In contrast to TBA, DFT calculations show that all types of narrow graphene nanoribbons (zigzag or armchair) have a finite energy band gap [3]. The electronic energy band gaps in AGNRs arise due to quantum confinement, whereas staggered sublattice potential produces the energy band gap in ZGNRs due to edge magnetization. In graphene nanoribbons, especially in ZGNRs, the correlation between the states present at the edges is responsible for the magnetization. Therefore, electronic correlations play a very significant role in governing the electronic properties of these nanoribbons. In these systems, the correlation effects can be simulated by Hubbard Hamiltonian [4],

$$H = \sum_{i,\sigma} \epsilon_i b_{i,\sigma}^{\dagger} b_{i,\sigma} + \sum_{\langle i,j\rangle,\sigma} t_{ij}\left(b_{i,\sigma}^{\dagger} b_{j,\sigma} + \text{H. c.}\right) + U\sum_{i} n_{i\uparrow} n_{i\downarrow}, \tag{4.1}$$

where ϵ_i and t_{ij} are the onsite energy and the nearest neighbor hopping integral, respectively. $b_{i,\sigma}^{\dagger}(b_{i,\sigma})$ annihilates (creates) an electron at ith site with spin σ. In the literature, the magnitude of t_{ij} varies from 2.4 to 3 eV to reproduce the experimental observations [2]. U is the on-site Coulomb interaction term. $n_{i\uparrow}(n_{i\downarrow})$ is the number operator. The value of U can be estimated as the difference between the first ionization energy and the electron affinity for carbon, which comes out to be 9.66 eV. The exact estimation of U has not been provided until now, although U/t varies from 1.7 to 5 eV [4].

In the absence of spin-polarization ($U = 0$), i.e., when spins are not considered, both TBA and DFT calculations show that ZGNRs are NM with metallic and doubly degenerate flat bands at the Fermi energy (see figure 4.4(d)) [2, 36]. The flat band corresponds to the states available at the edges of ZGNR and extends over 1/3 of the one-dimensional Brillouin zone from $-\pi/a$ to $-\pi/a$ (with a being the lattice constant of graphene with a value of 2.46 Å). The states present at these flat bands give rise to a sharp peak in density of states (DOS) around Fermi energy (see figure 4.4(a)). In the presence of Coulomb interactions, no zero-energy states appear in the finite-length ZGNRs. As a result, Coulomb interaction causes instability and raises the possibility of edge magnetism by producing a finite energy band gap at Fermi energy. However, with the increase in the length of ZGNRs, zero energy states appear when the length approaches infinity. The mean-field Hubbard model and spin-polarized DFT calculations have proven the ordering of these magnetic edge states. It is found that the ground state of ZGNRs shows AFM coupling between the magnetic edge states available at zigzag edges, and the correlation between these magnetic edge states results in a band gap opening (see figure 4.4(e)). Therefore, the ZGNRs are small band gap magnetic semiconductors with polarized electron spin reside at the opposite zigzag edges. Moreover, the AFM coupling makes the ZGNR semiconducting without breaking the spin degeneracy, resulting in net zero magnetic moments. The preservation of spin degeneracy and band gap opening of ZGNR is also confirmed by the DOS (see figure 4.4(b)). It has been observed that the spin degeneracy of the AFM ground state of ZGNR can be removed by applying external electric and magnetic fields. For instance, in the presence of an external magnetic field, AFM ground states of ZGNR can be converted into metastable FM states (see figure 4.4(f)). In contrast to the AFM ground state, the crossing in bands (at $k = 2\pi/3a$) structure makes the FM-ZGNRs metallic by breaking the spin degeneracy. Figure 4.4(c) shows the spin-up and spin-down DOS of FM-ZGNRs. Theoretically, it is also observed that FM-ZGNRs have

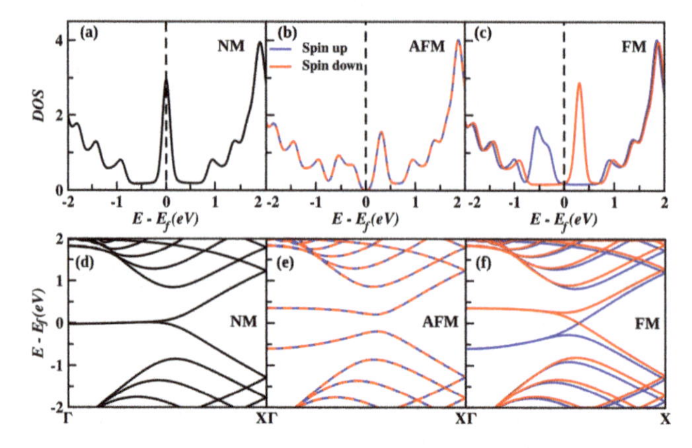

Figure 4.4. For H-terminated 12-AGNR, (a)–(c) DOS and (d)–(f) electronic band structures with NM, AFM, and FM states. Spin-up and spin-down states are denoted by blue and red colors, respectively. Adapted with permission from [36]. Copyright (2022) American Chemical Society.

a magnetic moment of 0.58 μ_B. However, bulk bands of NM, AFM, and FM cases of ZGNRs are negligibly affected by the magnetic effects [36].

It is important to note that although the ground state of AFM ZGNRs has spin-polarized carbon atoms (see figure 4.5(c)), both band structure and DOS plots suggest spin degenerate energy levels in AFM state, similar to that of NM-AGNRs (see figures 4.5(a) and (b)). Therefore, the presence of antiferromagnetically coupled localized edge states, which are only discovered for ZGNR, is the most obvious distinction between AGNR and ZGNR. It has been suggested that the occurrence of edge magnetic ordering at the ground state of ZGNR could serve as the foundation for new spintronics devices. However, the instability of the one-dimensional spin ordering at finite temperature limits the length of this magnetic ordering, or the spin correlation length [37]. It has been observed that the spin correlation length in a clean ZGNR is inversely related to temperature and its value is around 1 nm at ambient temperature. This temperature-related restriction can be overcome by using edge-terminated heavy metals or magnetic anisotropy resulting from substrate effects. It is, therefore, possible to produce spin correlation length for spintronic applications at a scale of many tens of nanometers at a fixed temperature [37].

In the presence of an external electric field applied in the direction of the cross-ribbon width, half-metallicity, a special characteristic of periodic ZGNRs with AFM ground states has been expected to appear. Half-metallic materials, unlike ordinary metals or semiconductors, have a zero band gap for electrons with one spin orientation while maintaining a semiconducting or insulating state for the other spin channel (see figure 4.6). This class of materials offers perfect control over the spin polarization of current with improved efficiency for magnetic memory storage due to their special band property, which only permits one kind of spin to flow when affected by an electric field. The intriguing magnetic properties of graphene nano-structures in spintronics were first discussed by Son *et al* [8]. It is observed that periodic ZGNRs become half-metallic over a specific magnitude of the electric field in the direction of the cross-ribbon width (figure 4.7(a)). Moreover, in the absence of

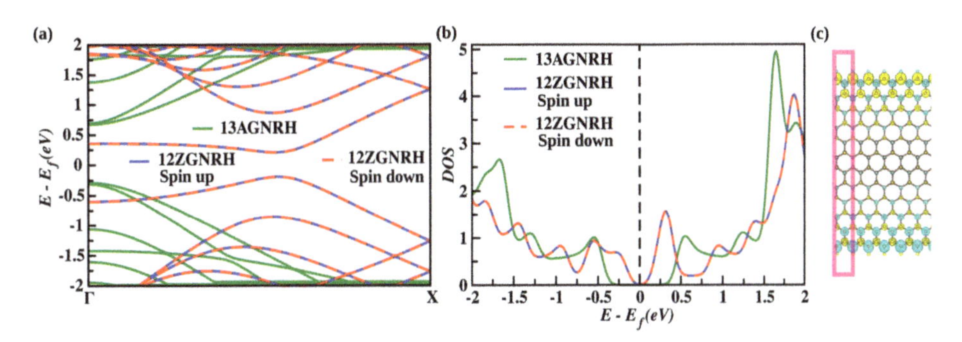

Figure 4.5. (a) Band structure of H-terminated pristine 13-AGNRs (13 dimer lines in width) with NM and H-terminated 12-ZGNRs (12 zigzag lines in width) with AFM state. (b) The corresponding DOS with Fermi level is set a zero energy (dotted black line). (c) Spin-up (cyan) and spin-down (yellow) densities of 12-ZGNR at isosurface value 0.01 μ_BÅ^3. Adapted with permission from [36]. Copyright (2022) American Chemical Society.

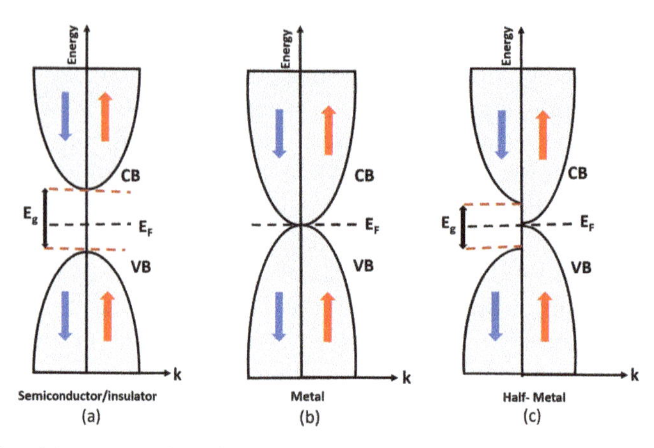

Figure 4.6. A pictorial representation of spin-polarized electronic band structure of (a) semiconductor/insulator, (b) metal, and (c) half-metal. The valence band (VB) and conduction band (CB) are separated by the Fermi energy (E_F), which is in between them. E_g denotes energy band gap. A direct band gap property of parabolic dispersion has been employed to streamline the characterization. Red and blue arrows denote the spin-up and spin-down electron states, respectively.

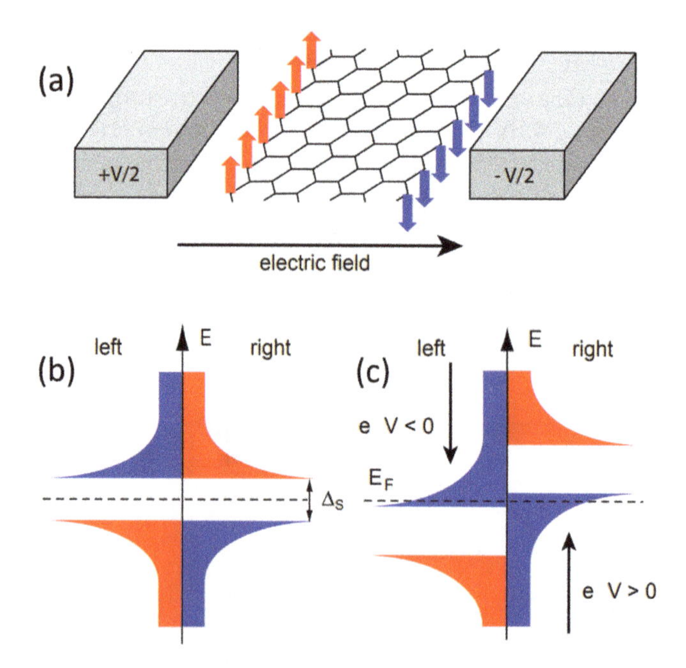

Figure 4.7. Schematic of half-metallicity in ZGNR caused by an applied electric field with a critical value of 3.0 V/width. (a) Schematic of the applied electric field across the nanoribbon from the left edge to the right edge. The red and blue arrows at the edges denote the spin-up and spin-down states, respectively. (b) Diagrammatic illustration of the spin-resolved local density at the opposite edges in the absence of an applied electric field. (c) Symmetry breaking of different spin states under the effect of an applied electric field. Reprinted with permission from [5]. Copyright (2014) Elsevier.

an electric field, the system exhibits an energy gap (Δ_s) due to the correlation effects of spin-polarized states at the opposite edges (see figure 4.7(b)). In the presence of an electric field, the degeneracy of energy levels for both spins is destroyed. The electric field has opposing effects on two distinct spin bands as a result of the localization of two opposite spins on either edge. With the gradual increase in the electric field, the gap for one kind of spin gradually increases while the other spin gap decreases. Therefore, an increase in the applied electric field closes the gap for one spin direction by breaking the spin symmetry (see figure 4.7(c)). Moreover, the direction of the applied electric field determines the spin channel with metallic conductivity [36].

In order to experimentally fabricate a spin-filtered graphene nanoribbon device, the transport properties of ZGNRs are theoretically predicted by combining the non-equilibrium Green's function theory with DFT [36, 38]. The interaction between a bulk system and a finite device is modeled using the non-equilibrium Green's function theory and a many-body concept called self-energy. The device Green's function can be written as [36]

$$G(E) = [ES - H - \Sigma_L - \Sigma_R]^{-1}, \tag{4.2}$$

where H and S are the Hamiltonian and overlap matrix of the device. The coupling of the left (right) lead to the scattering region is represented by $\Sigma_{L(R)}$. The transmission (T^σ) through the device can be calculated as [36]

$$T^\sigma(E) = \mathrm{Tr}(\Gamma_R^\sigma G^\sigma \Gamma_L^\sigma G^{\sigma\dagger}), \tag{4.3}$$

where $G^\sigma(G^{\dagger\sigma})$ retarded (advanced) Green's function with spin $\sigma = (\uparrow, \downarrow)$. $\Gamma_L(\Gamma_R)$ represents the level broadening due to coupling between the scattering region and the left (right) electrodes. Under a zero bias condition, the spin-filtered efficiency (η_{SFE}) can be calculated as

$$\eta_{SFE}(\%) = \frac{|T^\uparrow(E_f) - T^\downarrow(E_f)|}{|T^\uparrow(E_f) + T^\downarrow(E_f)|} \times 100. \tag{4.4}$$

The spin-filtered current through the device under the effect of bias voltage (V_b) can be defined with the help of Landauer's Büttiker approach as

$$I^\sigma(V_b) = \frac{e}{h} \int_{-\infty}^{\infty} dE\left[f(E - \mu_L) - f(E - \mu_R)\right] T^\sigma(E, V_b), \tag{4.5}$$

where f and $\mu_{L(R)}$ represent the Fermi function and electrochemical potential of the left (right) leads, respectively, such that $\mu_L - \mu_R = V_b$. Therefore, the total current through the device becomes

$$I = I^\uparrow + I^\downarrow. \tag{4.6}$$

Hence, under the effect of V_b, spin filtered efficiency can be obtained as

$$\eta_{SFE}(\%) = \frac{|I^\uparrow - I^\downarrow|}{|I^\uparrow + I^\downarrow|} \times 100. \tag{4.7}$$

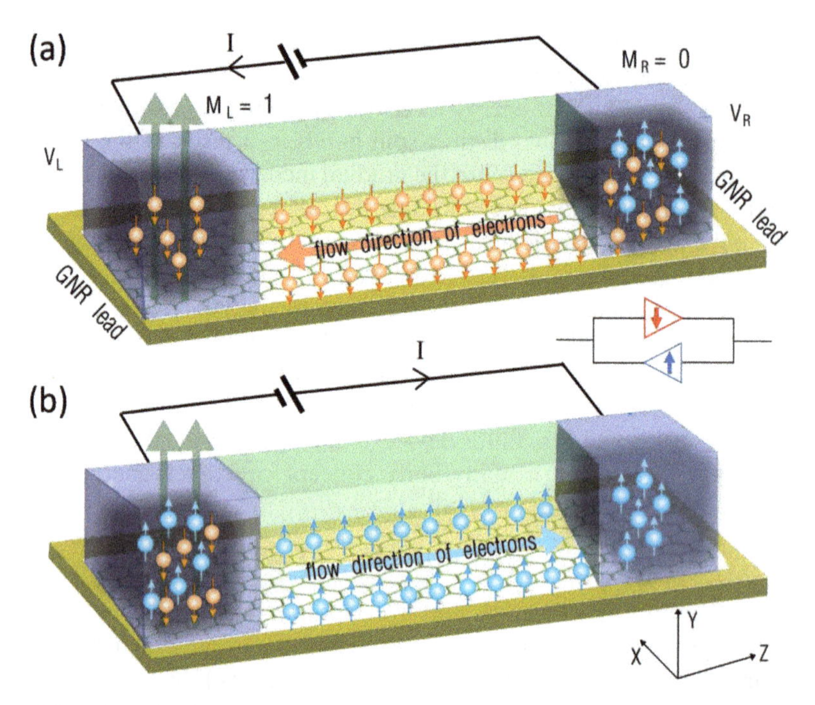

Figure 4.8. ZGNR-based bipolar spin diodes with magnetized leads. (a) Transport of spin-down current under positive bias and (b) transport of spin-up current under negative bias. Reprinted with permission from [38]. Copyright (2011) American Physical Society.

Therefore, spin-polarized current in ZGNR-based two-terminal devices can be tuned by varying the source to drain voltage (V_b), and the device functions as a bipolar spin diode. A symmetry selection rule makes this possible. Zeng *et al* proposed a theoretically ZGNR-based spin diode (see figures 4.8(a) and (b)), in which spin-dependent transport can be controlled by controlling the source to drain bias voltage or the magnetic configuration of the electrodes [38]. It was proposed that controls of the spin-polarized current in ZGNR-based spin diodes make it possible to use them in future spintronic devices [38].

4.4.2 MoS_2-based nanoribbons

Like graphene nanoribbons, MoS_2-based nanoribbons are also obtained by cutting the MoS_2 monolayer nanosheet. These nanoribbons also exhibit zigzag and armchair edge structures depending on the direction of edge termination. However, unlike graphene nanoribbons, the zigzag and armchair edge structures of these nanoribbons are further subdivided depending on the symmetries of the S- and Mo-atom layers. MANRs are divided into two categories: (i) symmetrical and (ii) asymmetrical. Similarly, MZNRs are categorized into six types by considering atoms at the edges and symmetry: one of the two edges can be ended with Mo atoms and the other by S atoms, or both of the two edges can be ended with Mo or S atoms [17]. Regardless of chirality and hydrogen termination, symmetrical nanoribbons are often more stable than their asymmetrical counterparts because of their relatively lower edge energy. The S atoms–terminated MZNRs are the most stable and should be easily formed in experiments even without hydrogen termination, according to the calculated edge energies and relaxed edge structures, while hydrogen termination should be required for the generation of other nanoribbons [17]. Similar to graphene nanoribbons, the magnetic properties of MoS_2-based nanoribbons can be quantified by the magnitude of the energy difference between the magnetic and NM states. In H-saturated or without H-saturated MANRs, the magnetic and NM states are equal and result in net zero magnetic moments for these structures. In addition to this, MANRs can possess net magnetic moments due to vacancy defects and the adsorption of foreign atoms. The presence of a MoS_2 triple vacancy in MANR results in a net magnetic moment of 2 μ_B. To use these materials in various applications, it is crucial to tune their physical properties using external fields. The application of the transverse electric field significantly decreases the energy band gap and results in a metal-insulator transition beyond some critical value of the electric field (see figures 4.9(a)–(c)). Moreover, Stoner ferromagnetism that can be controlled and even extinguished by an external electric field emerges when the Fermi level density at the vicinity of the metal-insulator transition is sufficiently high. The possibility of driving the system into magnetic instability is suggested by the presence of localized edge states that can be shifted to the Fermi level. Spin polarization calculations show that at a critical value of the electric field, a NM to magnetic transition occurs in MANRs. Moreover, when ribbon width increases, the value of the critical electric field decreases (see figure 4.9(c)).

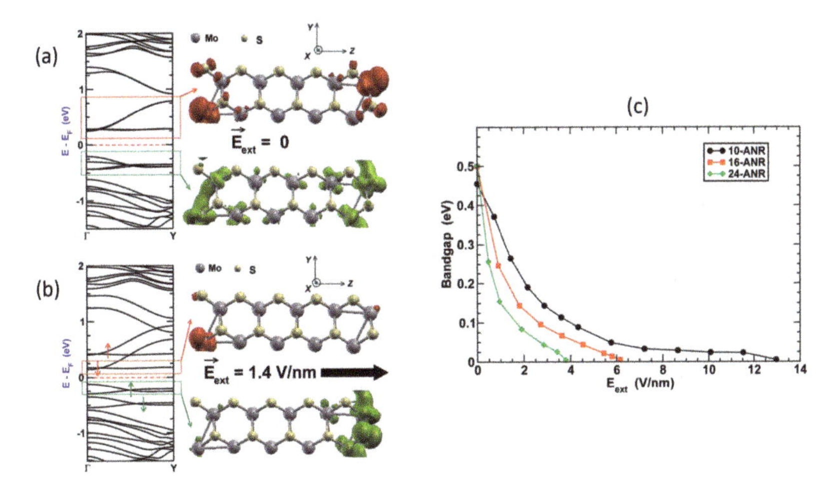

Figure 4.9. Electronic structure of MANRs in the presence and absence of external electric field (E_{ext}). (a) Band structure (left) and local density of states of the valence band (top) and conduction band (bottom) in the absence of an electric field. (b) Band structure and local density of states under the influence of electric field with strength 1.4 V nm^{-1}. The red and green rectangle above and below the Fermi energy represents the range of the local density of states. (c) The variation of the energy band gap of different-widths MANRs with the external field strength. Adapted with permission from [29]. Copyright (2012) American Chemical Society.

In contrast to MANRs, MZNRs are magnetic in nature because the magnetic states in these ribbons have lower energies than those of NM states, irrespective of the H-termination. The H-termination at the edges of MZNRs may improve the transition temperature and stabilize the magnetic moments by enhancing the magnitude of the energy difference between the NM and magnetic states of these ribbons. It is also investigated that, without spin polarization, both TBA and DFT calculations show metallic behavior of MZNR but with different numbers of metallic bands (see figures 4.10(a) and (b) (left panel)). The magnetism in MZNR arises due to the splitting of metallic edge states, and the study of these states is crucial for the proper description of the magnetic properties of these ribbons.

As described above, the ZGNRs show localized electron states at the edges, with FM coupling at each edge and AFM coupling between the edges. The magnetic moment calculations and the energy differences show that MZNRs are FM in nature. TBA and DFT calculations show that Stoner instability is caused by an increase in the DOS and finite Coulomb repulsion, which divides the partially occupied S atom band into a fully filled spin-down band and an almost empty spin-up band (see figures 4.10(a) and (b) (right panel)) [39]. The model proposed by Pan *et al* shows that in MZNRs, the spin orientations of Mo and S atoms at one edge are anti-parallel, and the magnetic moments of S atoms are greater than those of Mo atoms, leading to the FM coupling at this edge and parallel coupling between the edges [17]. Additionally, it is noted that although the magnetic moments and spin orientations at the edges can be changed by the H-termination, the FM coupling at each edge and between the edges maintains its stability. To sum up, irrespective of the H-termination at edges and structures, MZNRs are FM and more suitable for

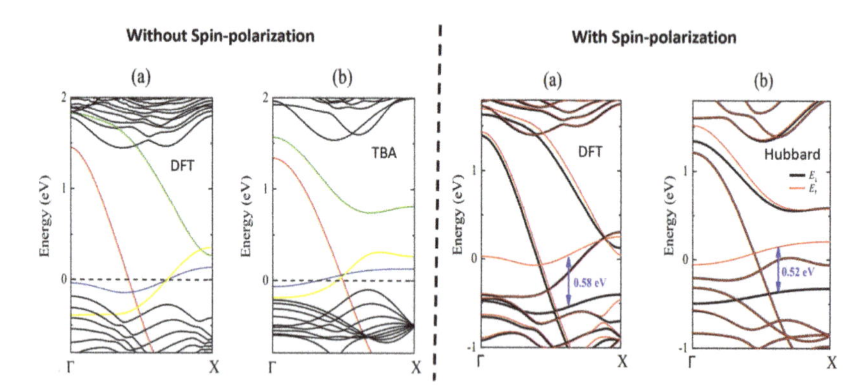

Figure 4.10. Without spin-polarization: band structure calculations of MZANR (a) by using DFT and (b) by using TBA with modified parameters for the edge atoms. When compared to narrower MZNR, midgap states (denoted by different colors) are the same, suggesting the negligible interaction between the Mo and S edge atoms. With spin-polarization: spin-polarized band structure calculation for narrower MZNR (a) by using DFT and (b) by using the Hubbard model with $U_S = 1.7$ eV and $U_{Mo} = 0.6$ eV. Here, U_S (U_{Mo}) represents the amplitude of the Hubbard interaction term for S (Mo) orbitals. Up and down spins are represented by red and black colors, respectively. The Fermi level is set at zero energy. Localized magnetic moments are formed along the S edge as a result of magnetization, which results from the splitting of the S atom's band. Reprinted with permission from [39]. Copyright (2019) American Physical Society.

spintronics-based nanodevices than ZGNRs because ZGNRs require an external electric or magnetic field to exhibit ferromagnetism. Moreover, the magnetic states in FM MZNRs are stronger than ZGNRs. The magnetic properties of carbon- and MoS_2-based nanoribbons show that control on the magnetism of these ribbons provides direct control on the electron spin. However, spin degrees of freedom of electrons can be controlled indirectly by controlling the SOC strength. SOC provides the ability to control the electron spin without any external field. Therefore, in designing the spintronic devices, the tunability of SOC in both types of nanoribbons becomes essential.

4.5 Impact of SOC on electronic properties of nanoribbons

SOC is a relativistic phenomenon. During the motion of an electron in a solid, it may encounter an electric field due to internal (from lattice) and/or external (from bias voltage) agencies. In the electron's rest frame of reference, this electric field manifests as an effective magnetic field. The momentum of the electron is proportional to this effective magnetic field. Because of this, any coupling between the effective magnetic field and the electron spin is equivalent to the coupling between the electron's momentum and spin, which is known as SOC. SOC offers an indirect means of controlling the electron spin. The potential to process, transport, and control data using an electron's spin rather than its charge has opened up fascinating new scientific and technological horizons.

In the rest frame of reference, an effective magnetic field ($\mathbf{B}_{\text{eff}}$) is experienced by the electron when it moves with a momentum ($\mathbf{p}$). This effective magnetic field is given by

$$\mathbf{B}_{\text{eff}} = -\frac{\mathbf{p}}{mc^2} \times \mathbf{E}, \tag{4.8}$$

with m and c being the mass and speed of light, respectively. In the non-relativistic limit, where the electron's rest frame analogue of the Lorentz force, the effective magnetic field gives rise to an energy term:

$$H \propto \mu_{\text{B}}\boldsymbol{\sigma}.\left(-\frac{\mathbf{p}}{mc^2} \times \mathbf{E}\right), \tag{4.9}$$

where μ_{B} and σ denote the Bohr magneton and Pauli spin matrices, respectively.

The SOC can be symmetry-dependent or symmetry-independent. Almost all crystal structures show symmetry-independent SOC because it arises due to spin–orbit interaction in atomic orbitals. This type of spin–orbit interaction is known as intrinsic SOC. In contrast to symmetry-independent SOC, symmetry-dependent spin–orbit interactions arise due to the spin–orbit fields. The inversion symmetry breaking is necessary for symmetry-dependent SOC because the spin–orbit field is odd in momentum. Depending on how the inversion symmetry breaks, the symmetry-dependent SOC can be divided into two categories: (i) Dresselhaus SOC (bulk induces symmetry breaks) and (ii) Rashba SOC (RSOC) (surface or interface induces symmetry breaks). If the electric field in equation (4.9) is externally applied or arises due to the system itself (from the presence of substrate), RSOC comes into the picture. However, in the absence of any inversion center, Dresselhaus SOC comes into play, i.e., caused by the breaking of bulk-induced symmetry. The impact of symmetry-dependent and symmetry-independent has been studied through spin transport measurements. However, in low-dimension materials such as graphene, the presence of symmetry-independent SOC identified by Kane and Mele has been shown to play an important role in the possibility of observing topologically nontrivial phases [40]. In the literature, the effect of intrinsic and RSOC on carbon- and MoS_2-based nanoribbons is widely studied for their application in spintronics.

4.5.1 Intrinsic spin–orbit interaction

The ground-breaking research of Kane and Mele sparked interest in novel quantum phases of matter as well as the SOC effect, which, despite being small in graphene, contributes to significant physics [40]. The quantum spin Hall (QSH) phase in particular has received a lot of attention. It is observed that the intrinsic SOC opens an energy gap and converts 2D semi-metallic graphene to an insulator with a quantized spin Hall effect. Moreover, geometric curvature and the presence of impurity atoms can enhance the strength of intrinsic SOC. The effect of spin–orbit interactions on monolayer as well as bilayer graphene is well known [2]. In graphene, the intrinsic SOC corresponds to an effective coupling of second order, while in buckled materials such as silicene, germanene, and stanene, it arises due to coupling of first order. In graphene, the intrinsic SOC is mostly contributed by the p_z orbitals of the carbon atoms, and it is very weak (approximately 10^{-6} eV). In second quantized form, the intrinsic spin–orbit interaction in the graphene system can be given as [40]

$$H_{\text{ISO}} = i\lambda_{\text{ISO}} \sum_{\langle\langle i,j\rangle\rangle,\alpha\beta} \nu_{ij}\sigma^{z}_{\alpha,\beta} b^{\dagger}_{i,\alpha} b_{j,\beta} + \text{H. c.}, \tag{4.10}$$

where sum is performed over all the second nearest neighbors denoted by $\langle\langle i, j\rangle\rangle$. $\nu_{ij} = \pm 1$, which connects the i and j sites via a common site k and are -1 (clockwise) or $+1$ (anticlockwise). $\sigma^{z}_{\alpha\beta}$ represents the corresponding Pauli spin matrices. Kane and Mele showed that in large-width ZGNRs, a new state of matter is associated with the QSH effect (QSHE). It is found that the presence of intrinsic SOC makes ZGNR exhibit insulating bulk states and conducting edge states [40]. Moreover, the odd Z_2 index for graphene manifests the spin-filtered edge states for the graphene nanoribbons, making these ribbons promising for spintronic applications. In the Kane–Mele study, the p_z orbitals are considered. However, by considering the s, p_x, p_y, and p_z orbitals of carbon atoms, Rhim *et al* showed that the QSHE in ZGNR does not occur because the σ edge bands located at Fermi energy lift up the energy of the spin-filtered chiral edge states at the zone boundary by warping the π-edge bands [41]. The system shows the QSHE by an increase in the carrier density within a certain range, and QSHE disappears when carrier density increases up to a certain value. This new topological insulator behavior in ZGNR arises due to the breaking of the particle-hole symmetry. Interestingly, ZGNRs recover the standard features suggested by Kane and Mele by H-passivation at the edges of nanoribbons. For AGNRs, it has been investigated that QSHE is mostly quite stable with or without H-passivation. The effect of SOC in ZGNRs (even number of zigzag chains in width) and AGNRs creates an energy band gap [41]. The results are confirmed using TBA and DFT calculations; however, conduction and valence bands remain degenerate at the Brillouin zone boundary. The effects of intrinsic SOC create an energy gap in ZGNRs with an even number of zigzag chains in their width and in metallic AGNRs. In semiconducting AGNRs, the effect of intrinsic SOC reduces the energy band gap and causes a band gap in ZGNRs with an odd number of zigzag chains of varying widths. To sum up, edge orientations of graphene nanoribbons and widths affect the strength of intrinsic SOC, and in narrower graphene nanoribbons, intrinsic SOC has stronger effects for AGNRs than ZGNRs [3]. The experimental realization of the topologically non-trivial insulating bands possessed by the Kane–Mele is still a challenge because of the very small magnitude of the intrinsic spin–orbit gap in graphene.

The Mo-heavy element causes a significant spin–orbit interaction in the pure MoS_2 monolayer. The valence bands have separated at K points of the Brillouin zone because the clean MoS_2 monolayer lacks inversion symmetry. In the magnetic field-free MoS_2 monolayer, the time-reversal symmetry results in bands with opposing spins at the K and $-K$ points. Therefore, pristine MoS_2 is favorable for its application in spintronics. Recently, Salami *et al* studied the behavior of n-type MoS_2-based nanoribbons in the presence of intrinsic and extrinsic SOC effects [42]. In the absence of any spin–orbit interactions, both MZNR and MANR exhibit n-type semiconductor behavior, which possesses a bandgap of about 1.56 and 1.04 eV, respectively (see figures 4.11(A)(b) and (B)(b)). In the presence of intrinsic SOC, both types of nanoribbons (MZNR and MANR) lift the band degeneracy, and spin-split sub-bands are produced (see figures 4.11(A)(c) and (B)(c)). This band splitting exists due to the simultaneous existence of both time-reversal and spatial inversion symmetries. The presence of intrinsic SOC also leads to the broadening of DOS to some extent due to lifting in spin degeneracy (see figure 4.11) [42].

4.5.2 Rashba spin–orbit interaction

In graphene, the extrinsic RSOC arises due to the mirror symmetry breaking of the graphene plane and can be very large compared to intrinsic SOC for graphene grown on a substrate. Experimentally, in pristine graphene, the value of RSOC strength has been reported up to 200 meV. The strength of the RSOC in graphene nanoribbons can be tuned by the external electric field applied perpendicular to the plane of nanoribbons. Therefore, the electric field is a good candidate to control the spins in graphene nanoribbons via Rashba spin–orbit interactions. In second quantized form, the Hamiltonian that describes the effect of RSOC can be described as [2]

$$H_{\mathrm{RSOC}} = i \sum_{\langle i,j \rangle, \alpha\beta} b_{i,\alpha}^{\dagger}(\boldsymbol{\sigma}.\ u_{ij})_{\alpha\beta} b_{j,\beta} + \mathrm{H.\ c.}\ , \tag{4.11}$$

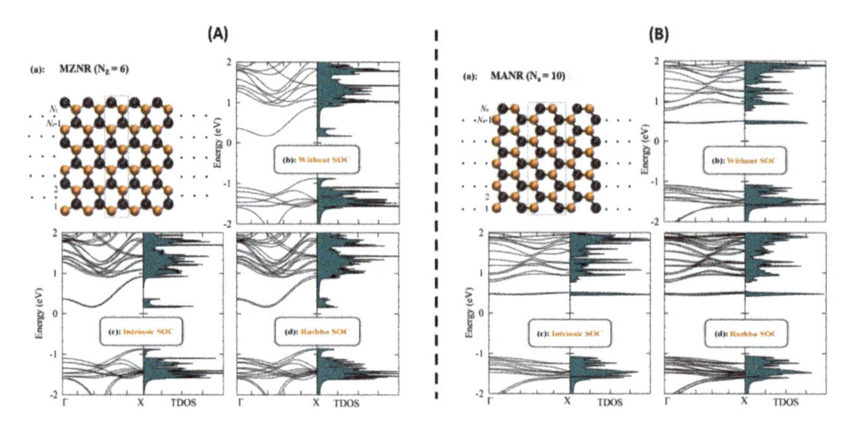

Figure 4.11. (A) Band structure and total density of states of MZNR ($N_z = 6$). (a) Schematic representation of MZNR with width = 1.457 nm. The total density of state (b) in the absence of any type of spin–orbit interactions, (c) in the presence of intrinsic SOC, and (d) in the presence of RSOC ($\lambda_R = 0.1$). (B) Band structure and total density of states of MANR ($N_a = 10$). (a) Schematic representation of MANR with width = 1.420 nm. The total density of state (b) in the absence of any type of spin–orbit interactions, (c) in the presence of intrinsic SOC, and (d) in the presence of RSOC ($\lambda_R = 0.1$). Reprinted with permission from [42]. Copyright (2017) Elsevier.

where $\boldsymbol{\sigma}$ is the vector of Pauli matrices, and

$$u_{ij} = \frac{e}{2me^2av_f}\mathbf{E} \times \delta_{ij} = -\frac{\lambda_R}{a}k \times \boldsymbol{\delta}_{ij}, \tag{4.12}$$

where λR is the spin–orbit parameter, v_f is the Fermi velocity, **E** is the electric field applied perpendicular to the plane of the system, and $\boldsymbol{\delta}_{ij}$ are the position vectors corresponding to the first nearest neighbors.

A number of studies have been done on the transport properties of graphene nanoribbons with RSOC. In the case of ZGNRs, spin-polarized states with high edge localization are produced by the RSOC. The spatial spin distribution of these states is highly dependent on the state under investigation and exhibits opposite polarization at opposite edges. As a result of the conservation of time-reversal symmetry under the RSOC interaction, the ribbon's net spin polarization remains zero in the absence of external fields. Both for AGNRs and ZGNRs, the RSOC modifies the conductance's step-like nature. Yet, compared to nanoribbons with zigzag edges, conductance is significantly altered in nanoribbons with an armchair edge. Around the zigzag edges of the ZGNRs, the RSOC generates strongly localized spin-polarized edge states [3]. In addition to this, the spin conductance of graphene nanoribbons depends on the width and length of these ribbons in the presence of large RSOC. Also, the presence of a spatially varying non-uniform magnetic field produces a large value of RSOC (≈10 meV) that results in helical modes existing in AGNRs that exhibit nearly perfect spin polarization and are robust against boundary defects. These results imply that in specific geometries, graphene nanoribbons could be employed as a spin-flip operator [3, 5].

In MoS_2 nanoribbons, RSOC caused by the application of a vertical electric field breaks the spin degeneracy of the MoS_2 monolayer at the Γ point. It is found that in response to a perpendicular electric field, the presence of edge states in MoS_2-based nanoribbons can induce spin-resolved currents. A vertical electric field can be used to control these spin currents. A numerical study of RSOC on both types of MoS_2-based nanoribbons demonstrates the spin splitting of energy bands (see figure 4.11(A)(d) and (B)(d)). It is observed that the low strength of RSOC ($\lambda_R = 0.1$) does not much affect the energy bands of MZNR or MANR. However, with an increase in the value of RSOC strength ($\lambda_R = 0.5$), the band splitting and broadening of the total density of states enhance (not shown here). This enhancement occurs due to the increment in spin-flip rates with an increase in the strength of RSOC. It is also observed that when a vertical electric field is applied in close proximity to an FM insulator substrate, the induced spin polarization increases such that the spin-polarization of the MZNR reaches 100% close to the Fermi level via considering the exchange field [42]. Consequently, the MoS_2 nanoribbons can offer tunable spin-dependent transport.

4.6 Electronic properties of strained nanoribbons

Nanoribbons may have a variety of potential uses in nanoelectronics if their physical properties can be effectively modified by external control. The deformations occur naturally and lead to changes in the electronic properties of nanoribbons when they grow on top of the substrate. There are several ways to put the system under strain, including thermal ripples, lattice mismatch ripples, piezoelectric substrate actuation, the interaction of thin films with the substrate, and bending and stretching of the flexible substrate. Therefore, the study of strain effects on nanoribbons becomes important from the nanodevice application point of view. Graphene has an extraordinary ability to tolerate reversible deformations up to a high value ($\approx 27\%$). This means that it could serve as a good candidate for using it to develop new strain devices. However, in graphene, the band gap opening does not occur for strain greater than 20%. Mechanical deformations also change the physical properties of monolayer-MoS_2. It is investigated that effect of strain has the ability to turn direct band gap monolayer-MoS_2 into the indirect band gap. Moreover, strain engineering on monolayer-MoS_2 leads to semiconductor–metal transitions. The variation in electronic properties with strain motivates researchers to study the effects on their nanostructures. In this section, we shall focus on the strain effects on the electronic properties of zigzag and armchair types of carbon- and MoS_2-based nanoribbons.

4.6.1 Carbon-based nanoribbons

Experimentally, by using Raman spectroscopy, the graphene lattice acquires the strain due to a lattice mismatch with the substrate. It is discovered that uniaxial strain below a threshold value only moves the Dirac point rather than creating a gap in the energy spectrum of graphene. However, in graphene nanoribbons (armchair or zigzag), the effect of tensile strain can open a finite band in their energy spectrum. In TBA, the impact of strain can be modeled as the modification to the nearest neighbor hopping integral. Both theoretically and experimentally, the effects of uniaxial and shear strain on the electrical characteristics of graphene nanoribbons have been studied. On the application of strain, a band gap is opened in AGNRs by shifting the Dirac points due to their unique permissible electronic states. When uniaxial strain is applied to AGNRs, it changes the electronic band gap, resulting in a vibrating transition between the metal and semiconducting states. In contrast, the band structure of AGNRs is hardly impacted by shear strain. In contrast to AGNRs, due to the presence of edge states in ZGNRs, shear and uniaxial strain are unable to produce an electronic energy band gap. Therefore, ZGNRs are a suitable contender for use as components in conductive graphene interconnects in nanoelectronics due to the stability of the electronic energy band gap against strain.

Current research in ZGNR has used continuum approximation and the TBA to examine how strain affects topological phases and associated edge modes. It is observed that a topological phase may change from one with opposite Chern numbers to another or become trivial due to a strain. Hence, the chiral and antichiral edge mode dispersions that depend on strain are formed. These modes might start to spread backward due to the imposed strain, which would eventually destroy them. This phenomenon may be used to generate strain-tunable edge currents in the case of topological insulators and 2D TMDs [43].

The presence of strain can also control the magnetism of graphene at room temperature. Yang *et al* numerically showed that in the presence of strain, weak Coulomb interactions may induce ferromagnetism in ZGNRs [44]. This induced magnetism in ZGNRs can be enhanced with an increase in strain applied along the zigzag edge of these ribbons, which may be helpful to spintronics and many other applications [44]. Azadparvar *et al* studied the effects of uniaxial or oblique strains on the non-linear transport properties of ZGNRs [45]. It has been observed that the current-voltage characteristics and negative differential resistance of ZGNR significantly vary with the type, strength, and direction of the applied strain. The on–off current ratio increases with tensile strain along the ribbon axis, while negative differential resistance gradually disappears in the presence of compressive strain. In contrast to this, a drastic decrease in the on-off current ratio occurs in the presence of oblique strain. This suggests that strained ZGNRs can perform nanoelectromechanical switching [45]. In ZGNRs, by using the non-equilibrium Green's function approach, Zuo *et al* draw attention to the fact that the generation of the 'snake states' demands the presence of reverse strains between the lower and half planes [46]. Moreover, for the experimental realization, out-of-plane strain effects such as strain folds are also studied in the case of graphene nanoribbons. It has been observed that along the direction of the strain fold, the local density of states enhances due to the localization of higher energy states and adds more conductance channels at lower energies [47].

4.6.2 MoS_2-based nanoribbons

As aforementioned, the semiconductor–metal transition takes place in monolayer MoS_2 via a direct-to-indirect band gap. This transition occurs at a tensile biaxial strain greater than 10% in the plane of 2D MoS_2 and can endure a maximum strain of 12%. In contrast to the biaxial strain, the uniaxial strains having a value of more than 4% can change monolayer MoS_2 from a direct to an indirect band gap material. In contrast to the biaxial strain, at a uniaxial strain of 12% semiconductor–metal transition does not take place. However, the direct-to-indirect band gap transition occurs at lower values of biaxial strains [48]. It is observed that the band gap roughly varies inversely to the uniaxial strain. In addition to this, the charge carriers (electron and hole) effective masses decrease with an increase in the value of uniaxial strains. A similar trend is observed for biaxial strains [48].

The increase in the application of uniaxial strain on MANR (width = 31.7 nm) decreases the band gap. It is found that the DOS near the band edge becomes smaller. Moreover, the DOS of the states at the edge of the ribbon does not change with an increase in the value of strain because the application of strain does not change the concentration of atoms at all. With the large value of strain, monolayer MoS_2 has an indirect band gap [48]. Yet, a direct bandgap at Γ valley is always present in the nanoribbon structure. This can be explained by the problem with the band folding. States in the K valley and the M valley will combine to form mixing bands. The folded K valley states begin to rise more quickly than the original Γ valley states, which eventually dominate the valence band characteristics with their large hole-effective mass. Moreover, the presence of uniaxial tensile strain in the transport direction or perpendicular to the transport direction reduces the carrier's effective mass due to stretch in the k-space [48].

In MZNR ($N_z = 8$), when uniaxial strain is applied along the ribbon growth direction, the net magnetic moment of ribbons increases. For MZNR ($N_z = 8$), the value of the magnetic moment is enhanced from 0.62 to 1.01 μ_B in the presence of 8% uniaxial strain. Under the effect of strain, the Mo–S bond length in each layer elongates, and bond energy decreases. This causes more electrons to become spin-polarized, which increases the magnetic moments [50]. The magnetic moment of MZNR becomes approximately three times its original value when strain reaches a value of 10%. In contrast, the compressive strain (beyond −4%) increases the interaction between Mo and S edge atoms and reduces the net magnetic moment to zero. However, the metallic behavior of narrower MZNRs remains in the presence of uniaxial strain [50]. DFT shows that the electronic and magnetic properties of MZNR are also sensitive to the combined effect of strain and electric field. An internal electric polarization and the competing covalent and ionic interactions cause energy-level shifts during tensile strain that result in the reversible manipulation of magnetic moments and electronic phase transitions between metallic, half-metallic, and semiconducting states (see figures 4.12(b)–(f)) [49]. Depending on how the electric field is directed in relation to the internal polarization, a simultaneously applied electric field can further increase or inhibit the strain-induced modulations. Therefore, to enhance the functionality of nano-devices based on these novel nanoscale materials, strained MZNRs' controllable magnetism and reversible electrical phase transitions offer a reliable and practical solution [49].

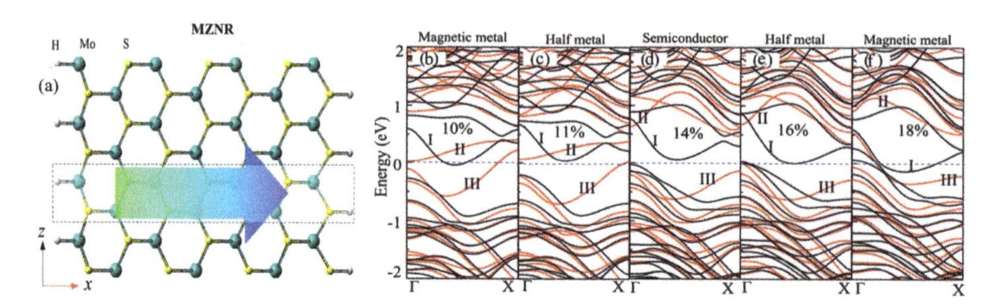

Figure 4.12. (a) Top view of the structural model of MZNR. (b)–(f) Typical spin-polarized band structure of MZNR ($N_z = 7$). Adapted with permission from [49]. Copyright (2012) American Chemical Society.

4.7 Summary

Starting from the very basic introduction, need, and recent synthesis techniques, this chapter provides a broad overview of the comparative study of carbon-based nanoribbons and MoS_2-based nanoribbons for spintronics-based devices. The physical properties of both carbon-based and MoS_2-based nanoribbons are highly sensitive to edge termination. The zigzag nanoribbons of both materials give rise to magnetic edge states. ZGNRs in the ground state behave as AFM semiconductors with net zero magnetic moments. The presence of an external electric field converts the ZGNR into half metals with only one type of spin transport channel near Fermi energy. Moreover, in the presence of a magnetic field, the semiconducting AFM states of ZGNR convert to the metallic FM state. In contrast to ZGNR, MZNR shows an FM ground state with a metallic character without the presence of external magnetic fields. On the basis of their FM properties, it is found that the MoS_2-based nanoribbons exhibit more stable magnetic moments compared to carbon-based nanoribbons. However, the armchair nanoribbons of both materials exhibit a NM character.

The presence of intrinsic SOC in carbon-based nanoribbons (ZGNR) results in insulating bulk states and a conductive edge state, while the presence of RSOC removes the spin degeneracy by breaking the mirror symmetry and providing highly localized spin-polarized edge states. However, the small value of intrinsic SOC limits the experimental realization of the topological behavior of carbon-based nanoribbons. The intrinsic and RSOCs in MoS_2-based nanoribbons result in the spin polarization of energy bands. This spin polarization is enhanced in the presence of an electric field in close proximity to an FM substrate.

The spin polarization in both types of materials varies with the type, direction, and strength of the mechanical deformations. AGNRs show periodic oscillations of the energy band gap with uniaxial strains. In contrast to this, the energy band gap of MANRs decreases with an increase in the strength of the uniaxial strain. The presence of uniaxial or shear strain barely affects the metallic nature of ZGNRs. However, in MZNRs, the phase transitions (magnetic metal—half metal—semiconductor—half metal—magnetic metal) occur with different strain strengths. Compared to carbon-based nanoribbons, MoS_2-based nanoribbons cannot endure large strains.

Based on the above-discussed properties, the future perspectives of both types of nanoribbons (carbon-based and MoS_2-based) in spintronics and straintronics are very promising. For instance, (i) armchair nanoribbons of both materials are semiconducting and could be used as an active channel in FETs; however, AGNRs provide more control over charge transport because the energy band gap varies inversely with width. (ii) The magnetic edge states in both types of nanoribbons could be used in spin filters and spin transistors that are more efficient and powerful than current devices. Moreover, due to spin-polarized states, both types of nanoribbons might be utilized in sensors that are sensitive to electron spin. (iii) The variation of the energy band gap with strain opens the possibility for both types of nanoribbons (carbon-based and MoS_2-based) to be utilized in strain sensors. Moreover, both materials are suitable for strain transistors which could be used in high-speed and flexible electronic devices.

There are still a number of challenges that need to be overcome in order to fully realize the potential of these nanoribbons. These challenges include: (i) The synthesis of these nanoribbons with controlled width, edge structure, and doping is still a challenge. (ii) The impurities frequently contaminate nanoribbons, which can cause their properties to deteriorate. Especially the carbon-based nanoribbons, which have dangling bonds at the edges, are highly influenced by the impurities. (iii) The large-scale production of these nanoribbons is still a challenge. The current synthesis methods are often complex and time-consuming, making it difficult to produce large quantities of these materials [3]. (iv) The experimental observation of spin-polarized edge states of zigzag nanoribbons is still a challenge [5]. Therefore, substantial collaboration from experimentalists is required for the verification of the theoretically predicted applications of these nanoribbons.

Acknowledgments

SP would like to thank Central University of Himachal Pradesh for providing the facilities during the course of this work. SK would like to thank University Grants Commission—New Delhi for his research fellowship.

References

[1] Geim A K and Novoselov K S 2007 The rise of graphene *Nat. Mater.* **6** 183–91

[2] Neto A C, Guinea F, Peres N M, Novoselov K S and Geim A K 2009 The electronic properties of graphene *Rev. Mod. Phys.* **81** 109

[3] Kumar S, Pratap S, Kumar V, Mishra R K, Gwag J S and Chakraborty B 2023 Electronic, transport, magnetic, and optical properties of graphene nanoribbons and their optical sensing applications: a comprehensive review *Luminescence* **38** 909–53

[4] Dutta S and Pati S K 2010 Novel properties of graphene nanoribbons: a review *J. Mater. Chem.* **20** 8207–23

[5] Kheirabadi N, Shafiekhani A and Fathipour M 2014 Review on graphene spintronic, new land for discovery *Superlattices and Microstruct* **74** 123–45

[6] Kan E J, Li Z, Yang J and Hou J 2008 Half-metallicity in edge-modified zigzag graphene nanoribbons *J. Am. Chem. Soc.* **130** 4224–5

[7] Wang D, Bao D L, Zheng Q, Wang C T, Wang S and Fan P *et al* 2023 Twisted bilayer zigzag-graphene nanoribbon junctions with tunable edge states *Nat. Comm.* **14** 1018

[8] Son Y W, Cohen M L and Louie S G 2006 Half-metallic graphene nanoribbons *Nature* **444** 347–9

[9] Dutta S, Manna A K and Pati S K 2009 Intrinsic half-metallicity in modified graphene nanoribbons *Phys. Rev. Lett.* **102** 096601

[10] Magda G Z, Jin X, Hagymási I, Vancsó P, Osváth Z and Nemes-Incze P *et al* 2014 Room-temperature magnetic order on zigzag edges of narrow graphene nanoribbons *Nature* **514** 608–11

[11] Wang Q H, Kalantar-Zadeh K, Kis A, Coleman J N and Strano M S 2012 Electronics and optoelectronics of two-dimensional transition metal dichalcogenides *Nat. Nanotechnol.* **7** 699–712

[12] Radisavljevic B, Radenovic A, Brivio J, Giacometti V and Kis A 2011 Single-layer MoS_2 transistors *Nat. Nanotechnol.* **6** 147–50

[13] Mak K F, Lee C, Hone J, Shan J and Heinz T F 2010 Atomically thin MoS_2: a new direct-gap semiconductor *Phys. Rev. Lett.* **105** 136805

[14] Wang Z, Li H, Liu Z, Shi Z, Lu J and Suenaga K *et al* 2010 Mixed low-dimensional nanomaterial: 2D ultranarrow MoS_2 inorganic nanoribbons encapsulated in quasi-1D carbon nanotubes *J. Am. Chem. Soc.* **132** 13840–7

[15] Botello-Méndez A R, Lopez-Urias F, Terrones M and Terrones H 2009 Metallic and ferromagnetic edges in molybdenum disulfide nanoribbons *Nanotechnology* **20** 325703

[16] Yu S and Zheng W 2016 Fundamental insights into the electronic structure of zigzag MoS_2 nanoribbons *Phys. Chem. Chem. Phys.* **18** 4675–83

[17] Pan H and Zhang Y W 2012 Edge-dependent structural, electronic and magnetic properties of MoS_2 nanoribbons *J. Mater. Chem.* **22** 7280–90

[18] Mack C A 2011 Fifty years of Mooreś law *IEEE Trans. Semicond. Manuf.* **24** 202–7

[19] Wang H, Wang H S, Ma C, Chen L, Jiang C and Chen C *et al* 2021 Graphene nanoribbons for quantum electronics *Nat. Rev. Phys.* **3** 791–802

[20] Houtsma R K, de la Rie J and Stöhr M 2021 Atomically precise graphene nanoribbons: interplay of structural and electronic properties *Chem. Soc. Rev.* **50** 6541–68

[21] Pratap S 2016 Transport properties of zigzag graphene nanoribbons in the confined region of potential well *Superlattices Microstruct.* **100** 673–82

[22] Pratap S 2020 Edge states in zigzag graphene nanoribbons in a finite potential well *AIP Conf. Proc.* **Vol. 2220** (New York: AIP Publishing) p 100011

[23] Pratap S 2017 Transmission and LDOS in case of ZGNR with and without magnetic field *Superlattices Microstruct.* **104** 540–6

[24] Bhalla P and Pratap S 2018 Aspects of electron transport in zigzag graphene nanoribbons *Int. J. Mod. Phys.* B **32** 1850148

[25] Han M Y, Ozyilmaz B, Zhang Y and Kim P 2007 Energy band-gap engineering of graphene nanoribbons *Phys. Rev. Lett.* **98206805**

[26] Pratap S, Kumar S and Singh R P 2022 Certain aspects of quantum transport in zigzag graphene nanoribbons *Front. Phys.* **10** 940586

[27] Yang C, Wang B, Xie Y, Zheng Y and Jin C 2019 Deriving MoS_2 nanoribbons from their flakes by chemical vapor deposition *Nanotechnology* **30** 255602

[28] Yagmurcukardes M, Peeters F M, Senger R T and Sahin H 2016 Nanoribbons: From fundamentals to state-of-the-art applications *Appl. Phys. Rev.* **3** 041302

[29] Dolui K, Pemmaraju C D and Sanvito S 2012 Electric field effects on armchair MoS_2 nanoribbons *ACS nano* **6** 4823–34
[30] Campos L C, Manfrinato V R, Sanchez-Yamagishi J D, Kong J and Jarillo-Herrero P 2009 Anisotropic etching and nanoribbon formation in single-layer graphene *Nano Lett.* **9** 2600–4
[31] Lee H J, Lim J, Cho S Y, Kim H, Lee C and Lee G Y *et al* 2019 Intact crystalline semiconducting graphene nanoribbons from unzipping nitrogen-doped carbon nanotubes *ACS Appl. Mater. Interfaces* **11** 38006–15
[32] Yano Y, Mitoma N, Ito H and Itami K 2019 A quest for structurally uniform graphene nanoribbons: synthesis, properties, and applications *J. Org. Chem.* **85** 4–33
[33] Kimouche A, Ervasti M M, Drost R, Halonen S, Harju A and Joensuu P M *et al* 2015 Ultra-narrow metallic armchair graphene nanoribbons *Nat. Comm.* **6** 10177
[34] Li Q, Newberg J, Walter E, Hemminger J and Penner R 2004 Polycrystalline molybdenum disulfide (2H-MoS $_2$) nano-and microribbons by electrochemical/chemical synthesis *Nano Lett.* **4** 277–81
[35] Li Z, Jiang Z, Zhou W, Chen M, Su M and Luo X *et al* 2021 MoS_2 Nanoribbons with a Prolonged Photoresponse Lifetime for Enhanced Visible Light Photoelectrocatalytic Hydrogen Evolution *Inorg. Chem.* **60** 1991–7
[36] Ali S, Bajaj A and Ali M E 2022 Quantum interference controlled spin-polarized electron transmission in graphene nanoribbons *J. Phys. Chem.* C **126** 14714–26
[37] Yazyev O V and Katsnelson M 2008 Magnetic correlations at graphene edges: basis for novel spintronics devices *Phys. Rev. Lett.* **100** 047209
[38] Zeng M, Shen L, Zhou M, Zhang C and Feng Y *et al* 2011 Graphene-based bipolar spin diode and spin transistor: rectification and amplification of spin-polarized current *Phys. Rev. B Condens. Matter* **83** 115427
[39] Vancsó P, Hagymási I, Castenetto P and Lambin P 2019 Stability of edge magnetism against disorder in zigzag MoS_2 nanoribbons *Phys. Rev. Mat.* **3** 094003
[40] Kane C L and Mele E J 2005 Quantum spin Hall effect in graphene *Phys. Rev. Lett.* **95** 226801
[41] Rhim J W and Moon K 2011 Quantum spin Hall effect in graphene nanoribbons: effect of edge geometry *Phys. Rev. B: Condens. Matter* **84** 035402
[42] Salami N and Shokri A 2017 Tunable spin polarization of MoS_2 nanoribbons without time-reversal breaking *Superlattices Microstruct.* **109** 605–18
[43] Mannaï M and Haddad S 2020 Strain tuned topology in the Haldane and the modified Haldane models *J. Phys.: Condens. Matter* **32** 225501
[44] Yang G, Li B, Zhang W, Ye M and Ma T 2017 Strain-tuning of edge magnetism in zigzag graphene nanoribbons *J. Phys.: Condens. Matter* **29** 365601
[45] Azadparvar M and Cheraghchi H 2021 Strain-induced switching in field effect transistor based on zigzag graphene nanoribbons *Physica B: Condens. Matter* **622** 413304
[46] Zuo C, Qi J, Lu T, Bao Z and Li Y 2022 Reverse strain-induced snake states in graphene nanoribbons *Phys. Rev. B: Condens. Matter* **105** 195420
[47] Carrillo-Bastos R, León C, Faria D, Latge A, Andrei E Y and Sandler N 2016 Strained fold-assisted transport in graphene systems *Phys. Rev. B Condens. Matter* **94** 125422
[48] Chen S F and Wu Y R 2017 Electronic properties of MoS_2 nanoribbon with strain using tight-binding method *Phys. Status Solidi* B **254** 1600565

[49] Kou L, Tang C, Zhang Y, Heine T, Chen C and Frauenheim T 2012 Tuning magnetism and electronic phase transitions by strain and electric field in zigzag MoS_2 nanoribbons *J. Phys. Chem. Lett.* **3** 2934–41

[50] Lu P, Wu X, Guo W and Zeng X C 2012 Strain-dependent electronic and magnetic properties of MoS_2 monolayer, bilayer, nanoribbons and nanotubes *Phys. Chem. Chem. Phys.* **14** 13035–40

[51] Dhanabalan S S, Arun T, Periyasamy G, Dineshbabu N, Chidhambaram N, Avaninathan S R and Carrasco M F 2022 Surface engineering of high-temperature PDMS substrate for flexible optoelectronic applications *Chem. Phys. Lett.* **800** 139692

[52] Sundar D S, Raja A S, Sanjeeviraja C and Jeyakumar D 2017 High temperature processable flexible polymer films *Int. J. Nanosci.* **16** 1650038

[53] Sundar D S, Sivanantharaja A, Sanjeeviraja C and Jeyakumar D 2016 Highly transparent flexible polydimethylsiloxane films-a promising candidate for optoelectronic devices *Polym. Int.* **65** 535–43

[54] Sundar D S, Sivanantharaja A, Sanjeeviraja C and Jeyakumar D 2016 Synthesis and characterization of transparent and flexible polymer clay substrate for OLEDs *Mater. Today: Proc.* **3** 2409–12

Part II

Fabrication—the art of realization

Chapter 5

Fabrication techniques for printed and wearable electronics

G R Raghav, M S Anoop, P C Jayadevan, R Ashok Kumar, K J Nagarajan and D Muthukrishnan

In the modern era, smart devices play an important role in day-to-day life, and their widespread applications are getting huge demand globally. Innovation in the field of the Internet of Things paves new possibilities for future endeavors of mankind. Printed electronics is a sustainable way for achieving the widespread popularity of smart devices around the world and this technology is in its nascent stage. In the current scenario, massive amounts of e-waste generated due to the digital revolution and its disposal become a greater challenge for sustainability. Printed electronics are composed in a process of registering thin functional material (ink) layer combinations on a low-cost substrate that will degrade naturally. This article discusses the possibilities of printed electronics and its ability to hurdle the limitations of traditional high-cost electronics, based on rigid silicon, and the production of different devices on flexible substrates. Efficient use of materials, optimized energy consumption both in production and utilization, reduction in hazardous substances, and enhanced recyclability are the several benefits associated with printed electronics technology. The additive manufacturing method is used in printed electronics technology and the rate of production is much improved as compared with other processes. The materials used for printed electronics like ink and substrates are derived from synthetic or natural polymers. The above-stated reasons make printed electronics a technology for the future digital revolution. This article discusses various fabrication techniques like lithographic process for the production of printed electronics and its application in a sustainable manner.

5.1 Introduction

Printed or wearable electronics have good potential to be utilized as eco-friendly and biodegradable electronics so as to reduce electronic wastes, which is also known as

doi:10.1088/978-0-7503-5492-9ch5

e-waste. Electronic wastes are due to the large number of electronic devices, which are disposed of every day [1, 2]. Another advantage of printed or wearable electronics is that it can be used in complex surfaces. Wearable devices thus help in improving the country's economic growth because of the sudden surge in printed or stretchable electronic devices. Now due to the advancement of artificial intelligence, electronic devices with artificial intelligence assistance have a great impact on the electronics industry. There are many conventional manufacturing processes to manufacture printed or wearable electronic devices. But due to the wastage of materials, and in order to avoid secondary operations such as etching and masking, the additive manufacturing process is most preferred in recent times. The printing process is also known as the additive method of manufacturing electronic applications by depositing electronic materials using functional inks along with the normal printing process [3]. As discussed earlier, this process thus eliminates the need for etching and masking and thereby involves environmentally friendly cleaner production compared to that of other traditional methods [4].

Nowadays a lot of sports and fitness equipment utilizes these wearable technologies in monitoring exercise and detecting the glucose level of diabetics [5]. Also, recently many printed and wearable devices have been made up of flexible or stretchable materials, which are used as sensors, that have close contact with human skin [6, 7].

The printing of electronic devices is classified into two types, one is contact type and the other is non-contact type. In the contact printing method, the die or pattern is immersed in a functional ink, and it is transferred onto the substrate by means of physical transfer. Screen printing, offset printing, flexography, and pad printing are the few types of contact printing [8].

On the other hand, in the non-contact printing process, the functional ink is sprayed via a nozzle onto the target substrate. There are two common types of non-contact printing processes, (i) inkjet printing and (ii) aerosol printing [9]. Normally the printing process is characterized by three steps or stages:

(i) Selection of materials
(ii) Printing process
(iii) Sintering/drying process

After transferring ink onto the substrate either by contact or non-contact mode, it is very important to sinter the printed surface so as to achieve the desired properties of ink and the substrate. Figure 5.1 shows the steps involved in printed electronics manufacturing. Even though producing flexible electronic devices with required properties and specification is difficult for mass production, current technology and materials development has shown a positive trend in both performance and biodegradable properties. Because of this development, printed or wearable devices are developed for varying applications such as Radio Frequency Identification devices (RFID), organic light emitting diodes, thin-film transistors (TFT), photovoltaic cells, energy devices such as batteries, and different types of sensor devices [10–12].

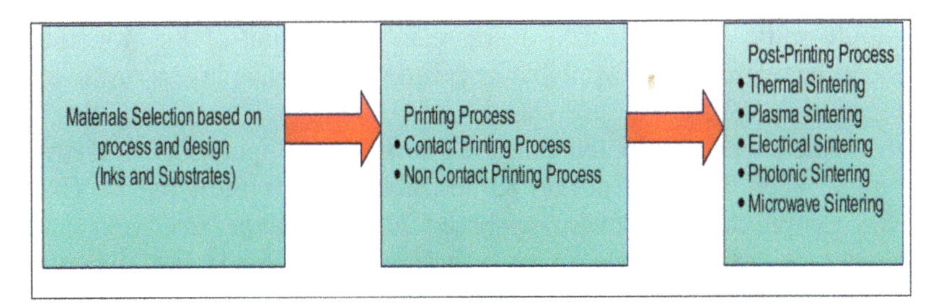

Figure 5.1. The steps involved in printed electronics manufacturing.

In this chapter, we provide the basic principle of various printing or fabrication techniques used for the development of wearable electronics. Further, this chapter will give clear insight into different materials and substrates employed.

5.2 Different types of flexible and printed electronic materials

In this part, we will explore the different types of materials used in flexible or printed electronics.

5.2.1 Inks used in printed electronics

In the production of wearable/printed electronics, inks play a vital role in creating structures or skeleton, which has a particular function. For printing of very complex electronic parts, different types of inks are used such as insulator/dielectric inks, semiconductor inks, and conducting inks. An important property of functional inks is the compatibility with other inks for simultaneous application, and it should be able to form uniform homogenous layers over the substrate. The inks used for wearable devices contain solvents, polymers, or resins. The inks can be made up of either organic or inorganic materials. Sometimes some additives will be added to improve the properties of the inks and avoid clogging of inks [13, 14].

5.2.1.1 Conducting ink materials

There are different types of conductive inks that are synthesized as nanoparticles. The nanoparticle inks are then dissolved onto the conductive polymer matrix [15]. Mostly metal nanoparticles are used as conductive ink material. Even though metal nanoparticle–dispersed inks have better properties, the synthesis of metal nanoparticles is very difficult and requires more time and labor. Moreover, stabilizers are necessary to prevent agglomeration in functional inks. The sintering process, which is the post-printing process, requires heating above 100 °C so as to cure macro particle–dispersed inks, whereas, for nanoparticle-dispersed inks, the sintering temperature can be less than 100 °C [16]. Silver nanoparticle inks are one type of conducting inks. The silver nanoparticle–dispersed inks are toxic due to the evolution of silver ions; hence, the application of nanosilver inks is limited [17]. On the other hand, metal-organic decomposition inks use metal particles as precursors and evaporable alcohols as solvents. Thus, the agglomeration of nanoparticles is prevented. But due to the evaporation of the solvents, there may be non-uniformity in the deposited patterns.

This non-uniformity will result in a decrease in conductivity [18]. Nonetheless, this drawback can be overcome by applying or printing successive layers one over another using the same ink [19].

The most widely used conductive inks are aluminum, copper, gold, and silver metal-dispersed inks. When talking about properties, silver-based conductive inks have better conductivity compared to that of copper-based inks. The silver-based inks have a good ability to resist oxidation. However, with the increase in oxidation, the conductivity decreases. The oxidation of metal nanoparticle–dispersed inks can be prevented by coating antioxidants over the nanoparticles or *in situ* synthesis in an organic solvent with a protective layer. Whereas the above-said methods are temporary solutions to control oxidation, new methods such as forming a bio-metallic core-shell or formation of a thick shell are made up of non-oxidizing conductive materials [15]. Gold is also one of the important conductive ink materials, which is eco-friendly and can be cured at very low temperatures. Whereas gold conductive inks are costly compared to other conductive inks. Another ink is aluminum-based ink, which can be synthesized using organic solvents, but aluminum inks are reactive in nature and tend to oxidize very quickly.

Apart from metal-dispersed inks, carbon-based inks such as carbon nanotubes (CNTs), graphene, and C60 can be altered and modified for applications such as conductive inks. The CNT's reinforced metallic conductive inks exhibit better stability, conductivity, and flexibility. The conductivity of CNT's reinforced metallic inks increases with the increase in thickness. The graphene and C60 also exhibit good light transmittance, flexibility, and conductivity. The light transmittance decreases when the number of graphene layers increases, whereas the conductivity increases with an increase in graphene layers [16, 20].

The recent development in conductive inks is the conducting polymers-based inks. They are very cheap, very light in weight, flexible in nature, and can be used in aqueous solvents as well as organic solvents. The main disadvantage of conductive polymer inks is their poor conductivity compared to metal inks, and production of conductive polymer inks is very difficult due to their processing difficulties, stability, and lesser solubility compared to metals. The polymer-based conductive inks are classified into the following types:

(i) Organic metal chelates
(ii) Conjugated polymers
(iii) Polymer electrolytes

The poly(3,4-ethylenedioxythiophene) polystyrene sulfonate is one type of conductive polymer that has good conductivity and decent temperature stability. There are a few other conductive polymers for the application of functional inks such as polypyrrole, polyacetylene, and polyacene [21]. Apart from the above materials, there are conducting ceramics, which are doped to improve conducting properties of ceramics. For example, aluminum-coated zinc oxide, indium-coated tin oxide (ITO), and gallium-coated zinc oxide. Among these, the ITO is most widely used for electronics applications, owing to its enhanced conductivity. But it should be noted that indium is a rare earth material, and hence, it is very costly [22]. Normally there are two types of ITO conductive inks. One type of ink is *in situ* sol–gel-based inks, and another type is nanoparticle-dispersed conductive ink. The sol–gel-derived ITO conductive inks have better conductivity compared to nanoparticle-dispersed ITO inks.

5.2.1.2 Dielectric ink materials

In printed or wearable electronics, the capacitor and insulator layers are made up of dielectric materials. In order to be a good insulator, the layers of dielectric materials should be thick and uniform. However, it is difficult to print dielectric materials as compared to conductive materials. There are some substrate materials that are dielectric in nature such as silk, gelatine, cellulose, etc [2]. There are many dielectric materials based on polymers, with less density, less toughness, and less curing temperature. The most widely used dielectric polymers are polydimethylsiloxane (PDMS), polylactic acid (PLA), polymethyl methacrylate (PMMA), and polyvinyl alcohol [22].

5.2.1.3 Semiconducting ink materials

The semiconducting material is most commonly used as an active layer. There are many semiconducting materials such as silicon, CNTs, and different derivatives of graphene owing to their mechanical and semiconducting properties. The multiple layers of graphene can improve the semiconducting properties of wearable devices. There are few ceramic materials that can be used as semiconductors. But semiconducting ceramic materials are very rare and expensive. They also require high sintering temperatures for curing the printed layers [23, 24].

In flexible and printed electronic applications, the semiconducting functional inks can also be prepared by dissolving polymers in specific solvents. Hence, these types of polymer-based semiconducting inks can be used either as p-type or n-type materials. Polymers like polyfluorenes are most widely used as semiconducting polymers. Poly(3-alkyl thiophene) is an example of a p-type semiconducting material that uses holes for charge transfer, whereas n-type conducting polymers such as poly(9,9-dioctyl-fluorene-co-bithiophene) (F8T2) uses electrons for charge transfer [2].

5.2.2 Substrate materials for printed electronics

The substrate forms the base for any printed electronic devices. This substrate can also act as an insulator. The conventional substrate materials are strong and can remain rigid for quite a long period. Even though these materials possess good rigidity, conventional substrates are more brittle; hence, it is very difficult to machine conventional substrates for flexible or wearable devices. However, the development of flexible, biodegradable, and light polymer substrates has resulted in the rapid improvement of wearable devices with a long service life [25]. The substrate can be made up of natural or synthetic materials. These substrates are highly flexible, heat resistant, thin, low weight, and low cost [21]. Another important step in printed electronics is post-treatment, also known as the sintering process.

This sintering process might damage the surface of the substrate. So, materials with good thermal resistance should be considered for substrates.

5.2.2.1 Biodegradable polymeric substrates

Nowadays paper-based substrates are replacing conventional printed electronics because of their flexibility, biodegradability, and low cost [26]. However, the paper substrates have disadvantages such as porosity, surface roughness, and poor resistance to moisture. But the properties of paper-based substrates can be improved by metallic or ceramic coating as per the application [27]. Nano cellulose is another important contender for substrates because of its heat resistance, better surface smoothness, transparent nature, and good mechanical properties [21, 28]. Similarly, starch, silk, and shellac were also considered for the fabrication of substrates. Silk is a biodegradable material with better properties. Shellac is also a naturally available resin that can be used for preparing substrates in printed electronics. These materials have good surface smoothness, and further, these materials are relatively cheaper [28].

5.2.2.2 Synthetic polymer substrates

Polymer-based substrates are most widely used for printed electronic substrates. Polyethylene terephthalate (PET), polycarbonate, polyethylene naphthalate (PEN), and polyimide are the most commonly uses synthetic polymers in flexible electronic applications [2]. Because of its high flexibility, transparency, and resistance to solvent, PET is the most preferable and commonly used substrate material in flexible electronic devices. Polycarbonate substrates have high rigidity, are light in weight, and possess good mechanical properties. Although PEN has good transparency, it is very costly [29].

There are a few synthetic biodegradable polymers such as PLA, polyvinyl alcohol, PDMS, and polyethylene glycol, which can be utilized for the fabrication of substrates. The PLA is stiff in nature but possesses very poor heat resistance. Whereas PDMS can be used as substrates in flexible or stretchable electronic applications owing to their elastic nature. The need for producing flexible devices increases day by day. The research based on the direct printing of flexible or wearable electronics on polymer and fabric substrates is getting widespread acceptance.

5.3 Fabrication methods for printed electronics

Figure 5.2 shows the different types of fabrication methods in developing printed or wearable electronic devices. It Includes inkjet printing, offset printing, gravure printing, screen printing, and flexography. These methods are also classified into two types: contact and non-contact printing processes [8]. The inkjet printing and aerosol printing are contactless printing processes. There are some methods that involve both printing and deposition or coating techniques. The basic purpose of this printing or deposition is to develop multiple layers of structures that can be a conductive layer, semiconducting structure, or insulating structure for printed

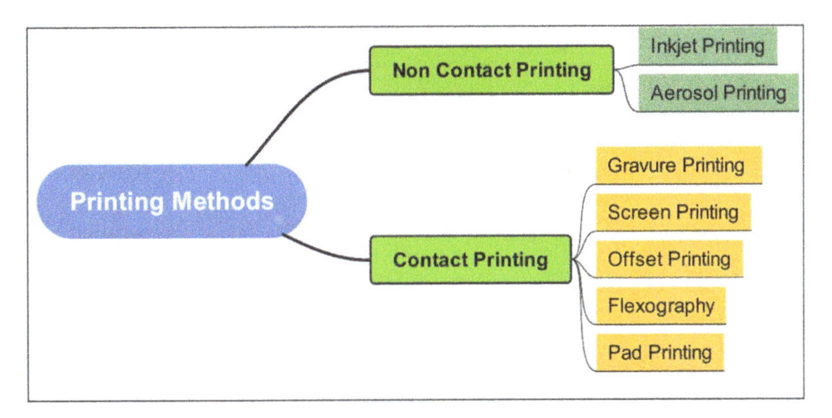

Figure 5.2. The different types of fabrication methods in developing printed or wearable electronic devices.

electronic devices. After the coating or deposition process, the functional inks will change their phase and orientation. This change in phase and flow rate on the coated substrate are mainly due to the viscosity of the ink material. It is also to be noted that the higher the viscosity, the higher the efficiency. The density of the ink is also important property for developing effective wearable devices. The surface tension is the property by which the ink can stick to the surface of the substrate. The evaporation rate and sintering temperature of the ink also play a vital role in the quality of flexible electronic devices [30]. The most commonly available printed electronic device is RFID stickers or flexible antennae, where antennas can be printed. The printing of RFID on flexible or printed substrates is a very efficient method, since it is a lighter, smaller, and cheaper device compared to conventional RFID antennas. The high-volume production techniques for printed electronics are offset printing, gravure method, and flexographic method. These techniques are utilized for the mass production of solar cells, sensors, etc. The organic or inorganic conducting materials can be printed using flexography and offset methods. The organic semiconductors and insulators are coated using the gravure printing method.

Recently, many new novel printing methods are identified and employed in the manufacturing of flexible or wearable devices. Devaraj *et al* developed a method known as the form-fuse method. In this method, silver nanoparticles are coated on polymer films by using an aerosol jet with mask and without mask. The entire process is carried out in a vacuum to develop desired shape and pattern. The sintering process is carried out to reduce the resistivity of printed materials [31]. Constante *et al* [32] also developed a new 4D coating method employing a 3D extrusion process along with melt-electro writing, which proves to be a good potential method for the development of flexible devices. Table 5.1 shows the important parameters of different printing methods. The inkjet method is more suited for high-quality research applications. The screen printing method is most useful for printing multiple layers, whereas the flexographic and gravure methods

Table 5.1. Parameters of different printing methods.

S. No.	Printing methods	Throughputs $m^2\ s^{-1}$	Resolution lines cm^{-1}	Operating speed $m\ min^{-1}$
1	Gravure printing process	3–60	20–400	100–1000
2	Screen printing process	2–3	50	10–15
3	Offset printing process	3–30	100–200	100–900
4	Flexography printing process	3–30	60	100–700
5	Inkjet printing process	0.01–0.5	60–250	15–500

are useful for mass production. The output rate of the inkjet printing method is $0.5\ m^2\ s^{-1}$, which is much less compared to other methods.

5.3.1 Contact printing of flexible electronics

In contact printing, the functional ink is transferred onto the substrate directly. This method is also known as roll-to-roll printing or transfer printing. The roll is used to transfer the ink to the substrate. The major disadvantage of this method is the large time consumption and high initial cost of the equipment. But the production cost is low and has good reproducibility, which makes them favorites for mass production [33, 34].

5.3.1.1 Gravure printing method

Gravure printing is the process in which the design to be printed on the substrate is first engraved on the printing cylinder, also known as the gravure cylinder. The doctor blade, which is made of steel, is used to remove the excess ink present in the printing cylinder before the ink is transferred to the impression cylinder from where the design is transferred onto the surface of the substrate as shown in figure 5.3. The printing cylinder is made up of rubber. The printing or gravure cylinder is made up of steel coated with copper. This process utilizes inks with low viscosity and possesses good efficiency. This method of printing proved to be more economical with good-quality printing. The quality of printing can be improved by using electrostatic forces for transferring ink onto the substrate [9].

It is to be noted that many devices such as antennas, TFTs, pressure sensors, surface-enhanced Raman scattering (SERS), and electrochemical sensors are developed using the gravure printing method [35]. Recently a wrinkle-structured solvent-excluded surface substrate developed for the detection of drugs like cocaine was fabricated by Maddipatla *et al* as shown in figure 5.4(A). In this study, the wrinkle-shaped structures were developed on the thermoplastic polyurethane (TPU) substrates by varying the proportions and printing silver ink of 150 nm particle size onto the TPU substrate using the gravure process [36]. Another recent study also utilizes the gravure printing process for fabricating a novel RFID antenna made up of paper substrate as shown in figure 5.4(B). In this work, Zhu *et al* manufactured a

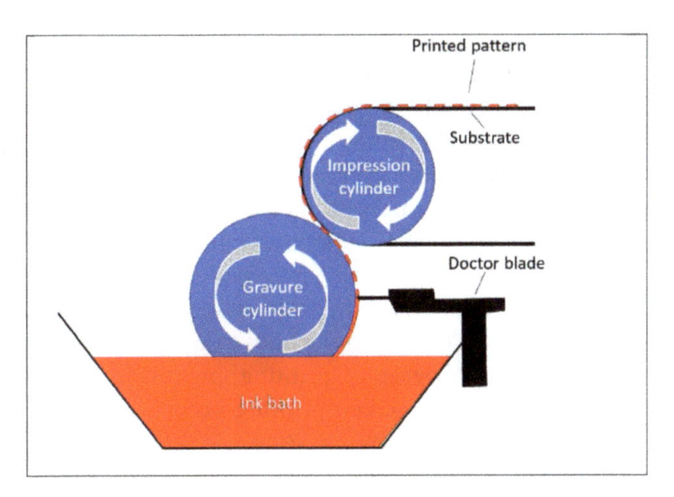

Figure 5.3. The Schematic representation of the Gravure printing process, reprinted from [13] with permission from MDPI, Copyright (2021).

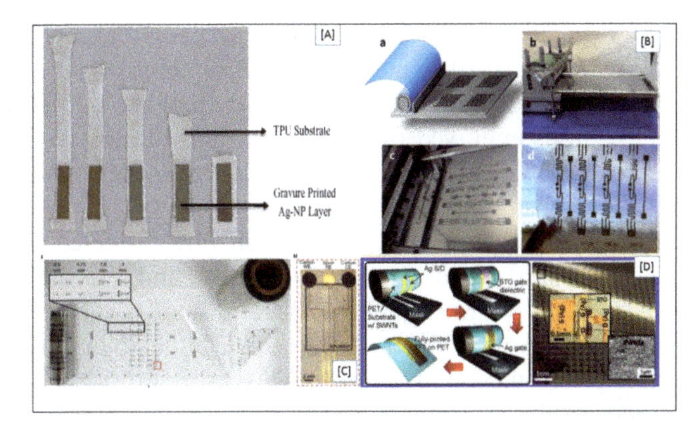

Figure 5.4. (A) Wrinkle SERS substrate, reprinted from [36] with permission from Elsevier, Copyright (2019). (B) RFID antenna made up of nano-paper, reprinted from [37] with permission from RSC, Copyright (2014). (C) PET-based electrochemical sensors for the detection of heavy metals, ions, and metabolites, reprinted from [38] with permission from ACS, Copyright (2018). (D) High-performance TFT made up of CNTs, reprinted from [39] with permission from ACS, Copyright (2013).

nano-structured paper by employing cellulose nanofibers and ultra high frequency (UHIF) UHIF RFID tag, also known as squiggle, fabricated by depositing silver ink onto the paper substrate [37]. Bariya *et al* developed electrochemical sensors using PET for the detection of heavy metals, ions, and metabolites as shown in figure 5.4(C). Here the working and counter electrode is the carbon electrode, and silver is used as the reference electrode [38]. Lau *et al* fabricated CNT-based TFTs through the gravure printing process as shown in figure 5.4(D). In this work, silver ink is deposited as drain, gate, and source electrodes on walled carbon nanotube (WCNT)-coated PET substrates. The nano barium titanate is printed as insulator layers. This gated TFT exhibits better performance, high flexibility, and minimal hysteresis [39]. In the process of gravure printing, the gravure cylinder, also known as the printing cylinder, contains patterns that are very expensive, and further during the process of printing, a small percentage of the inks gets clogged due to evaporation and dries out in the printing cylinder, thereby reducing the quality of succeeding printings. The availability of functional ink is very low in the case of both gravure and flexography printing processes. These two processes are widely adopted in all graphics and packaging applications because of their ability for mass production. The major disadvantage or task involved in these two printing processes is that the development of functional ink, which requires hours of research and development, incurs more cost. Owing to the above details, the research undertaken based on the gravure and flexography printing process is very minimal when compared to that of the inkjet and screen printing process.

5.3.1.2 Screen printing

The screen printing process is also known as push-through method, which uses an ink of sticky nature. The ink is transferred onto the substrate through a screen, which may be made up of wire, plastic, and metals. This method can be done using bare hands or by using fully or semi-automatic systems. The coating machine consists of the following parts: (i) stencil, (ii) squeegee, and (iii) screen as shown in figure 5.5. The squeegee is the material that is made up of rubber. The design to be printed on the substrate is engraved on the screen, and the ink is allowed to pass through the screen either by means of pushing or squeezing the screen. Thus, the ink is transferred onto the substrate [40].

In this process, the quality of printing depends on wire diameter, the thickness of the emulsion, the mesh count of the screen, offset height, and screen deflection angle. This process has a good output rate at a low cost, and there is very minimal wastage of materials. The screen printing is very famous for its flexibility. There are many flexible or wearable electronics such as wearable sensors fabricated using a screen printing process and exhibiting similar properties to that of conventionally manufactured electronic devices. A piezoelectric touch sensor was fabricated using screen printing by Emamian *et al* [41]. They fabricated the sensor using polyvinylidene

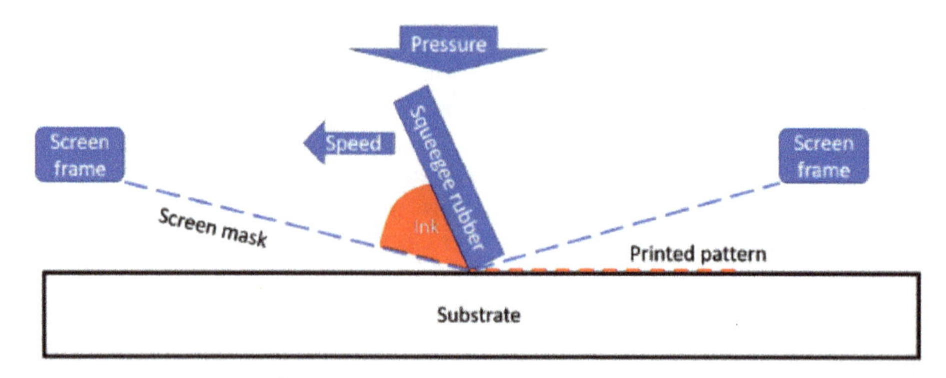

Figure 5.5. Schematic representation of screen printing, reprinted from [13] with permission from MDPI, Copyright (2021).

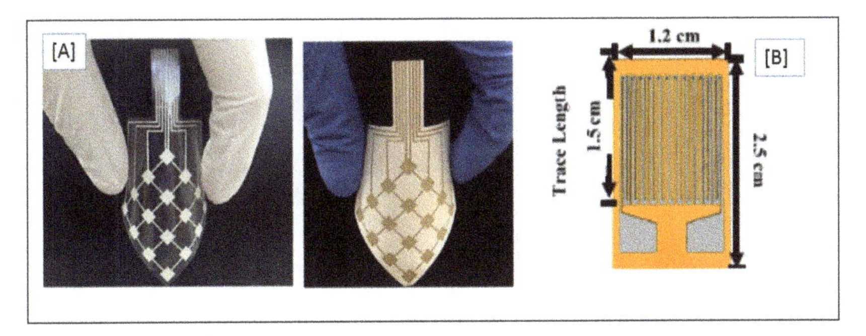

Figure 5.6. (A) Touch sensor based on polyvinylidene fluoride on PED and paper substrates, reprinted from [41] with permission from Elsevier, Copyright (2017). (B) Schematic representation of the RTD sensor, reprinted from [42] with permission from IEEE, Copyright (2019).

fluoride–based piezoelectric layer covered at the top and bottom by silver layers. These layers are coated using a screen printing process on PET and paper substrate as shown in figure 5.6(A). This PET substrate sensor has an observed sensitivity of 1.2 V N^{-1}, and the paper substrate sensors exhibit 0.3 V N^{-1} sensitivity. Hence, these sensors have good potential in applications such as robotics and sensors in automobiles. A sensor for temperature detection was developed by Turkani *et al.* They fabricated Ni-coated polyimide substrate sensors for detecting a wide range of resistance temperatures starting from −60 °C to 180 °C. This flexible resistance temperature detector (RTD) exhibits good repeatability and stability at all temperatures [42]. Figure 5.6(B) represents the RTD sensors. A flexible and stretchable sensor was fabricated by Bose *et al* using the screen printing process. The ink used for printing is silver ink, and the substrate is made up of TPU. The outcomes of the above work are that 20% of strain was detected by the wavy configured sensor, and they also exhibited excellent flexibility compared to conventional sensors. There are many studies that show that the screen printing process is the most viable and cost efficient [43–46]. This process can be utilized for the fabrication of flexible, wearable, and stretchable electronics [47, 48].

5.3.1.3 Offset printing process

This method is also known as the indirect printing method, because the ink is transferred from the initial or printing cylinder to the intermediate or blanket cylinder. From the blanket cylinder, the ink is transferred onto the substrate as shown in figure 5.7. The water roller will apply a small amount of water to the undesired part of the pattern so as to remove the ink. Surface engineering is an important aspect of offset printing. The image or desired areas accept the presence of ink but reject water, whereas undesired or non-image areas repel ink and accept water. The spreading of ink can be controlled by the surface energy. The ink is transferred to the paper substrate at pressure. In this process, the multiple layers of inks are coated simultaneously without an intermediate drying process; hence, this process is also called a wet-on-wet printing process. The coated inks are dried due to evaporation, absorption, and polymerization [49, 50].

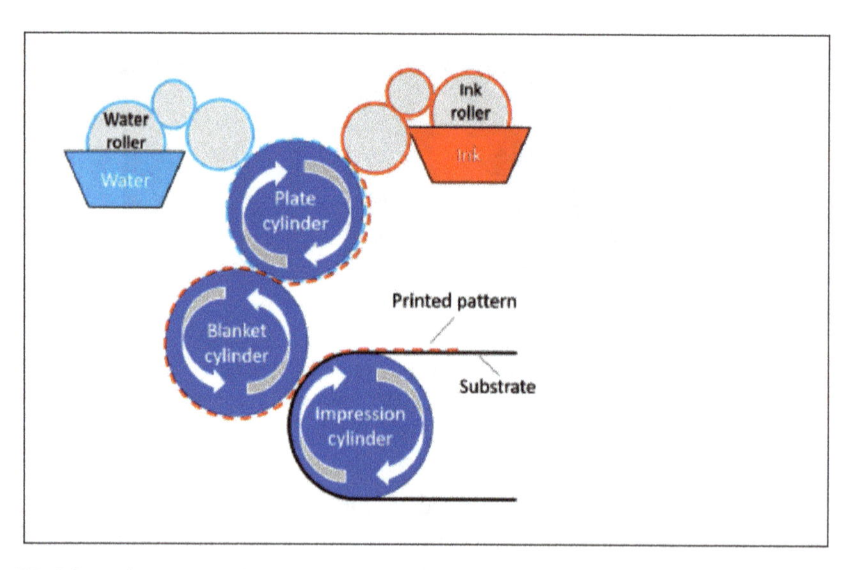

Figure 5.7. Schematic representation of the offset printing process, reprinted from [13] with permission from MDPI, Copyright (2021).

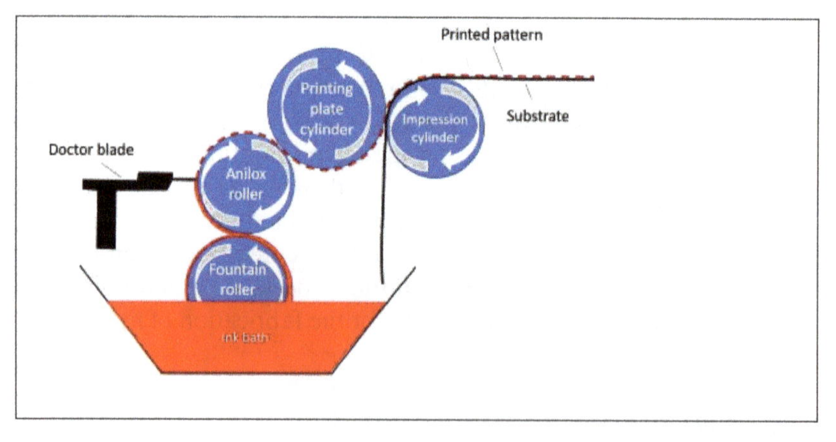

Figure 5.8. Schematic representation of the flexography printing process, reprinted from [13] with permission from MDPI, Copyright (2021).

5.3.1.4 Flexography printing process

The flexographic printing process is a kind of roll-to-roll printing process that has high throughput. Flexographic printing is also known as the rotational printing process. This method is an indirect contact-based printing method that has the ability to print ink of varying thicknesses with good resolution. As shown in figure 5.8, the flexography printing machine consists of the following parts [40]:

(i) Printing cylinder
(ii) Anilox cylinder

(iii) Impression cylinder
(iv) Ink reservoir
(v) Doctor blade

A cylinder made of steel, which is otherwise called an anilox cylinder is used to transfer ink from the reservoir. The cylinder is engraved with a design on its surface, normally the anilox cylinder is made up of ceramics or chromium. Then the ink is transferred onto the printing cylinder, which is also known as a printing plate. Then from the printing plate, the ink is transferred to the target substrate. When compared to screen printing and inkjet printing, the flexographic process has high processing speed. This method is most widely used in printing graphics and in printing operations in packaging industries. There are only very few studies in flexible electronics manufacturing through the flexographic printing process [51, 52]. CNT-based flexible TFTs were developed by Higuchi *et al* as shown in figures 5.9(A) and (B). They deposited nano silver ink, resist ink, and polyimide ink as source, gate and drain gates electrodes, CNT patterner, and insulator over a thin film made up of PEN fabricated through a flexographic printing process. The CNTs synthesized using chemical vapor deposition were coated out of the TFT electrodes. The TFT developed using flexographic printing has exhibited good stability and mobility. A paper-based sensor (strain) was developed by Maddipatla *et al.* They developed silver ink–coated strain gauges through a flexographic printing process. The strain gauges are of different lengths. This sensor has the ability to detect even small displacements of the order of 1 mm with good repeatability [53].

5.3.1.5 Pad printing process

Pad printing is a method in which a 2D pattern or cliche is printed on a 3D substrate or object, as shown in figure 5.10. This process utilizes an indirect offset printing method, in which the ink is transferred from the cliché or a stereotype using a silicon pad. This process is widely used as a replacement for the screen printing process. The products such as transistor electrodes are manufactured using this printing process [55].

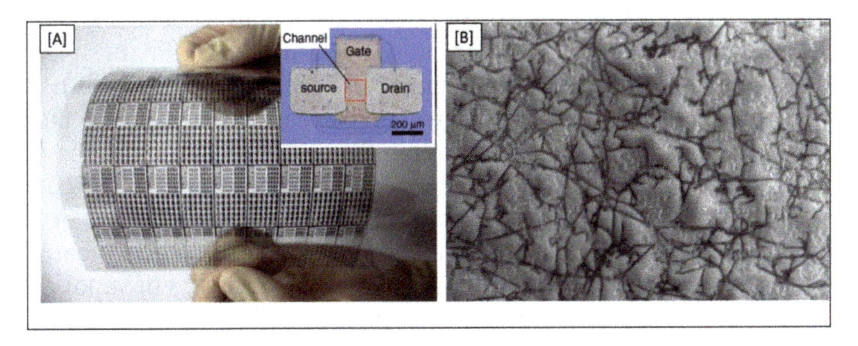

Figure 5.9. (A) TFT sensor based on CNTs fabricated on PEN substrate using flexography process. (B) Scanning electron microscopy image of CNT film [54], reprinted from [54] with permission from IOP Publishing, Copyright (2013).

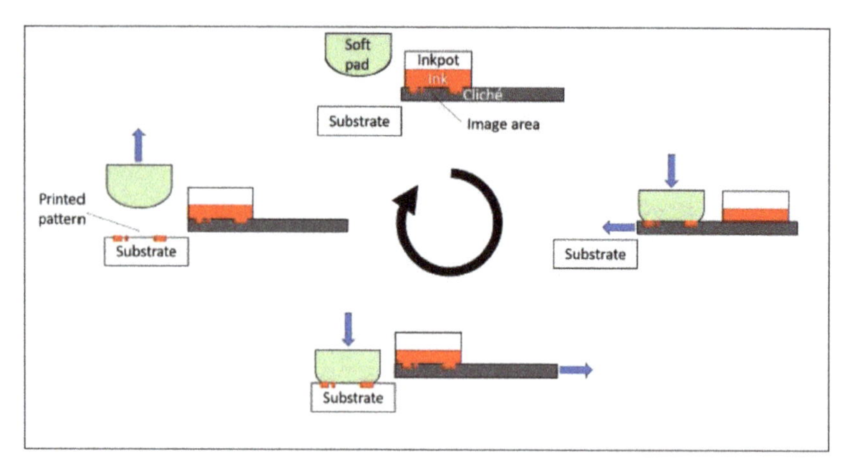

Figure 5.10. Schematic representation of the pad printing process, reprinted from [13] with permission from MDPI, Copyright (2021).

5.3.2 Non-contact printing methods

The non-contact printing process is one in which the nozzles are utilized to spray the ink onto the substrate without any contact with the substrate. The advantage of non-contact printing is that the life of the nozzle is longer because of less contamination, although the non-contact printing process is slow compared to the roll-to-roll manufacturing process. However, the efficiency of the non-contact mode is higher compared to the contact printing process, and it is possible to print computer-aided design (CAD) models, which is not feasible in the contact printing process. Hence, contact printing can be used to manufacture prototypes or used for highly demanded products [56].

5.3.2.1 Inkjet printing process

The inkjet printing process is a type of additive manufacturing process, which transfers ink onto the substrate based on digital CADs and does not utilize any physical patterns. Normally, the inkjet printing process uses inks with low viscosities so that deposition on ink will be easily compared to highly viscous ink. The inkjet printing process is of two types: (i) drop-on-demand and (ii) continuous printing. A voltage source is used to maintain a continuous flow of ink on the substrate. The inkjet printers, which use a thermal source, are very widely used in the packaging and graphic designing industries. In drop-on-demand inject printing, the inks are forced toward the substrate based on the digital signal received from the computers [40, 57]. The droplets can be generated either by means of piezoelectric or thermal methods, which is shown in figures 5.11(A) and (B). In thermal energy–based inkjet printers, the inks are enforced out of the printer nozzle by means of vaporization of ink. The volume of the printer nozzle is compressed or decompressed based on the digital input from the computers and the inks are dispersed out of the nozzle in piezoelectric printers. However, the drop-on-demand–based inkjet printing process was most widely used in flexible or wearable electronic fabrication processes owing

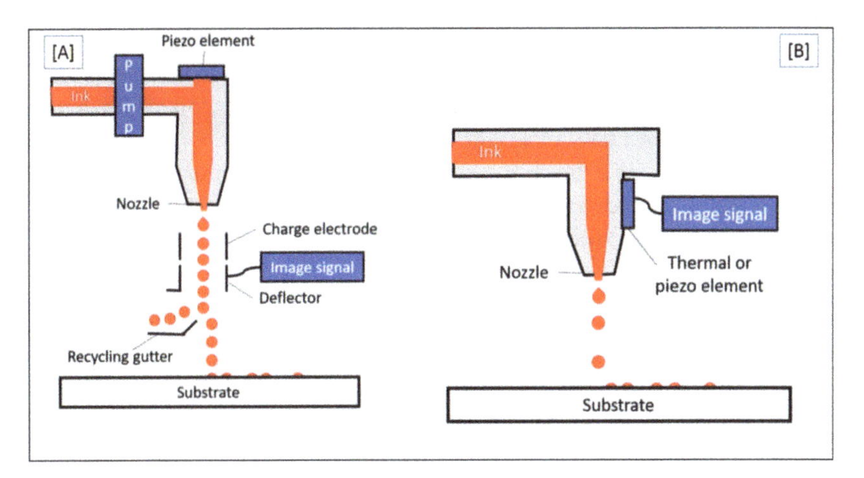

Figure 5.11. Schematic representation of (A) continuous printing and (B) drop-on-demand printing, reprinted from [13] with permission from MDPI, Copyright (2021).

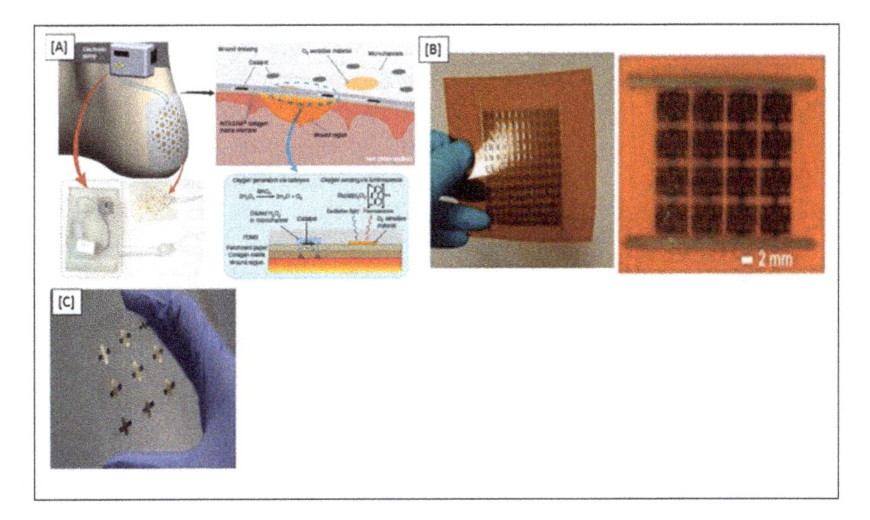

Figure 5.12. (A) Represents the bandage made up of flexible paper for sensing and delivery of oxygen in the treatment of chronic wounds, reprinted from [59] with permission from SPIE, Copyright (2018). (B) Supercapacitors based on graphene used as electrodes, reprinted from [65] with permission from ACS, Copyright (2017). (C) Metal–insulator–metal capacitor fabricated using inkjet printing, reprinted from [61] with permission from Nature, Copyright (2019).

to its advantages such as cost efficiency, high resolution, and easy fabrication without masks [58].

There are various types of flexible or wearable devices that are manufactured through inkjet printing methods. A paper-based oxygen-sensing and delivery sensor in bandage form was fabricated using an inkjet printing process by Ochea *et al.* This bandage was intended to treat chronic wounds as shown in figure 5.12(A). In this work, parchment paper was employed as the substrate over which manganese oxide and ruthenium inks are deposited, which has the ability to generate and measure oxygen in the wounded area. It is also reported that by varying the thickness of the manganese oxide layer, the oxygen concentration was controlled. The ruthenium coated on the substrate facilitates contactless measuring of oxygen at the wounded region owing to the fluorescence nature of ruthenium ink. These smart bandages also possess good mechanical properties and flexibility compared to that of conventional bandages but also possess additional properties such as oxygen generation, delivery, and therapeutics [59].

The wounds are not the same in nature; it varies according to the type of injuries, location of injuries, and the depth of injuries. Hence, the different concentration of oxygen generation and different therapeutics is necessary. The fabrication of smart bandages that should be done using the inkjet printing process is very important to carry out further research and to customize bandages for mass production. Hence, the importance of this work is in the treatment of wounds [60].

A graphene based super micro capacitors manufactured using exfoliation of graphene using electrochemical process for the application of electrodes and collectors of current and poly(4-styrenesulfonic acid) as ink. The image of the supercapacitors is shown in figure 5.12(B). A metal–insulator–metal capacitor fabricated by Mikolajek *et al* using inkjet printers as shown in figure 5.12(C). The silver ink is coated as metal electrodes and the PMMA/BST ($Ba_{0.6}Sr_{0.4}TiO_3$) is coated as an insulator on PET substrate using inkjet printing process. By using this process, thin, homogeneous, and smooth layers along with better resolution and fewer defects can be produced. This PMMA/BST composite insulator layer shows a better dielectric constant when compared to that of pure PMMA [61]. A flexible microfluidic sensor was designed and fabricated by Narakathu *et al* using inkjet printing. This silver-coated sensor has the capacity to detect different hazardous chemicals such as cadmium sulfide, molybdenum disulphate, and mercury sulfide by electrochemical impedance spectroscopy [62]. Because of the attractive characteristics of the inkjet printing process, which includes high resolution and ease of fabrication, a lot of researchers are attracted to develop devices such as SERS substrate sensors for detecting gases, antennas, and bandages [63]. In the inkjet printing process, the clogging of nozzles happens due to the faster rate of evaporation and agglomeration of ink particles, which in turn reduces the efficiency of inkjet printing. Hence, frequent cleaning of the nozzle is a big challenge. Also, the nozzle cartridges for one-time use are usually expensive, owing to the inkjet printing process speed, which is very slow, and the use of the nozzle increases the production time compared to other contact printing processes [64].

5.3.2.2 Aerosol printing process

In the aerosol printing process, the ink is atomized so as to reduce the size of the ink droplets in the range of 1–5 μm in diameter as shown in figure 5.13. The ultrasonic technique or pneumatic method can be utilized to atomize the ink droplets. This system is entirely maintained in vacuum condition, and the ink is directed toward the ceramic nozzle by means of nitrogen gas and transferred onto the substrate under high pressure. It is possible to print on conformal as well as plane surfaces. However,

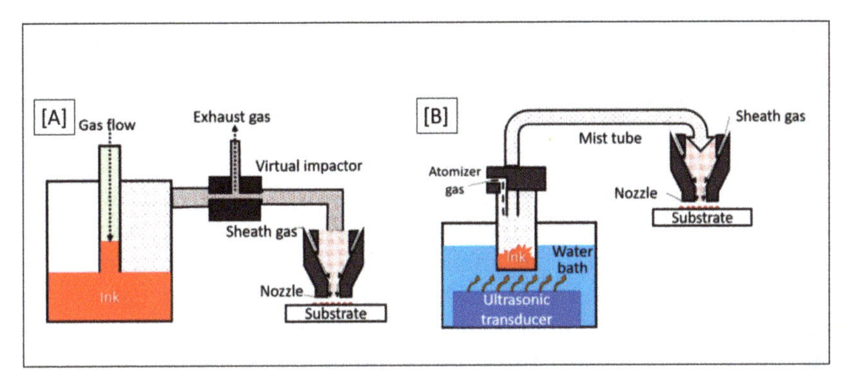

Figure 5.13. Schematic representation of (A) pneumatic aerosol printing and (B) ultrasonic aerosol printing, reprinted from [13] with permission from MDPI, Copyright (2021).

in order to print complex designs, it is necessary to control the beam. Thus, by varying the size of the nozzle, it is possible to control the beam. It is also to be noted that the distance between the nozzle and the substrate should be less than 10 mm or above 1 mm so as to achieve good accuracy because if these boundary conditions are crossed, then there is a possibility of overspray defects in the pattern printed on the substrate. The aerosol printing process also does not need any physical pattern, digital models can be printed using this technique. The following are the advantages of the aerosol printing process: (i) there is no contamination or clogging of the nozzle, and (ii) even small sizes such as 10 μm can be fabricated with high resolution using this process. Conversely, the main shortcoming of the aerosol printing process is the speed. The machining rate is 12 m min^{-1}, which is slow compared to other processes. Hence, this process is not considered for mass production [56, 66, 67].

5.4 Conclusions

The advancement of flexible or wearable electronic devices via different additive manufacturing or printing processes is rising exponentially owing to apparent reasons such as low cost, fast fabrication, light weight, and need for thin devices. In this chapter, the importance of flexible or wearable electronics and the processes to be carried out before and after printing, along with many recently developed flexible electronic devices utilizing different processes, is explored. It is to be noted that there are many scientific difficulties that still exist in manufacturing flexible or wearable electronic devices, which needs to be focused to perform research in a better way, and in adopting suitable advanced printing methods for the fabrication of flexible or wearable devices. One such challenge is the wet film thickness from the screen printing process, which results in more spreading of ink, and the resolution of the printing will be less if the wet ink is not sintered immediately.

Additionally, the evaporation of solvent present in the ink leads to a reduction in the quality of the mesh. This is due to the exposure of the printed surface to the atmosphere for a longer duration during the printing process [68].

In addition, the drawbacks like drying of inks, the viscosity of inks, the quantity of inks, an incorrect volume of anilox roller, and the working speed of the printers result in a squashed ink look at the exterior edges of the patterns printed using the flexographic printing process.

Because of many different issues in fabrication processes, there are enormous inconsistencies in the development of flexible or wearable devices, which leads to unreliability and unsteady performance. To overcome the above-said drawbacks in flexible or wearable devices, the improvisation and standardization of various printing parameters are necessary to develop devices with better reliability, stability, and repeatability. So as to maintain standards and to improve flexible and wearable electronics research to the next level, standardization of all parameters that have a great impact on flexible electronics fabrication is necessary. The parameters include CAD designs, characterization, printing parameters such as deposition time, humidity, and temperatures, post-printing processes such as sintering, mechanical testing, etc. Researchers around the world are working to develop advanced manufacturing systems to develop the research area of flexible or wearable devices to the next level. Hence, implementing these new additive printing methods for flexible or wearable electronic applications would possibly lead to cost-efficient and reliable production. Thus, it is possible to revolutionize the application of flexible or wearable devices in many areas such as agriculture, health, automobile, defense, and food industries.

References

[1] Zeng X, Yang C, Chiang J F and Li J 2017 Innovating e-waste management: from macroscopic to microscopic scales *Sci. Total Environ.* **575** 1–5

[2] Tan M J, Owh C, Chee P L, Kyaw A K K, Kai D and Loh X J 2016 Biodegradable electronics: cornerstone for sustainable electronics and transient applications *J. Mater. Chem.* C **4** 5531–58

[3] Maddipatla D, Narakathu B B and Atashbar M 2020 Recent progress in manufacturing techniques of printed and flexible sensors: a review *Biosensors* **10** 199

[4] Dizon J R C, Espera A H, Chen Q and Advincula R C 2018 Mechanical characterization of 3D-printed polymers *Addit. Manuf.* **20** 44–67

[5] Liu Y, Wang H, Zhao W, Zhang M, Qin H and Xie Y 2018 Flexible, stretchable sensors for wearable health monitoring: sensing mechanisms, materials, fabrication strategies and features *Sensors* **18** 645

[6] Yao S, Swetha P and Zhu Y 2018 Nanomaterial-enabled wearable sensors for healthcare *Adv. Healthcare Mater.* **7** 1700889

[7] Nakata S, Arie T, Akita S and Takei K 2017 Wearable, flexible, and multifunctional healthcare device with an ISFET chemical sensor for simultaneous sweat ph and skin temperature monitoring *ACS Sens.* **2** 443–8

[8] Kipphan H 2001 Printing technologies with permanent printing master *Handbook of Print Media: Technologies and Production Methods* ed H Kipphan (Berlin: Springer) pp 203–448

[9] Kipphan H 2001 Printing technologies without a printing plate (NIP technologies) *Handbook of Print Media: Technologies and Production Methods* ed H Kipphan (Berlin: Springer) pp 675–758

[10] Chang J S, Facchetti A F and Reuss R 2017 A circuits and systems perspective of organic/printed electronics: review, challenges, and contemporary and emerging design approaches *IEEE J. Emerg. Sel. Top. Curcuits Syst.* **7** 7–26
[11] Hashmi S G, Özkan M, Halme J, Zakeeruddin S M, Paltakari J and Grätzel M *et al* 2016 Dye-sensitized solar cells with inkjet-printed dyes *Energy Environ. Sci.* **9** 2453–62
[12] Kjar A and Huang Y 2019 Application of micro-scale 3D printing in pharmaceutics *Pharmaceutics* **11** 390
[13] Wiklund J, Karakoç A, Palko T, Yiğitler H, Ruttik K and Jäntti R *et al* 2021 A review on printed electronics: fabrication methods, inks, substrates, applications and environmental impacts *J. Manuf. Mater. Process.* **5** 89
[14] Magdassi S 2009 *The Chemistry of Inkjet Inks* (Singapore: World Scientific)
[15] Kamyshny A and Magdassi S 2014 Conductive nanomaterials for printed electronics *Small* **10** 3515–35
[16] Izdebska J 2016 Aging and degradation of printed materials *Printing on Polymers: Fundamentals and Applications* (Oxford, UK: Elsevier) PDL Handbook Series 353–70
[17] Martin D P, Melby N L, Jordan S M, Bednar A J, Kennedy A J and Negrete M E *et al* 2016 Nanosilver conductive ink: a case study for evaluating the potential risk of nanotechnology under hypothetical use scenarios *Chemosphere* **162** 222–7
[18] Valentine A D, Busbee T A, Boley J W, Raney J R, Chortos A and Kotikian A *et al* 2017 Hybrid 3D printing of soft electronics *Adv. Mater.* **29** 1703817
[19] Choi Y, dong S K and Piao Y 2019 Metal–organic decomposition ink for printed electronics *Adv. Mater. Interfaces* **6** 1901002
[20] Janczak D, Słoma M, Wróblewski G, Młożniak A and Jakubowska M 2014 Screen-printed resistive pressure sensors containing graphene nanoplatelets and carbon nanotubes *Sensors* **14** 17304–12
[21] Suganuma K 2014 Introduction *Introduction to Printed Electronics* (New York: Springer) pp 1–22
[22] Cui Z 2016 Introduction *Printed Electronics* (New York: Wiley) pp 1–20
[23] Ji T, Sun M and Han P 2014 A review of the preparation and applications of graphene/semiconductor composites *Carbon* **70** 319
[24] Kim M, Safron N S, Han E, Arnold M S and Gopalan P 2010 Fabrication and characterization of large-area, semiconducting nanoperforated graphene materials *Nano Lett.* **10** 1125–31
[25] Hwang S W, Tao H, Kim D H, Cheng H, Song J K and Rill E *et al* 2012 A physically transient form of silicon electronics *Science* **337** 1640–4
[26] Kim S 2020 Inkjet-printed electronics on paper for RF identification (RFID) and sensing *Electronics* **9** 1636
[27] Agate S, Joyce M, Lucia L and Pal L 2018 Cellulose and nanocellulose-based flexible-hybrid printed electronics and conductive composites—a review *Carbohydrate Polym.* **198** 249–60
[28] Välimäki M K, Sokka L I, Peltola H B, Ihme S S, Rokkonen T M J and Kurkela T J *et al* 2020 Printed and hybrid integrated electronics using bio-based andrecycled materials—increasing sustainability with greener materials andtechnologies *Int. J. Adv. Manuf. Technol.* **111** 325–39
[29] Fischer T, Rühling J, Wetzold N, Zillger T, Weissbach T and Göschel T *et al* 2018 Roll-to-roll printed carbon nanotubes on textile substrates as a heating layer in fiber-reinforced epoxy composites *J. Appl. Polym. Sci.* **135** 45950

[30] Torrisi F, Hasan T, Wu W, Sun Z, Lombardo A and Kulmala T S *et al* 2012 Inkjet-printed graphene electronics *ACS Nano* **6** 2992–3006
[31] Devaraj H and Malhotra R 2019 Scalable forming and flash light sintering of polymer-supported interconnects for surface-conformal electronics *J. Manuf. Sci. Eng.* **141** 041014
[32] Constante G, Apsite I, Alkhamis H, Dulle M, Schwarzer M and Caspari A *et al* 2021 4D biofabrication using a combination of 3d printing and melt-electrowriting of shape-morphing polymers *ACS Appl. Mater. Interfaces* **13** 12767–76
[33] Khan S, Lorenzelli L and Dahiya R S 2015 Technologies for printing sensors and electronics over large flexible substrates: a review *IEEE Sens. J.* **15** 3164–85
[34] Saengchairat N, Tran T and Chua C K 2017 A review: additive manufacturing for active electronic components *Virtual Phys. Prototyp.* **12** 31–46
[35] Reddy A S G, Narakathu B B, Atashbar M Z, Rebros M, Rebrosova E and Joyce M K 2011 Fully printed flexible humidity sensor *Procedia Eng.* **25** 120–3
[36] Maddipatla D, Janabi F, Narakathu B B, Ali S, Turkani V S and Bazuin B J *et al* 2019 Development of a novel wrinkle-structure based SERS substrate for drug detection applications *Sens. Bio-Sens. Res.* **24** 100281
[37] Zhu H, Narakathu B B, Fang Z, Tausif Aijazi A, Joyce M and Atashbar M *et al* 2014 A gravure printed antenna on shape-stable transparent nanopaper *Nanoscale* **6** 9110–5
[38] Bariya M, Shahpar Z, Park H, Sun J, Jung Y and Gao W *et al* 2018 Roll-to-roll gravure printed electrochemical sensors for wearable and medical devices *ACS Nano* **12** 6978–87
[39] Lau P H, Takei K, Wang C, Ju Y, Kim J and Yu Z *et al* 2013 Fully printed, high performance carbon nanotube thin-film transistors on flexible substrates *Nano Lett.* **13** 3864–9
[40] Kipphan H (ed) 2001 *Handbook of Print Media: Technologies and Production Methods* (Berlin: Springer) p 1207
[41] Emamian S, Narakathu B B, Chlaihawi A A, Bazuin B J and Atashbar M Z 2017 Screen printing of flexible piezoelectric based device on polyethylene terephthalate (PET) and paper for touch and force sensing applications *Sensors Actuators* A **263** 639–47
[42] Turkani V S, Maddipatla D, Narakathu B B, Altay B N, Fleming P D and Bazuin B J *et al* 2019 Nickel based RTD fabricated via additive screen printing process for flexible electronics *IEEE Access* **7** 37518–27
[43] Dhanabalan S S, Arun T, Periyasamy G, N D, N C and Avaninathan S R *et al* 2022 Surface engineering of high-temperature PDMS substrate for flexible optoelectronic applications *Chem. Phys. Lett.* **800** 139692
[44] Sundar D S, Raja A S, Sanjeeviraja C and Jeyakumar D 2017 High temperature processable flexible polymer films *Int. J. Nanosci.* **16** 1650038
[45] Sundar D S, Sivanantharaja A, Sanjeeviraja C and Jeyakumar D 2016 Synthesis and characterization of transparent and flexible polymer clay substrate for OLEDs *Mater. Today: Proc.* **3** 2409–12
[46] Shanmuga sundar D, Sivanantha Raja A, Sanjeeviraja C and Jeyakumar D 2016 Highly transparent flexible polydimethylsiloxane films—a promising candidate for optoelectronic devices *Polym. Int.* **65** 535–43
[47] Cinti S and Arduini F 2017 Graphene-based screen-printed electrochemical (bio)sensors and their applications: efforts and criticisms *Biosens. Bioelectron.* **89** 107–22
[48] Cao R, Pu X, Du X, Yang W, Wang J and Guo H *et al* 2018 Screen-printed washable electronic textiles as self-powered touch/gesture tribo-sensors for intelligent human–machine interaction *ACS Nano* **12** 5190–6

[49] Hakola E 2009 Principles of conventional printing ed P Oittinen and H Saarelma *Papermaking Science and Technology* (Finland: Finnish Paper Engineers' Association) pp 40–87

[50] Dhanabalan S S, R S, Madurakavi K, Thirumurugan A, M R and Avaninathan S R *et al* 2022 Flexible compact system for wearable health monitoring applications *Comput. Electr. Eng.* **102** 108130

[51] Shrestha S, Yerramilli R and Karmakar N C 2019 Microwave performance of flexo-printed chipless RFID tags *Flex. Print. Electron.* **4** 045003

[52] Fung C M, Lloyd J S, Samavat S, Deganello D and Teng K S 2017 Facile fabrication of electrochemical ZnO nanowire glucose biosensor using roll to roll printing technique *Sensors Actuators* B **247** 807–13

[53] Maddipatla D, Narakathu B B, Avuthu S G R, Emamian S, Eshkeiti A and Chlaihawi A A *et al* 2015 A novel flexographic printed strain gauge on paper platform *IEEE Sensors* **2015** 1–4

[54] Higuchi K, Kishimoto S, Nakajima Y, Tomura T, Takesue M and Hata K *et al* 2013 High-mobility, flexible carbon nanotube thin-film transistors fabricated by transfer and high-speed flexographic printing techniques *Appl. Phys. Express* **6** 085101

[55] Knobloch A, Bernds A and Clemens W 2001 Printed polymer transistors *1st Int. IEEE Conf. on Polymers and Adhesives in Microelectronics and Photonics Incorporating POLY, PEP & Adhesives in Electronics Proc. (Cat No01TH8592)* pp 84–90

[56] Saengchairat N, Tran T and Chua C K 2017 A review: additive manufacturing for active electronic components *Virtual Phys. Prototyp.* **12** 31–46

[57] Izdebska-Podsiadły J and Thomas S 2015 *Printing on Polymers: Fundamentals and Applications* (Amsterdam: Elsevier)

[58] Corzo D, Almasabi K, Bihar E, Macphee S, Rosas-Villalva D and Gasparini N *et al* 2019 Digital inkjet printing of high-efficiency large-area nonfullerene organic solar cells *Adv. Mater. Technol.* **4** 1900040

[59] Ochoa M, Rahimi R, Zhou J, Jiang H, Yoon C K and Oscai M *et al* 2018 A manufacturable smart dressing with oxygen delivery and sensing capability for chronic wound management *Micro- and Nanotechnology Sensors, Systems, and Applications X* 10639; T George, A K Dutta and M S Islam (Bellingham, WA: SPIE) p 106391C

[60] Maddipatla D, Narakathu B B, Ochoa M, Rahimi R, Zhou J and Yoon C K *et al* 2019 Rapid prototyping of a novel and flexible paper based oxygen sensing patch via additive inkjet printing process *RSC Adv.* **9** 22695–704

[61] Mikolajek M, Reinheimer T, Bohn N, Kohler C, Hoffmann M J and Binder J r 2019 Fabrication and characterization of fully inkjet printed capacitors based on ceramic/polymer composite dielectrics on flexible substrates *Sci. Rep.* **9** 13324

[62] Narakathu B B, Avuthu S G R, Eshkeiti A, Emamian S and Atashbar M Z 2015 Development of a microfluidic sensing platform by integrating PCB technology and inkjet printing process *IEEE Sens. J.* **15** 6374–80

[63] Abutarboush H F and Shamim A 2018 A reconfigurable inkjet-printed antenna on paper substrate for wireless applications *IEEE Antennas Wirel. Propag. Lett.* **17** 1648–51

[64] Martin G D, Hoath S D and Hutchings I M 2008 Inkjet printing—the physics of manipulating liquid jets and drops *J. Phys. Conf. Ser.* **105** 012001

[65] Li J, Sollami Delekta S, Zhang P, Yang S, Lohe M R and Zhuang X *et al* 2017 Scalable fabrication and integration of graphene microsupercapacitors through full inkjet printing *ACS Nano* **11** 8249–56

[66] Dimitriou E and Michailidis N 2021 Printable conductive inks used for the fabrication of electronics: an overview *Nanotechnology* **32** 502009
[67] Chen Y D, Nagarajan V, Rosen D W, Yu W and Huang S Y 2020 Aerosol jet printing on paper substrate with conductive silver nano material *J. Manuf. Processes* **58** 55–66
[68] Joannou G 1988 Screen inks ed R H Leach, C Armstrong, J F Brown, M J Mackenzie, L Randall and H G Smith *The Printing Ink Manual* (Boston, MA: Springer) pp 481–514

Chapter 6

Laser patterning in fabrication of flexible perovskite solar cells

Samuel Paul David, Sahaya Dennish Babu and Ananthakumar Soosaimanickam

Organic–inorganic hybrid perovskite-based solar cells have shown incredible efficiency owing to their outstanding structural and optical properties. Progress in the fabrication approaches of perovskite solar cells has led to maximum efficiency of over 25% within a short period. For the fabrication of high-efficiency perovskite solar cells, different approaches are followed starting from preparation of precursors, incorporation of additives, variation of solvents for the coating, deposition etc. To fabricate perovskite solar cells as a module, lasers play an important role in patterning the substrate and deposited layers. Through laser patterning, highly efficient perovskite solar cells are achieved by carefully modifying fabrication parameters. Nature of laser pulse (nanosecond or femtosecond), wavelength and focal spot are severely affecting the removal of different layers in the solar cell and their interconnections. Specifically, in order to apply for the indoor and outdoor applications, all these benefits are concentrated towards the industrial production of flexible perovskite solar cell (FPSC) modules. Carefully ablated layers through P1, P2 and P3 pattern steps and their impact on the electrical properties of perovskite solar cells clearly envisages how lasers are essential for the fabrication of highly efficient FPSCs. In this chapter, the principle of laser patterning in solar cells and the current developments achieved in the case of FPSCs using the same have been discussed.

6.1 Introduction to solar cells

Photovoltaic (PV) technology is the most promising technology to address the ever-growing global energy demand. With zero carbon emissions, solar PV technology is a renewable technology and is based on the principle of extracting electrical energy from the photon energy of sunlight (Doumon *et al* 2022). Solar cells are made of organic, inorganic and organic–inorganic hybrid semiconducting materials, with an

doi:10.1088/978-0-7503-5492-9ch6

ultimate goal of achieving large-area modules with high power and power conversion efficiency (PCE). Recently, LONGi from China has claimed to have achieved a record PCE of 26.81% for any silicon-based solar cells (Green *et al* 2023). Based on III–V compound semiconductor quantum well nanostructures, France *et al* from the National Renewable Energy Laboratory, United States of America, created a triple-junction tandem solar cell with the highest record efficiency of 39.5% of any solar cell type measured under standard 1 sun condition (France *et al* 2022). At a concentration of 665 suns, researchers from the Fraunhofer Institute for Solar Energy Systems achieved a global milestone efficiency of 47.6% using III–V compound semiconductor materials (Green *et al* 2023). Figure 6.1 shows 30 years' progress in efficiency of solar cells and modules with a minimum area of 1 cm^2. The graph shows both 'one-sun' cells and also concentrator cells. Green *et al* provides a constant update on the highest confirmed efficiency of solar cells and modules every six months with a recent report on large-area solar cells (Green *et al* 2023). However, the production cost of large-area commercial solar cells are still quite high. Considering the goal of addressing global electricity demand, crucial efforts are being made to improve the overall device performance with an enhanced PCE. This requires an improvement in both active and other device materials used in the module and also modification in cell architecture. Various approaches have been undertaken by researchers to achieve environmentally friendly solar modules at low cost to cover large surface areas with excellent flexibility. Some of the approaches

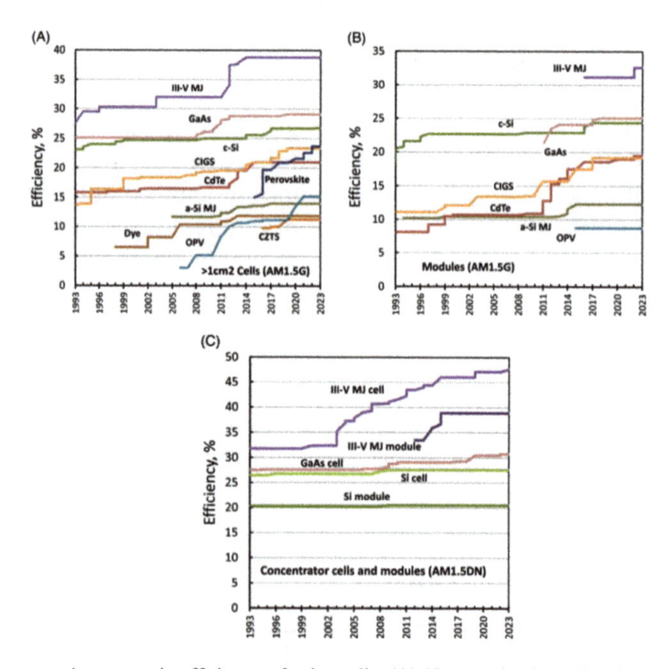

Figure 6.1. Thirty years' progress in efficiency of solar cells. (A) 'One-sun' solar cells with a minimum area of 1 cm^2. (B) Progress of most popular copper indium gallium selenide (CIGS) and cadmium telluride (CdTe) solar cells. (C) Efficiency of concentrator cells and modules. Reproduced from Green *et al* (2023) with permission from John Wiley & Sons.

are (i) chemical compositional change and tuning, (ii) fabrication of thin-film solar modules and nanostructures, (iii) interface control and (iv) band alignment. Thin-film solar modules are attractive because they require significantly less material and low manufacturing cost, thereby reducing the overall device cost.

6.2 Laser patterning of thin-film solar cells

Thin-film PV modules consist of a thin absorber layer having a thickness of a subnanometer to a few microns, sandwiched between top and bottom conducting electrodes. The top conducting electrode (e.g. indium tin oxide, ITO) needs to be transparent to permit the incident solar light to reach the absorber, whereas the bottom electrode can be either transparent or opaque. Absorber layers are semi-conducting active layers typically deposited on a stable substrate such as glass having surface areas ranging from a few square centimeters to several square meters (Stegemann *et al* 2012). High-power thin-film solar modules are generally made of several individual cells electrically connected in series. In such cases, individual cells need to be electrically isolated but connected in series with the electrodes present at the top and bottom of the PV module. Another primary advantage of such thin-film modules is the possibility of monolithic cell integration in which the electrical series interconnection is made within the structure itself. This requires an important additional process of patterning or scribing in sequence with the deposition of conductive, absorber and buffer layers. At least three steps of patterning are needed to electrically isolate the cells while connecting them in series (Jamaatisomarin *et al* 2023). Figure 6.2 shows step-by-step patterning of a monolithic copper indium

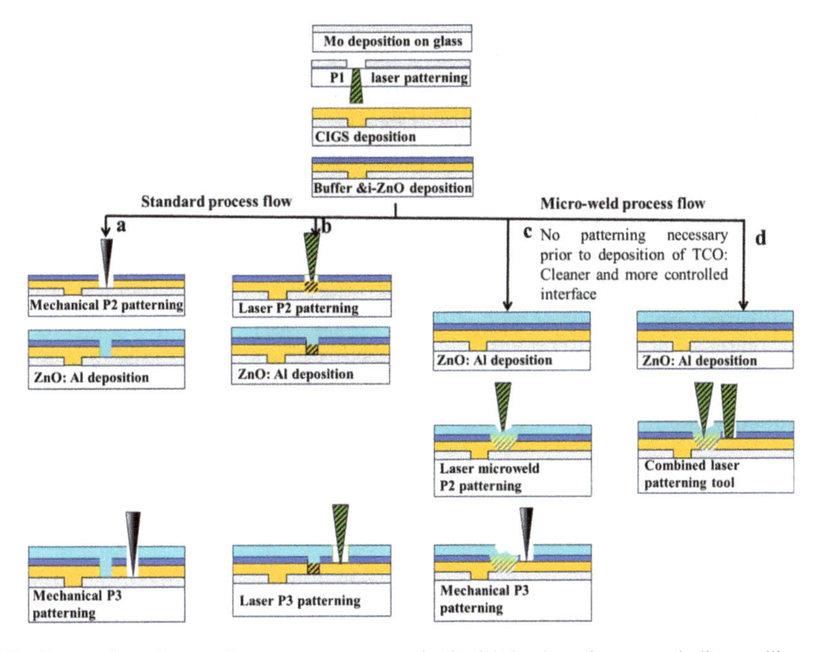

Figure 6.2. Various deposition and patterning processes in the fabrication of a copper indium gallium selenide (CIGS) PV module. Reprinted from Westin *et al* (2011) with permission from Elsevier, Copyright (2011).

gallium selenide (CIGS) solar module using mechanical and laser scribing techniques (Westin *et al* 2011).

The bottom conducting electrode of PV module is deposited on a substrate such as glass or organic material. The first patterning step (P1) separates the bottom layers, isolating them as separate cells. An optimum size of a cell is decided based on the maximum achievable photocurrent per solar cell with a minimum resistance loss. This step is followed by deposition of the absorber and electron transport layers (ETLs). The second patterning step (P2) selectively removes deposited absorber/buffer layers, creating a channel in order to create electrical contact between the bottom and top conducting layers. This step is followed by the deposition of a conducting layer that acts as the top electrode. The final patterning step (P3) is performed to electrically isolate the top contact layer by removing a thin line of deposited material. In this way, monolithic integration of thin-film solar modules is achieved by patterning technique.

Chemical methods such as polymer-based photolithography and mechanical methods such as stylus patterning are some of the techniques used to perform the above-mentioned scribing steps. However, both of these methods suffer from serious limitations, which are not ideal for next-generation solar cells. Comparatively, patterning-using lasers are attractive due to their non-contact in nature with substrate, lack of chemicals, high precision, repeatability, flexibility, and less demanding maintenance. Even though nanosecond pulsed lasers are widely used in patterning processes, this pulse duration can still be long enough to alter properties of absorber material in some solar cells, especially cells made of chalcopyrites such as copper-indium-gallium-diselenide. Therefore, ultrashort laser sources with picosecond and femtosecond laser pulses are preferred in thin-film PV modules due to their high precision and higher production rates (few meters per second). During the past few years, highly stable ultrashort solid-state lasers are commercially available at a reasonable price. This provides opportunities to readily employ ultrashort lasers for patterning processes due to ablation. Ablation occurs when the material absorbs incident laser beam and evaporates. With a high degree of coherence and short pulse duration, lasers perform selective removal of absorber thin films and electrode layers by ablation. Low divergence laser beams, when focused tightly, develop very narrow scribing lines, which make them very attractive in device fabrication (Stegemann *et al* 2012). Laser wavelength, pulse energy, fluence and beam shape are carefully chosen based on the materials and patterning parameters. Since this chapter aims to discuss laser patterning of flexible solar cells, we introduce some aspects of flexible substrates and the role of laser patterning on the same.

6.3 Flexible substrates

Solar cells can be classified into four generations. The first generation solar cell was based on crystalline silicon (c-Si) wafers. Thin-film technology was adopted during the second generation of solar cells, thereby significantly reducing the quantity of usable material. Even though thin-film solar cells drastically reduce the cost of the cell, their efficiency was still limited. Efficiency is significantly improved by third generation solar cells that are based on novel materials and concentrated energy capture technique. Third generation solar cells are multijunction PV solar cells based on organic materials, polymers, perovskites, quantum dots and so on (Khatibi *et al* 2019). However, they suffer from high cost, which led to the fourth generation solar cells. Fourth generation solar cells are based on polymer thin films in combination with nanostructure of inorganic materials. They exhibit high efficiency, flexibility and stability at a cheaper price (Shanmuga Sundar *et al* 2016a, Iqbal *et al* 2022). Flexible solar cells are third generation solar cells and have gained significant attention in recent times because of their flexibility, light weight, portability, less demanding maintenance, easy installation and easy integration with any surfaces with irregular shapes (Pagliaro *et al* 2008). They can also be integrated with flexible energy storage devices such as super capacitors and batteries. Such lightweight solar modules with good mechanical strength are attractive for wearable devices, smart phones, and outdoor applications, especially in large-scale industrial roofs and so on. With the best power to mass ratio and less weight, flexible solar cells can be utilized as solar arrays for space missions. Use of brittle metal oxide conducting layers such as ITO, which require high processing temperature, prevented the fabrication of flexible substrates in the first two generations of solar cells. After years of intense research, flexible solar cells are now commercially available. Recently Saravanapavanantham *et al* from Massachusetts Institute of Technology demonstrated ~50 μm ultrathin PV systems that can be easily integrated with composite fabric (Saravanapavanantham *et al* 2023). The device performed without any significant drop in PCE even after several hundreds of roll-up cycles as shown in figure 6.3. Organic and amorphous semiconducting materials with low synthesis temperature along with flexible substrates are crucial to achieve flexible solar cells.

Glass plates are conventionally used as substrates for the fabrication of solar cells. These glass substrates contribute to more than 90% of the cell weight and are

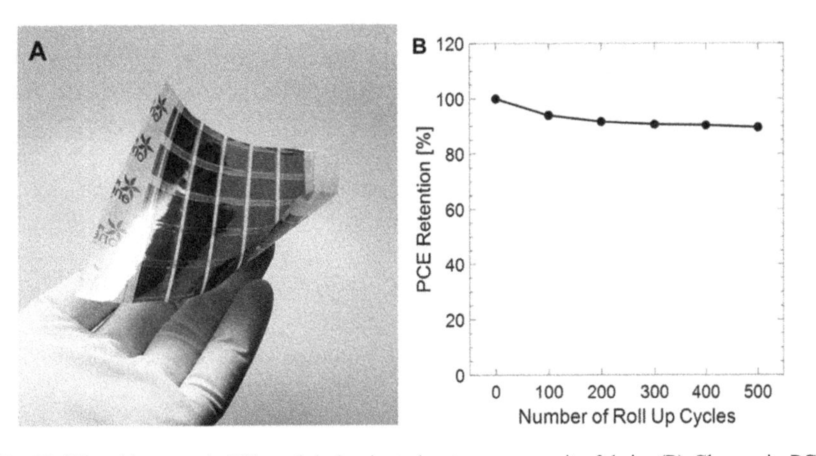

Figure 6.3. (A) Ultrathin organic PV module laminated onto a composite fabric. (B) Change in PCE as a function of roll-up cycles around a 6.4 mm cylindrical rod. Reprinted from Saravanapavanantham *et al* (2023) with permission from John Wiley & Sons.

replaced with flexible substrates for flexible solar cells. Flexible substrates need to be optically transparent to allow incoming sunlight and should possess high conductivity, good flexibility, good thermal stability and mechanical and robust bending durability. Flexible substrates are based on three categories of materials: metal thin foils, ultrathin flexible glasses and polymer materials. Metal thin films (Mo, Al, Ti and so on) have good mechanical stability, conductivity and a good chemical resistance; however, they are limited by their optical transparency. On the other hand, ultrathin flexible glasses exhibit good transparency but are limited by their weight and cost. Polymers such as polyimide (PI), poly(ethylene terephthalate) (PET) and poly(ethylene naphthalate) (PEN) have become more attractive substrates for flexible electronics due to their low cost, less weight, roll-to-roll (R2R) processability and flexibility (Sundar *et al* 2017). However, the processing temperature of the module cannot exceed ~150°C. Polymer substrates are widely chosen as flexible substrates for various flexible electronic devices (Sundar and Sivanantharaja 2013, Shanmuga Sundar *et al* 2016b). Flexible substrates are primarily developed by low temperature solution and R2R processing methods. Other methods, such as vapor deposition, drop casting and screen printing, are also employed to achieve flexible substrates.

To prepare flexible solar modules, flexible bottom electrodes are deposited on top of the substrates. The electrodes should be transparent with high electrical conductivity and are based on transparent conducting oxides (TCO). With a low sheet resistance, ITO is commonly used as flexible and transparent conducting electrodes (TCE). Nanomaterials such as metal nanowires, carbon nanotubes, graphene and oxide-based multilayers are also used as TCE materials especially for perovskite solar cells as a replacement to ITO (Jung *et al* 2019). Deposition of TCE is followed by the deposition of thin films of semiconducting active materials (e.g. Si, CIGS, cadmium telluride (CdTe), perovskites, hybrid materials) and ETLs (e.g. ZnO, TiO_2, SnO_2). Deposited active layers undergo an annealing process to induce crystallization of thin-film layers, which enhances the overall performance of the module. However, annealing temperature and duration time are limited to the maximum processing temperature of plastic substrates (PET, PEN) to avoid any thermal damage. Subsequent deposition of the top electrode (e.g. Au and Ag) results in a flexible PV module (Wang *et al* 2018, Xu 2022). The incoming light enters through the transparent bottom flexible electrode and is absorbed by the active layers to generate charges. These photogenerated charges are then collected by ETLs. Nanolayers of metal, such as Au, Ag or Al, are deposited by vacuum deposition over the substrates, and they act as back contacts.

Several flexible solar cells are developed based on both organic, inorganic and hybrid active layers. Using c-Si wafers, scientists have recently achieved high-performing foldable solar cells with 24% efficiency even after 1 000 times of side-to-side bending cycles exhibiting an excellent bending durability (Liu *et al* 2023). With a conversion efficiency of more than 25%, perovskites are promising active layers for flexible solar cells and are extensively investigated for large-scale commercialization (Xu *et al* 2022). Perovskites possess a high absorption coefficient, high carrier mobility and high tolerance to defects. In addition, perovskites can be deposited as thin films using the solution process, which makes perovskite solar cells preferable in large-area PV modules at low cost and high-power conversion efficiency (Kim *et al* 2020). Based on performance, FPSCs are used in various wearable and portable devices with an easy integration with other electronic products (Liang *et al* 2021, Dhanabalan *et al* 2022).

6.4 Fabrication of perovskite solar cells in flexible substrates

The basic structure of hybrid halide perovskites is ABX_3, where A is either an organic or inorganic cation, such as methylammonium $[CH_3NH_3]^+$, formamidinium $[CH(NH_2)_2]^+$, caesium (Cs), rubidium (Rb), or a combination of them. B is a cation of a metal, often Pb, and X is a halogen (I, Cl, Br or F). As discussed in the previous section, because of the advantages of halide perovskites, such as high absorption coefficient, tunable bandgap with respect to composition and low exciton binding energy, fabrication of FPSCs is extensively focused in the recent times. Polymeric plastic substrates have several advantages such as light weight, flexibility under different environments, low cost and transparency (Dhanabalan *et al* 2022a). There are two kinds of deposition process used to fabricate perovskite layers (for example: $CH_3NH_3PbI_3$) on the flexible substrate: namely, the one-step method and two-step method. In the one-step method, precursors are dissolved in solvents such as dimethyl sulfoxide or dimethyl formamide. This precursor is then deposited on a substrate, and crystallization takes place under annealing. In the two-step method, the already deposited dried lead halide (PbX_2) film is exposed to a methylammonium halide (Cl, Br, I) source to form a perovskite film. Regarding perovskite solar cell structure, if the perovskite active layer is deposited on the ETL, this is called an n-i-p (substrate/TCO/ETL/perovskite/HTM (hole-transport material)/electrode) or regular structure, and if the same is deposited on the hole-transport layer (HTL), it is classified as p-i-n (substrate/TCO/HTM/perovskite/ETL/electrode) or an inverted structure. Metal oxides such as TiO_2 and SnO_2 are the most often used electron-transport materials (ETMs), whereas 2,2′,7,7′-Tetrakis-(*N*,*N*-di-4-methoxyphenylamino)-9,9′-spirobifluorene (Spiro-OMeTAD) is the most regularly employed HTM. For the coating process, a precursor ink is prepared with the appropriate formulation in the suitable solvent(s) with the required viscosity. In general, the efficiency of perovskite solar cells based on glass/ITO is higher than the same with ITO/PET substrate, owing to the lower sheet resistance of ITO/glass. However, efforts are undertaken to fabricate FPSCs, owing to the promising indoor applications. Concerning the stability of the substrate, most of the plastic polymeric substrates are stable only up to 150°C, which restricts their use for high temperature processing. In other words, the low coefficient of thermal expansion of plastic substrates prohibits the use of plastic substrates at high temperatures. Although PI has high temperature stability (glass transition temperature over 200°C), it is less transparent than PET and PEN, which limits its usage. For this reason, other than polymeric (PET, PEN etc) substrates, FPSCs are fabricated using thin flexible metallic substrates. Especially, titanium (Ti) foil is utilized for this purpose. Through low-temperature processing approaches, around 8%–9% of efficiency has been

Table 6.1. Physical properties of polymeric substrates used for the fabrication of FPSCs. Adopted from Wei Zi *et al* (2018).

Property	PET	PEN	PI
Coefficient of thermal expansion	15 ppm K^{-1}	13 ppm K^{-1}	17 ppm K^{-1}
Heat resistance	150°C	200°C	350°C
Transparency	>85%	>85%	Yellow
Barrier property	Poor	Poor	Poor

achieved in the FPSCs. The best performing perovskite solar cell using polymeric substrate has delivered around 15%–16% efficiency, whereas for the flexible metallic substrates, it is over 20%. The lower efficiency in this case is attributed with several factors including processing temperature of metal oxide ETL. For example, titanium dioxide (TiO_2) is a widely used ETL and is usually annealed at higher temperatures in order to achieve a highly mesoporous layer. This, however, cannot be carried out when TiO_2 is deposited on a flexible substrate due to the temperature limitation. Besides, the glass substrates generally have higher transmittance than flexible substrates, which limits their efficiency. Also, maintaining the stability of FPSCs is a challenging task compared with glass-based perovskite modules because of the practical difficulties with encapsulation. Compared with TiO_2, zinc oxide (ZnO) has high electron mobility values, and therefore, it us used as an ETL in FPSCs delivering high efficiency. The extremely developed solvent and ligand chemistry in preparing highly crystallized perovskite layers is helpful to understand the effective ways to fabricate high quality perovskite films. The fabrication of large-area perovskite modules consists of three important steps: namely, (i) development of coating techniques for a large scale, (ii) patterned deposition or post-patterning methodologies and (iii) optimization in the cell fabrication and interconnection methodology (Di Giacomo *et al* 2016). The first report on the fabrication of FPSCs was by Di Giacomo *et al*, who achieved 3.1% efficiency (Di Giacomo *et al* 2015). Here, the layers were scribed through a CO_2 laser. Since then, progress has been made to fabricate highly efficient multilayered perovskite solar cells by different deposition approaches (table 6.1).

6.5 Fundamental concepts of laser patterning in FPSCs

6.5.1 Basic principle of laser patterning in FPSCs

Because it is quick, noncontact (preventing mechanical interaction that could cause stress and fractures to the substrates), highly accurate, wavelength selective (allowing a material that is highly soaked in a particular wavelength range to be eliminated from the lesser absorbing substrates beneath without ruining it), regional, highly computerized, and economically practical, laser processing has attracted increasing interest in the fabrication of PV devices. Perovskite solar cells are thin-film PV devices that utilize a hybrid organic–inorganic lead halide perovskite material as the light-absorbing layer (Turan *et al* 2017). This material exhibits excellent optoelectronic properties such as a high light absorption coefficient, long carrier diffusion length, and low defect density. These characteristics make perovskite solar cells highly efficient in converting sunlight into electricity. FPSCs have emerged as a promising technology for next generation PVs due to their high efficiency, low-cost fabrication, and potential for integration into various applications. To maximize the performance and functionality of these devices, precise patterning of the perovskite layer is essential. Laser patterning has gained significant attention as a versatile and efficient method for achieving high-resolution patterns on FPSCs. In this section, the principle of laser patterning and its application in the fabrication of FPSCs will be explored. In perovskite solar cells, various layers, including the transparent electrode, perovskite absorber, and ETLs/HTLs, are stacked to form a complete device. In perovskite-based modules, laser patterning can minimise the dead zones that are utilised for contacts. To produce the requisite spacing/isolation for the electrodes of neighbouring subcells (P1–P3) and to isolate and clean the contact regions (P3), patterning can be done in three consecutive phases (P1–P2–P3). Patterning these layers allows for the selective placement of these components, thereby optimizing the cell's efficiency and functionality (Moon *et al* 2015). Additionally, patterning can enable the integration of solar cells into flexible substrates, opening up possibilities for lightweight, flexible, and portable solar devices. While laser patterning offers several advantages, it can also damage solar cells, reducing their efficiency. Additionally, the process has been used to be expensive and time-consuming. However, recent developments help to achieve large scale laser patterning to be very quick and also at lower cost.

6.5.2 Importance and benefits of laser patterning in FPSCs

This section explores the importance of laser patterning in FPSCs and highlights its key benefits and challenges (Matteocci *et al* 2014, Rai *et al* 2021).

- **Enhanced light absorption:** One of the primary advantages of laser patterning in perovskite solar cells is the ability to create complex and optimized patterns on the active layer. By precisely tailoring the surface morphology and patterning the perovskite films, laser techniques enable increased light absorption. The controlled patterning improves the light-trapping capability of the solar cells, maximizing the utilization of incident photons and enhancing the overall device efficiency.
- **Improved stability:** Perovskite materials are known to be sensitive to moisture, oxygen, and other environmental factors, which can degrade their performance over time. Laser patterning allows for the precise removal of degraded or damaged regions, enabling selective repair and rejuvenation of the perovskite layer. By removing localized defects and preserving the active areas, laser patterning significantly improves the long-term stability and reliability of FPSCs.
- **Increased flexibility:** FPSCs have garnered significant attention due to their potential for integration into various applications, including wearable electronics and curved surfaces. Laser patterning plays a crucial role in achieving

flexibility by enabling selective removal of unwanted materials while maintaining the structural integrity of the solar cell. This technique facilitates the fabrication of lightweight, bendable, and conformable devices, opening up new possibilities for efficient energy harvesting in unconventional form factors.
- **Enhanced electrical performance:** Laser patterning techniques also offer benefits in terms of optimizing the electrical performance of perovskite solar cells. By selectively removing excess or undesired material, laser patterning helps to improve the charge carrier transport and minimize energy losses. Moreover, it allows for the precise placement of contacts and interconnections, reducing resistive losses and enhancing the overall electrical conductivity of the device.
- **Scalability and cost-effectiveness:** Scalability is a critical factor for the commercialization of any technology. Laser patterning enables high-throughput processing and can be readily integrated into existing manufacturing processes, making it suitable for large-scale production of FPSCs. Additionally, laser patterning offers cost advantages by reducing material waste and simplifying the fabrication steps, thus improving the overall cost-effectiveness of the technology.
- **Selective patterning:** Laser patterning allows for precise and selective removal or modification of specific regions of the perovskite layer. This capability is essential for defining the device architecture, such as patterning the active area, electrode contacts, or interconnections. Selective patterning helps to optimize the device performance and enhances overall efficiency.
- **Enhanced efficiency:** Laser patterning enables the creation of complex and optimized device architectures including novel electrode designs and light trapping structures. By tailoring the perovskite layer's morphology and optimizing the light absorption and charge collection pathways, laser patterning can significantly enhance the efficiency of perovskite solar cells.
- **Rapid prototyping and design iteration:** Laser patterning offers a rapid prototyping capability, allowing for quick iteration and optimization of device designs. Researchers can easily modify and test different patterns, architectures, and materials, accelerating the development of efficient and stable perovskite solar cell technologies.

The advantages related to laser patterning in perovskite solar cells are schematically represented in figure 6.4.

In general, laser patterning is of paramount importance in perovskite solar cells due to its ability to selectively pattern the perovskite layer, enhance device efficiency, improve stability, enable scalability, provide design flexibility, and expedite the development process. It is a versatile and powerful technique that contributes to advancing perovskite solar cell technology towards practical applications (Rai *et al* 2021). While laser patterning offers numerous advantages, there are some challenges that need to be addressed. Optimization of laser parameters, such as pulse duration and energy density, is essential to ensure precise and controlled material removal

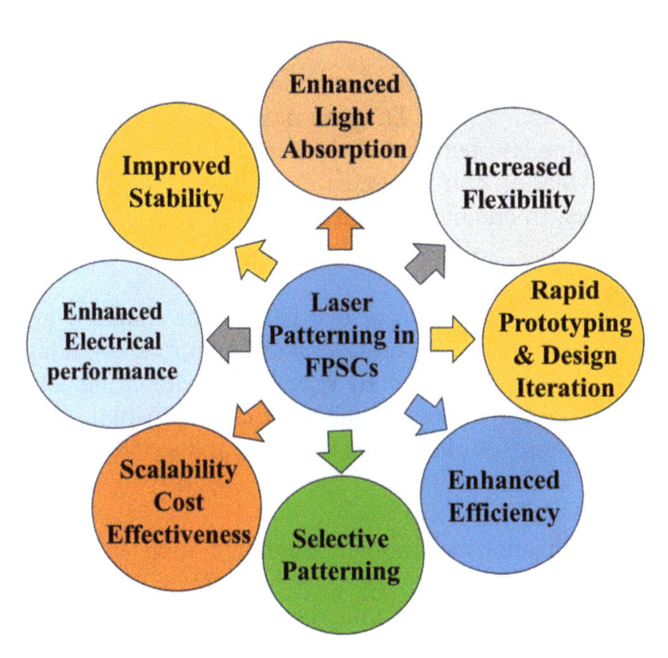

Figure 6.4. Schematic representation of key advantages of laser patterning in FPSCs.

without damaging the underlying layers. Moreover, further research is needed to explore the scalability and long-term stability of laser-patterned perovskite solar cells (Turan *et al* 2017, Rai *et al* 2021).

Laser patterning plays a pivotal role in the fabrication of FPSCs. Its ability to enhance light absorption, improve stability, increase flexibility, enhance electrical performance, and enable scalability makes it an indispensable technique for the advancement of perovskite-based PVs. Continued research and development in laser patterning will contribute to the realization of efficient and reliable FPSCs, paving the way for sustainable and cost-effective energy generation.

6.5.3 Methodology of laser patterning in perovskite solar cells

A straightforward solution technique, such as spin coating, may be used to create high-quality perovskite films with a modest area (25 cm^2). It is generally acknowledged that there are restrictions on the extension of the process to large-area modules (25 cm^2 or larger), despite a few reports on the fabrication of large-area PSCs using spin coating. This is because it is challenging to control the morphology and kinetics of perovskite crystals over a large area without pinholes (Bayer *et al* 2017). Laser patterning involves using a laser beam to selectively remove or modify specific regions of the perovskite layer. The principle relies on the interaction between the laser light and the perovskite material. When a laser beam is incident on the perovskite layer, it can induce various physical and chemical processes including ablation, scribing, and crystallization.

6.5.3.1 Ablation

Laser ablation involves the removal of material from the perovskite layer through localized heating and vaporization. The laser pulse energy and duration are carefully controlled to achieve precise removal of the perovskite layer without damaging the underlying layers. Ablation can be used to create patterns, such as interdigitated electrodes or bus bars, to enhance charge collection and minimize resistive losses.

6.5.3.2 Scribing

Laser scribing refers to the creation of fine grooves or channels in the perovskite layer. This process facilitates the separation of different functional layers such as the perovskite absorber and ETLs/HTLs. By selectively scribing the perovskite layer, it is possible to define the active area of the solar cell and isolate individual cells or modules, leading to improved device performance and scalability.

6.5.3.3 Crystallization

Laser-induced crystallization is another technique used in laser patterning of perovskite solar cells. By exposing the perovskite layer to laser irradiation, the amorphous perovskite material can undergo a phase transition, resulting in the formation of well-defined crystalline structures. Controlled crystallization can enhance the material's charge transport properties and reduce defects, leading to improved device performance.

6.5.4 Advantages of laser patterning in FPSCs

Laser patterning offers several advantages over conventional patterning techniques in the fabrication of FPSCs (Di Giacomo *et al* 2020, Razza *et al* 2021):

(i) Laser systems can provide sub-micrometer resolution, enabling the fabrication of intricate patterns with precise dimensions. This precision ensures accurate placement of functional components and maximizes the utilization of active materials, leading to improved device performance.
(ii) Laser patterning is a noncontact method that eliminates the need for physical masks or direct contact with the device. This feature reduces the risk of contamination and damage to the sensitive layers, resulting in higher device reliability and yield.
(iii) Laser systems offer versatility in terms of wavelength, power, and pulse duration, allowing optimization for different perovskite materials and device architectures. This flexibility enables the fabrication of various patterns, such as electrodes, bus bars, and interconnects, to meet specific design requirements.
(iv) Laser patterning is a scalable technique suitable for both laboratory-scale research and industrial production. The high-speed processing capability of lasers allows for efficient large-area patterning, making it compatible with the high-throughput requirements of commercial manufacturing.

The principle of laser patterning has revolutionized the fabrication of FPSCs. By leveraging the unique interaction between laser light and perovskite materials, precise patterns can be created, enhancing the performance and functionality of the devices. Laser patterning offers high precision, contactless processing, versatility, and scalability, making it a promising technique for the mass production of FPSCs and the realization of efficient, lightweight, and customizable solar energy solutions (Di Giacomo *et al* 2018).

6.5.5 Parameters influencing laser patterning in perovskite solar cells

Laser patterning is a technique commonly used in the fabrication of FPSCs to define and optimize the device's architecture and performance (Bayer *et al* 2017, Lin *et al* 2023). Several parameters can influence the outcome of laser patterning, and understanding their impact is crucial for achieving desired results. Here are some key parameters and their influences:

- **Laser power:** The laser power determines the intensity of the laser beam and affects the ablation or modification of the perovskite material. Higher laser power can result in faster material removal but may also cause thermal damage or undesired effects like melting. Optimizing the laser power ensures precise and controlled patterning without compromising the structural integrity of the perovskite layer.
- **Pulse duration:** The pulse duration of the laser determines the temporal length of the laser beam's interaction with the perovskite material. Shorter pulse durations are preferred for precise patterning, as they minimize heat transfer and reduce the risk of thermal damage. Ultrashort pulse lasers, such as femtosecond lasers, are commonly used for perovskite patterning due to their high precision.
- **Wavelength**: The choice of laser wavelength is critical, as it determines the absorption characteristics of the perovskite layer. The laser wavelength should match the absorption peak of the perovskite material to ensure efficient energy coupling and precise patterning. The commonly used wavelength range for perovskite patterning is in the near-infrared to visible spectrum.
- **Scanning speed:** The scanning speed of the laser beam across the flexible substrate affects the pattern resolution and quality. Higher scanning speeds can lead to reduced pattern fidelity, while slower speeds can increase processing time. Optimizing the scanning speed ensures a balance between pattern quality and fabrication efficiency.
- **Spot size:** The spot size of the laser beam determines the minimum achievable feature size and pattern resolution. Smaller spot sizes result in higher resolution but require more scanning passes to cover larger areas, which can increase processing time. The spot size should be chosen based on the desired pattern dimensions and fabrication efficiency.
- **Substrate material:** The choice of flexible substrate material can influence laser patterning. Different substrates have different optical and thermal

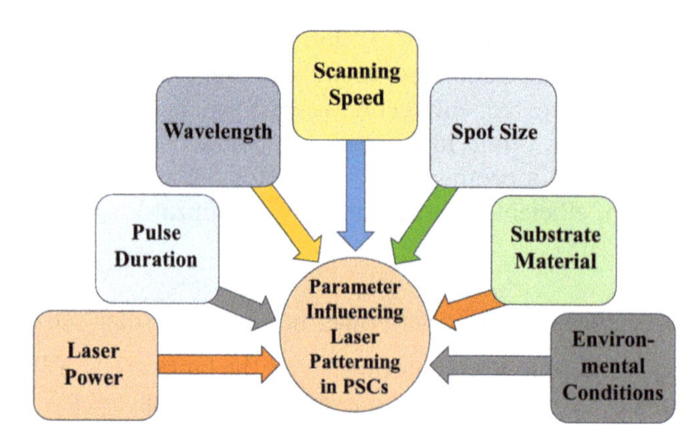

Figure 6.5. Schematic representation of parameters influencing laser patterning in perovskite solar cells.

properties, which can affect laser–material interactions and energy absorption. It is important to select a substrate that is compatible with the laser parameters to achieve the desired patterning results without causing substrate damage or delamination.

- **Environmental conditions:** Environmental conditions such as temperature, humidity, and air quality can affect the laser patterning process. These factors can influence the material properties, laser–material interactions, and overall device performance. Maintaining controlled environmental conditions can help ensure consistent and reproducible patterning results.

Overall, the influence of these parameters in laser patterning of FPSCs is interconnected and requires careful optimization to achieve precise patterning, maintain device performance, and ensure the structural integrity of the perovskite layer. The parameters influencing the laser patterning of perovskite solar cells are schematically given in figure 6.5.

6.6 Role of laser patterning in fabrication and analysis of FPSCs

6.6.1 Patterning strategy and instrumentation of laser patterning in perovskite solar cells

Large-area perovskite solar modules are built by interconnecting several smaller solar strips called subcells or submodules rather than a large single solar cell. This is due to an increase in internal electrical resistance for large-area cells, which reduces the overall output. As discussed in the earlier section, several steps of layer deposition and scribing are carried out in sequence to define the optimum cell width and interconnections. With smaller subcells, charge transport distances are reduced, thereby reducing the overall parasitic resistance losses (Kothandaraman *et al* 2022). However, an increase in the number of subcells and interconnections reduces the module active area compared to the overall aperture area, i.e. a lower geometric fill factor (GFF). Area with no contribution to output power is termed as 'dead area' and needs to be significantly minimized. Therefore, an optimized module width is

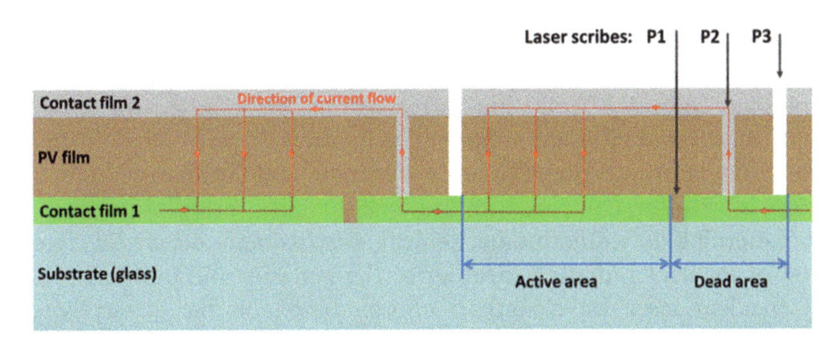

Figure 6.6. Schematic representation of all three patterning steps in PV fabrication and the direction of current flow. Reprinted with permission from Jamaatisomarin *et al* (2023). CC BY 4.0.

chosen to have a high GFF and low resistance loss. Mechanical, chemical and laser-based scribing/patterning techniques are adopted for module fabrication with a high GFF. Figure 6.6 shows a schematic representation of three patterning steps in a monolithic serial interconnection (Jamaatisomarin *et al* 2023).

Use of toxic chemicals and expensive exposure masks in photolithographic etching process limit the use of chemical-based patterning techniques for next generation solar cells. Similarly, the wear and tear of scribing needles require constant maintenance in mechanical scribing methods. Mechanical scribing also produces inhomogeneous patterning and material debris. On the other hand, the laser scribing process overcomes all the challenges faced by mechanical and chemical scribing methods. They can be used to selectively remove regions of bottom contact, active layer, ETL and bottom contact by laser-controlled ablation. The laser is a promising tool with high precision and processing speed in patterning submicron layers for large-area devices (Gebhardt *et al* 2013). Laser scribing is also an environmentally friendly technique with an excellent repeatability and low manufacturing cost, leaving much less material debris (Razza *et al* 2021).

Pulsed lasers having pulse widths of nanoseconds, picoseconds and femtoseconds are employed for scribing processes. Other laser parameters such as wavelength, fluence, focal area and peak power play a crucial role in scribing. These parameters vary with materials that are to be patterned. For instance, ultraviolet (UV) lasers with good beam quality were employed to obtain narrow scribe width on microcrystalline silicon thin films with an integrated width of less than 500 μm (Kuo *et al* 2010). The role of laser pulse duration has been extensively investigated on various materials. Lin *et al* reported bifacial semi-transparent perovskite solar cells by scribing all three steps (P1, P2 and P3) using an Nd:YAG pulsed laser emitting at 532 nm with a pulse duration of 7 ns at 7 500 Hz. PCE of 12.5% was achieved using the fabricated module at low cost and high stability (Lin *et al* 2023). Lauzurica and Molpeceres demonstrated selective ablation of deposited layers (TCO front contact, a-Si:H active layer and back contact) in a-Si:H solar cells using nanosecond and picosecond UV laser beams (Lauzurica and Molpeceres 2010). Compared to nanosecond laser pulses, use of picosecond or femtosecond laser beams reduces thermal damage of the substrate and heat affected zone (HAZ), resulting in very fine narrow grooves. Bayer *et al* investigated laser ablation and patterning of perovskite thin films using nanosecond, picosecond and femtosecond laser pulses at wavelengths ranging from 248–2 500 nm (Bayer *et al* 2017). Based on incident laser parameters, they found different ablation mechanisms of perovskite film scribing. It was found that laser wavelength, pulse duration and irradiation direction play crucial roles in ablation compared to laser energy and overlap of pulses.

Laser patterning is a thermomechanical process and occurs in three steps: absorption of incident radiation, heating of the material and material removal by ablation. Incident laser wavelength is chosen based on the absorptivity of the deposited material. For instance, due to strong band gap absorption in the UV wavelength range, lasers emitting at UV wavelengths are employed for ZnO deposited layers rather than infrared wavelength laser beams. Scribing using ultrashort lasers can be achieved even at lower pulse energy, thereby producing only a smaller ablation area and HAZ. Thermal effects are drastically reduced for picosecond and femtosecond laser beams. This is because electron heat conduction takes much longer than the interaction time between the laser pulse and the material under femtosecond laser irradiation. Femtosecond scribing occurs through nonlinear processes that result in rapid ablation from solid to vapor without any heat accumulation.

6.6.2 Practical methodology of laser patterning in perovskite solar cells

Flexible substrates with higher conductivity and transparency are often employed for the fabrication of perovskite solar cells. The transparent conducting metal oxides ITO and fluorine-doped tin oxide are generally used as the working electrode. On this, ETLs that consist of metal oxides such as TiO_2 are applied by the solution-phase deposition approach. A perovskite solar cell has many layers with few hundred nanometers' range, and careful removal of the layers in the prescribed format would help to interconnect them in a series connection. Because of the difference in the light reflection and robustness of the plastic films under the interaction of high-energy photons, the laser patterning is different between glass and flexible substrates. Here, removal of each layer is cautiously undertaken in order to not to affect the other layers, and hence the interconnection is assured. Laser patterning in FPSCs has gained much interest owing to its importance in industrial scale production. The mechanism of laser ablation can be done either through ablation or delamination lift-off or the combination of these two processes. At first, a laser is used to scribe the bottom layer (conducting metal oxide layer) of the substrate. This is called the P1 pattern, and it helps to fabricate a subcell in the solar panel. Secondly, after depositing the active layer materials and HTL, the P2 pattern is carried out. As a final step, the P3 pattern is carried out on the deposited top electrodes in order to divide them into several subcells. The interconnection between each cell allows electron transport in the device, which results in energy conversion. This P1, P2 and P3 scheme of scribing is a generally accepted pattern to fabricate large-area perovskite solar cell modules in glass or flexible substrates. Especially the R2R fabrication of perovskite solar cells has been accelerated using this approach, and several promising results predict promising industrial roles for this method. With respect to the layer, different kinds of laser sources are used. For example, for the conducting electrode (bottom electrode), generally, the P1 scribe is carried out using a 1 064 nm laser (Nd:YVO_4), and P2 and P3 are carried out using a 532 nm laser. Owing to the low absorption of transparent conducting oxides and high spectral absorption of perovskite materials in the visible region, a green laser is used to eliminate the perovskite layer and also the polymeric HTL. Usually, a nonlinear optical crystal is used to convert the wavelength from 1 064 to 532 nm. For the P1 scribing, either solid-state or gas lasers are utilized to scribe the conducting oxide layer from the substrate. Here, the laser should not damage or affect the substrate, and hence usually a linewidth lower than 10 mm, without damaging the PET substrate, is preferred. For the removal of the nc-TiO_2 layer, a high fluence rate is normally followed, but sometimes it may still remove the underlayer (TCO) from the substrate. Hence, utmost care should be taken to identify the fluence rate to remove each layer and also the corresponding laser wavelength. In summary, the laser patterning process helps to integrate the smaller solar cells into a single line, which is useful to fabricate a solar module. A typical FPSC with the laser patterned substrate and corresponding *I–V* curve and correlation of PCE with respect to time are represented in the figure 6.7.

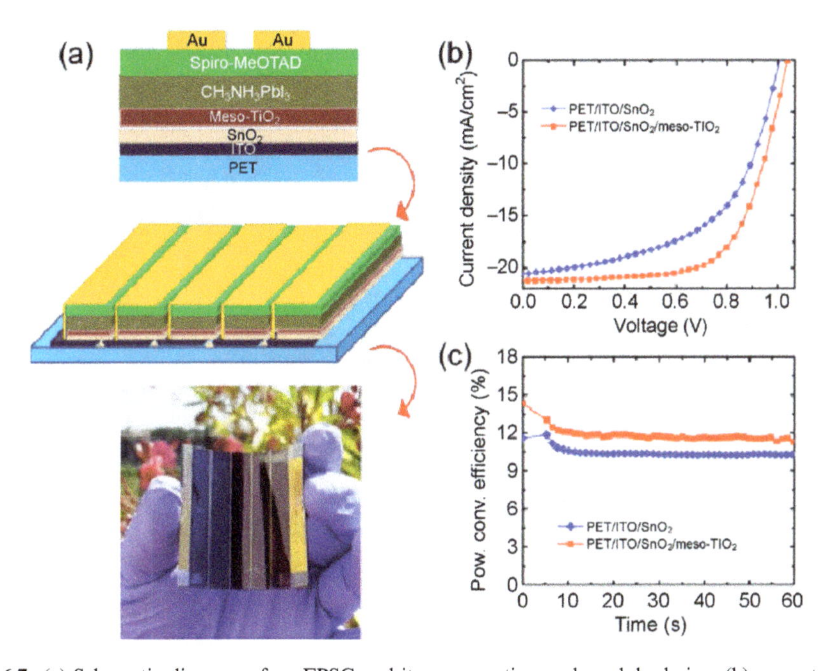

Figure 6.7. (a) Schematic diagram of an FPSC and its cross section and module design, (b) current–voltage (*I–V*) curve of solar cells with and without a mesoporous-TiO_2 layer, and (c) evolution of power conversion efficiency of the cells with respect to time. Reprinted from Dagar *et al* (2018) with permission from Springer Nature, Copyright (2018).

6.6.3 Experimental evidence and factor influences in laser patterning in flexible solar cells

Under a laser, irradiation organic materials can be more easily volatized than the inorganic materials. Inorganic materials are bonded by strong covalent or ionic bonds, whereas organic materials are bonded by van der Waals or hydrogen bonds that can be easily deformed even with less energy (Zhao *et al* 2022). Laser ablation and achieving optimal interconnections on flexible substrates are challenging and differ drastically compared to that of rigid glass substrates. Taheri *et al* performed laser scribing on SnO_2-based large-area FPSCs fabricated on flexible plastic substrate using a spray drying technique. Figure 6.8 shows the schematic sketch of the interconnection structure of a flexible perovskite module with all three patterns.

Under optimum conditions, a heterogeneous stack of active layers was selectively removed by a UV nanosecond laser without any degradation of the bottom ITO layer. With a maximum PCE of 15.3%, the research results prove the potentials of spray-dried and laser-scribed commercial large-area flexible perovskite solar modules (Taheri *et al* 2021). Bian *et al* reported selective ablation of metallic thin film on flexible polymer substrate by performing laser patterning of 800–850 nm thick molybdenum layers deposited on 25 μm PI flexible substrate for CIGS thin-film solar cells (Bian *et al* 2011). Using an 800 nm laser system operating at ~60 fs pulse width and frequency of 1 kHz, groove geometry and morphology of Mo thin films were studied at different laser fluences, pulse widths and laser scanning speeds. It was found that the ablation threshold increased from 0.08–0.1 J cm^{-2} when the pulse duration was varied from 60–600 fs. The groove width was found to increase by four times from 3–13 μm when the laser fluence was increased by three times from 0.34–0.9 J cm^{-2}. While groove width and depth depend on laser pulse energy, their morphology and structure vary with laser scanning speed. Here, we will briefly discuss flexible perovskite solar modules and the role of laser scribing in next generation commercial solar cells.

As discussed previously, the removal of the active layer, HTLs, and ETLs using P2 scribing is challenging owing to the low absorption selectivity of the TiO_2 layer

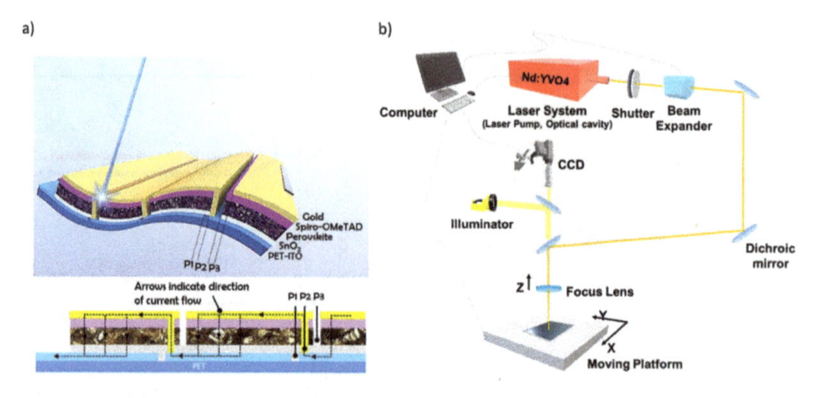

Figure 6.8. (a) Schematic sketch of flexible perovskite solar module with all three patterns scribed using a laser. (b) The laser scribing setup used for perovskite solar devices. Reprinted from Taheri *et al* (2021) with permission from ACS, Copyright (2021).

and ITO/PET substrate. Hence, partial elimination of the TiO_2 layer often results in contact resistance, which will be a hurdle to achieving interconnectivity in the module. However, these problems can be rectified through optimized laser conditions. The laser fluence rate is an important parameter to eliminate the layers through P2 scribing, and this eases the interconnection problems of cells in a module. Taheri *et al* have experimentally observed that laser power over 46 mW damages the bottom ITO layer, whereas laser power less than 23 mW was found to be insufficient to remove the first layer (Taheri *et al* 2021). Here, the authors varied the laser power from 31–52 mW to fabricate the solar cell structure PET/ITO/SnO_2/$Cs_{0.05}FA_{0.80}MA_{0.15}Pb(I_{0.85}Br_{0.15})_3$/spiro-OMeTAD. Removal of the $CH_3NH_3PbI_3$ perovskite layer through the P2 pattern should be carried out cautiously, since a nanosecond laser pulse ablation may leave PbI_2 after scribing. However, experimental observations indicate that use of a picosecond pulse laser could considerably reduce this effect. Despite these observations, it was not clear whether the P3 scribe could damage the unscribed perovskite layer. The laser scribe process may induce a degradation of the perovskite layer owing to its high power. Also, since halide perovskites are prone to high energy electrons, this effect is also observed with laser treatment. A recent study describes that when a picosecond laser is used to ablate in the P3 step, this could produce a significant change in the composition of the perovskite layer by forming a needle-like morphology of PbI_2 (Kosasih *et al* 2019). In this experiment, the authors used a UV laser with the wavelength $\lambda = 355$ nm, with the pulse duration 10 ps. By varying the laser fluence in the P2 scribe, Christof Schultz *et al* have revealed an interesting observation (Christof Schultz *et al* 2020). The authors observed that shorter picosecond pulses are useful for the complete removal of the perovskite layer compared with the nanosecond pulses. This efficiency of the picosecond pulse is due to the fact that lower heat output during the scribing process leads to the complete ablation, whereas an incomplete ablation by nanosecond pulses leads to the formation of PbI_2 as residue. The mechanism behind nanosecond and picosecond pulse laser ablation (532 nm) of P2 scribing and corresponding scanning electron microscopy images are given in figure 6.9.

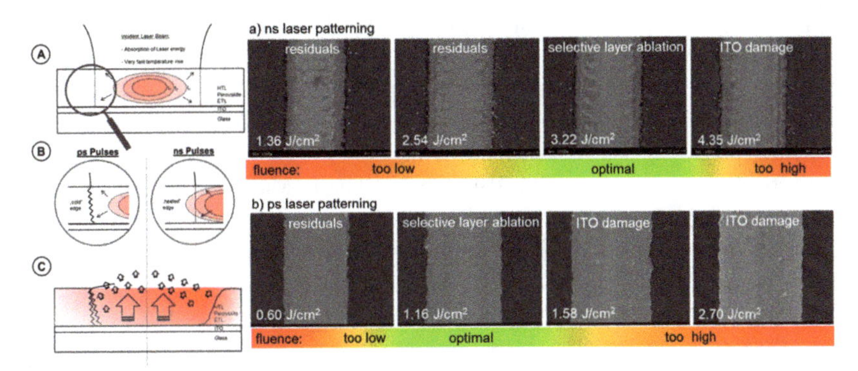

Figure 6.9. (a–c) Schematic representation of laser ablation mechanism using nanosecond and picosecond laser pulses. (d and e) Scanning electron microscopy images of the nanosecond (ns) and picosecond (ps) laser patterning under different influences. Reprinted from Schultz *et al* (2020) with permission from Elsevier, Copyright (2020).

These investigations are envisaging the crucial role of laser pulses in fabricating flexible perovskite modules.

6.7 Summary and future aspects of laser patterning

Based on the discussion on recent developments in laser patterning, it is imperative to note the need of laser scribing in developing next generation solar cells. With experimental and theoretical understanding of this process, researchers have revolutionized this technique to achieve flexible solar cells. Recent results prove the great potential of laser-based thin-film solar cell production with simplicity and wider flexibility. Apart from optimized laser power, pulse duration and wavelength, homogeneous laser scribing requires uniform laser intensity, which in turn is decided by laser beam shape. More studies need to be performed on laser beams with elliptical and top hat shapes for different applications. Additionally, more focus needs to be given to nanomaterial-based solar cells using laser patterning. Since transparent solar cells are gaining more attention, it will be better to apply laser patterning in such devices. Large-scale industrial production of transparent and flexible solar cells with good PCE and high power should be the primary goal of researchers. Laser patterning can play a major role in such future solar cells, which require more intensive research.

Acknowledgments

Samuel Paul David acknowledges the management of the Vellore Institute of Technology for providing the funding (RGEMS—Sanction Order No.—SG20220119) during this work. Dennish Babu acknowledges the management of Chettinadu College of Engineering and Technology for their extended support during this work. Ananthakumar Soosaimanickam sincerely acknowledges the management of Intercomet S.L. for their constant encouragement during this work.

References

Bayer L, Ehrhardt M, Lorenz P, Pisoni S, Buecheler S, Tiwari A N and Zimmer K 2017 *Appl. Surf. Sci.* **416** 112–7

Bayer L, Ye X, Lorenz P and Zimmer K 2017 *Appl. Phys.* A **123** 61

Bian Q, Yu X, Zhao B, Chang Z and Lei S 2011 *30th Int. Congress on Laser Materials Processing, Laser Microprocessing and Nanomanufacturing*

Dagar J, Castro-Hermosa , Gasbarri M, Palma A L, Clina L, Matteocci F, Calabro E, Di Carlo A and Brown T M 2018 *Nano Res.* **11** 2669–81

Dhanabalan S S, Madurakavi R S, Thirumurugan K, Avaninathan S R and Carrasco M F 2022 *Comput. Electr. Eng.* **102** 108130

Dhanabalan S S, Arun T, Periyasamy G, N D, N C, Avaninathan S R and Carrasco M F 2022a *Chem. Phys. Lett.* **800** 139692

Di Giacomo F, Castriotta L A, Kosasih F U, Di Girolamo D, Ducati C and Di Carlo A 2020 *Micromachines* **11** 1127

Di Giacomo F, Fakharuddin A, Jose R and Brown T M 2016 *Energy Environ. Sci.* **9** 3007–35

Di Giacomo F *et al* 2018 *Sol. Energy Mater. Sol. Cells* **181** 53–9

Di Giacomo F *et al* 2015 *Adv. Eng. Mater* **5** 1401808

Doumon N Y, Yang L and Rosei F 2022 *Nano Energy* **94** 106915

France R M, Geisz J F, Song T, Olavarria W, Young M, Kibbler A and Steiner M A 2022 *Joule* **6** 1121

Gebhardt M, Hänel J, Allenstein F, Scholz C and Clair M 2013 *Laser Tech. J.* **10** 25–8

Green M A, Ewan D D, Siefer G, Yoshita M, Kopidakis N, Bothe K and Hao X 2023 *Prog. Photovolt. Res. Appl.* **31** 3–16

Green M A, Ewan D D, Yoshita M, Kopidakis N, Bothe K, Siefer G and Hao X 2023 *Solar Cell Efficiency Tables (Version 62)* **31** 651

Iqbal M A, Malik M, Shahid W, Ud DinS Z, Anwar N, Ikram M and Idrees F 2022 *Thin Films Photovoltaics* (Rijeka: IntechOpen)

Jamaatisomarin F, Chen R, Hosseini-Zavareh S and Lei S 2023 *J. Manuf. Mater. Process.* **7** 94

Jung H S, Han G S, Park N G and Ko M J 2019 *Joule* **3** 1850–80

Khatibi A, Astraea F R and Ahmadi M H 2019 *Energy Sci. Eng.* **7** 305–22

Kim Y Y, Yang T Y, Suhonen R, Kemppainen A, Hwang K, Jeon N J and Seo J 2020 *Nat. Commun.* **11** 5146

Kosasih F U, Rakocevic L, Aernouts T, Poortmans J and Ducati C 2019 *ACS Appl. Mater. Interfaces* **11** 45646–55

Kothandaraman R K *et al* 2022 *RRL Solar* **6** 2200392

Kuo C F J, Tu H M, Liang S W and Tsai W L 2015 *J. Intell. Manuf.* **26** 677–90

Kuo C F J *et al* 2010 Optimization of Microcrystalline Silicon Thin Film Solar Cell Isolation Processing Parameters Using Ultraviolet Laser *Opt. Laser Technol.* **42** 945–55

Lauzurica S and Molpeceres C 2010 *Phys. Proc.* **5** 277–84

Liang X, Chuangye G, Fang Q, Deng W, Dey S, Lin H, Zhang Y, Zhang X, Zhu Q and Hu H 2021 *Front. Mater.* **8** 634353

Lin B Q, Huang C P, Tian K Y, Lee P H, Su W F and Xu L 2023 *Int. J. Precis. Eng. Manuf.-Green Technol.* **10** 123–39

Liu W, Liu Y and Yang Z *et al* 2023 *Nature* **617** 717

Matteocci F, Razza S, Di Giacomo F, Casaluci S, Mincuzzi G, Brown T M, D'Epifanio A, Licoccia S and Di Carlo A 2014 *Phys. Chem. Chem. Phys.* **16** 3918–23

Moon S-J, Yum J-H, Lofgren L, Walter A, Sansonnens L and Benkhaira M 2015 *IEEE J. Photovolt.* **5** 1087–92

Pagliaro M, Ciriminna R and Palmisano G 2008 *Flexible Solar Cells* (Heidelberg: Wiley)

Razza S, Pescetelli S, Agresti A and Di Carlo A 2021 *Energies* **14** 1069

Rai M, Yuan Z, Sadhu A, Leow S W, Etgar L, Magdassi S and Wong L H 2021 *Adv. En. Mater* **11** 2102276

Saravanapavanantham M, Mwaura J and Bulović V 2023 *Small Methods* **7** 2200940

Stegemann B, Fink F, Endert H, Schüle M, Schultz C, Volker , Quaschning V, Niederhofer J and Pahl H 2012 *Laser Technik J.* **9** 25

Schultz C, Fenske M, Dagar J, Zeiser A, Bartelt A, Schlatmann R, Unger E and Stegemann B 2020 *Sol. Energy* **198** 410–8

Shanmuga Sundar D, Sivanantharaja A, Sanjeeviraja C and Jeyakumar D 2016a *Mater. Today Proc.* **3** 2409–12

Sundar D S, Raja A S, Sanjeeviraja C and Jeyakumar D 2017 *Int. J. Nanosci.* **16** 1650038

Shanmuga Sundar D and Sivanantharaja A 2013 *Opt. Quantum Electron.* **45** 2397–403

Shanmuga Sundar D, Sivanantha Raja A, Sanjeeviraja C and Jeyakumar D 2016b *Polym. Int.* **65** 535–43

Taheri B, Rossi F D, Lucarelli G, Castriotta L A, Carlo A D, Brown T M and Brunetti F 2021 *Appl. Energy Mater.* **4** 4507–18

Turan B, Huuskonen A, Kuhn I, Kirchartz T and Haas S 2017 *Sol. RRL* **1** 1700003

Wang H, Cao Y, Feng J, Du M, Zhang D, Wang K, Qin W and Liu S 2018 *Strateg. Study Chinese Acad. Eng.* **20** 3

Westin P O, Zimmermann U, Ruth M and Edoff M 2011 *Sol. Energy Mater. Sol. Cells* **95** 1062

Xu Y, Lin Z, Wei W, Hao Y, Liu S, Ouyang J and Chang J 2022 *Nano-Micro Lett.* **14** 117

Zhao J, Chai N, Chen X, Yue Y, Cheng Y B, Qiu J and Wang X 2022 *Nanophotonics* **11** 987–93

Zi W, Jin Z, Liu S and Xu B 2018 *J. Energy Chem.* **27** 971–89

Part III

Applications—pioneering new horizons

IOP Publishing

Advances in Flexible and Printed Electronics
Materials, fabrication, and applications
Shanmuga Sundar Dhanabalan and Arun Thirumurugan

Chapter 7

Flexible electrochemical sensors for biomedical applications

Kashmira Harpale, Shweta Jagtap and Chandrashekhar Rout

In the last three decades, electrochemical sensors have gained major advancement, which has opened up the way for wearable electrochemical-sensing systems for real-time chemical monitoring. Electrochemical sensors are widely employed in the diverse disciplines of biosensing, electrochemical analysis, and drug administration because of their high sensitivity, selectivity, and cycle stability. Recently, electrochemical technology offers a promising framework for life healthcare by mimicking human tissue biocompatible to report electrical signals, potentially enabling timely disorder prediction through non-invasive real-time and simultaneous health monitoring. Stepping forward from traditional rigid electrodes, recent advancements in non-rigid electrochemical sensors offer new and exciting opportunities for various biomedical applications. This chapter details the wide range of developments in the field of flexible electrochemical devices for life healthcare, including their manufacture, analytical performance, and biomedical applications.

7.1 Introduction

The need for flexible and wearable electronic gadgets has dramatically expanded in the age of modern technology. Wearable and flexible electrochemical sensors have become primarily important for monitoring and detecting numerous analytes for real-time monitoring. The introduction of flexible sensors, fabricated using various materials, resulted in innovative applications that are expected to satisfy the demands of the next generation of affordable, foldable, portable medical devices. These sensors provide an exceptional blend of adaptability, mobility, and high sensitivity that opens the door for ground-breaking uses in healthcare. The exponential growth in availability of flexible smartphones, watches, rings, and bracelets as a result of the digital revolution has attracted a lot of attention for the development of numerous practical portable accessory devices. Due to its convenience, portability, and great potential for

doi:10.1088/978-0-7503-5492-9ch7

tracking an ideal state of health and fitness, flexible sensors have drawn a lot of interest among these [1, 2]. Additionally, flexible electrochemical-sensing systems provide one of the few opportunities to provide comprehensive chemical information, such as the presence or concentration of glucose, lactate, dopamine, cortisol, drugs, etc. Flexible and stretchable electrochemical sensors, as opposed to conventional rigid electrochemical sensors, have the unique ability to interact with human skin, which greatly enhances detection performances for monitoring electrolytes, biomarkers, drugs, or various biomolecules. Electrochemical sensors with mechanical compliance have considerable potential for wearable and implantable devices. Soft electrochemical sensors may provide new avenues for investigating in-depth biomedical research [3–5].

Wearable sensors are important, yet research and development in this field have progressed unevenly. Early research efforts were on developing wearable sensors that can assess bodily mobility, temperature, and the electrocardiogram. Wearable sensors currently in use routinely monitor a user's physical activity and vital signs (like heart rate). Individuals' physiological status can be continuously monitored to provide a unique health profile for the person's well-being. A major objective from the perspective of biosensors is the accurate and early detection of analytes, which can be achieved and improved through the fabrication of wearable devices that can be used for individualized healthcare monitoring. According to this perspective, the fundamental necessity for wearable biosensors has drawn more attention to the incorporation of flexibility into biosensors [6, 7].

Skin-worn electrochemical-sensing platforms have received particular focus because they mark a significant shift from conventional laboratory-based electroanalytical systems. Electrochemical sensors are particularly appealing when compared to other sensors because of their outstanding detecting capacity, ease of experimentation, and low price. They hold a leading place among the current crop of commercially viable sensors that have a wide range of significant uses in the sectors of clinical, industrial, and biological analyses. Recent advances in chemically modified electrodes, microelectrodes, and some electrochemical techniques (such as adsorptive stripping voltammetry and potentiometric stripping analysis) have made it possible to use electrochemical sensors to investigate a wide range of chemical species (figure 7.1) [8, 9].

Planar electrochemical sensors have been made from conventional semiconductor materials (such as Si or Ge), solid electrolytes, insulators, metals, and catalytic materials. The most common production methods for planar electrochemical sensors include thick-film and thin-film technologies, photolithography, and silicon technology [10, 11]. The sophisticated functionalized nanomaterial-based electrochemical sensors are sensitive and selective and are unquestionably crucial in several analytical sciences and the field of biomedicine [12]. Systems for electrochemical sensing are modular, elastic, and made up of numerous components. Real-world biomedical or soft robotic applications cannot be fulfilled by isolated flexible and elastic electrodes but need integrated systems. The basic components of flexible electrochemical-sensing systems are the sensor, electrochemical transducer, energy source, and signal processing unit. The first difficulty is a lack of materials that can

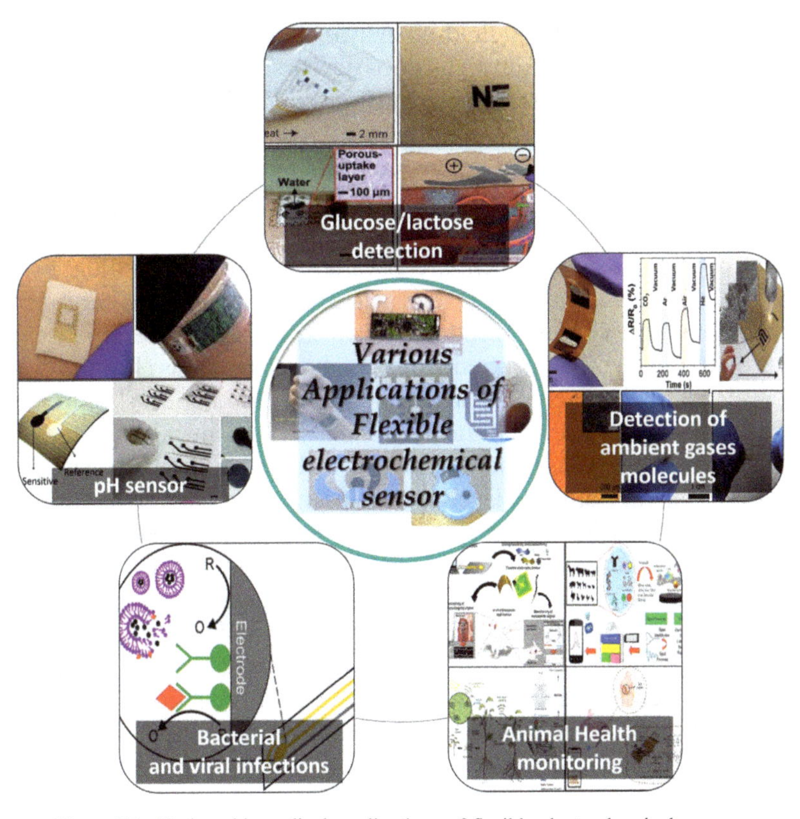

Figure 7.1. Various biomedical applications of flexible electrochemical sensors.

maintain conductivity in the electrodes and supporting substrates. Electrodes on soft substrates must be suitable electrochemical transducers that can operate as platforms for the functionalization of active layers, such as biorecognitions, mediators, or catalysts, in addition to being sturdy and physically resilient [3, 13].

Flexible electrochemical bioelectronics uses biological components in electrochemical reactions to address human biomolecular information. It takes advantage of the developments in analytical electrochemistry, microelectronics, and materials technology, as well as the expanding knowledge of human biology and biomedical research [14]. In order to quantify the concentration of the target analyte, the biorecognition element interacts with the target analyte by exchanging or transferring ions and electrons. An electrical current or potential can be measured by the electrochemical signal transducer, which transforms such contact. An analytical tool called a signal acquisition system enables data collection, processing, analysis, and display. To enable the biosensing measurement, it may include basic electronic components with signal processors or sophisticated electronic circuits, depending on the sensing modalities and intended applications. The in-body quantification of chemical components is made possible by flexible biosensors. In contrast to traditional analytical approaches, this represents a promising alternative for probing biomarkers in the human body. The detection of analytes, in a variety of biofluids, particularly blood, cerebrospinal fluid, interstitial fluid, sweat, saliva, and tears, has been the focus of recent breakthroughs in biosensors. Understanding current target biological constituents and their significance in biomedical research is essential to comprehending recent research developments in flexible electrochemical bioelectronics. Blood has been used to obtain data on physiological processes, medical diagnoses, and therapeutic intervention. Both blood and cerebrospinal fluid are invasively obtained through venipuncture or using implanted devices. Alternatively, biofluids for non-invasive assessment are gaining popularity. Interstitial fluid, which surrounds the microenvironment of cells, acts as a transport medium for nutrients, signaling molecules, and waste products between cells and the blood capillaries. Interstitial fluid is currently most frequently used in continuous monitoring of blood glucose levels due to its close relationship with blood glucose concentrations. Further, saliva, which an ultrafiltrate of plasma used in metabolomics, is another intriguing example. Saliva can be collected in better quantities. It was discovered that blood and salivary ethanol levels were closely connected. Tears are one more candidate for non-invasive assessment because they contain several proteins, salt, metabolites, and immunoglobin to maintain a healthy eye surface. Tear analysis is a different area of study that is growing due to its rich chemical compositions, but it is not as developed as other studies because of its challenging wearability issues [14–16]. Sweat is a newly discovered non-invasive biofluid that is easily retrieved through physical activity, heat stress, and a chemical induction process known as iontophoresis. Although largely in trace amounts, it is made up of ions, metabolites, amino acids, and proteins. Due to its accessibility, wearable bioelectronics for sweat sensing have recently developed at a rapid rate [2, 17].

7.2 Fabrication of the flexible electrochemical sensors

A suitable and efficient way of fabricating flexible devices for medical monitoring and diagnosis can offer several advantages for various technologies [18–20]. Fabrication of flexible electrochemical sensors involves the integration of flexible substrates (curved or irregular shapes), electrode materials, and sensing elements [21]. Flexible electrochemical devices could be readily prepared using a number of fabrication methods. While choosing the fabrication technology, substrate, electrode material, size of device, uniformity, durability, efficiency and cost, etc., need to be considered. Printable electrodes are one of the widely used materials for developing the flexible sensors [22–24].

Screen-printing technologies can easily provide high specificity and sensitivity. It is one of the well-known methods as it is easily accessible, low-cost, and has excellent operability. Screen printing involves efficient use of resources and the feasibility of production, which is employed to prepare mechanically stable, flexible, cost-efficient, lightweight printed electronic products [25, 26]. Electrochemical sensors fabricated by screen printing have the ability to bridge the gap between laboratory experiments and on-field implementation [27]. The flexibility of screen-printed electrodes is also very helpful in research; the ability to quickly modify the electrodes through various commercially available inks for the reference, counter, and working electrodes allows for the production of highly specific and finely calibrated electrodes for particular target analytes. Thixotropic fluid, which may contain carbon black, graphite, solvents, and polymeric binders, can be used for screen printing using mesh to define proper electrode placement [19]. Inks used in the printing process are highly viscous and undergo thinning, defining final shape and design when forced through the screen mesh by the squeegee blade. The thickness of the film can be controlled by designing of the mesh or stencil used. Substrate used for the printing purpose can be plastic or ceramic, which is decided by the intended application. Aside from the huge cost savings, the substitution of big cells and heavy electrodes allowed for experiments with significantly reduced sample quantities. Despite the fact that conventional printing techniques have been around for a while, the early versions could not endure the powerful mechanical stresses that the human body is subjected to. Because screen-printed electrodes must compromise their mechanical and electrical properties, adding more conductive filler to the ink (for increased conductivity) makes these products stiffer and less stretchable. The creation of soft/curvilinear human skin-compatible electrical devices is required for the construction of epidermal sensing platforms [26].

Inkjet printing is a digital, non-contact printing process that is increasingly being employed in the production of electrochemical sensors. This is a printing process used to fabricate a structure by depositing thin material layers from precursor inks in specified locations and in specific patterns by precision layering. Because of its widespread use in home and workplace environments around the world, inkjet printing is a well-known deposition technology. It is a very low-cost process that lends itself to the manufacturing of paper-based electrochemical sensors with particular advantages as a result of its popularity. Unlike screen printing, inkjet printing does not require the pre-fabrication of a template or stencil. Instead, the pattern is created with CAD software and delivered straight to the printer, which can deposit very small ink droplets row by row to make the required two-dimensional (2D) shape. Furthermore, because of the nature of the deposition procedure, inkjet printing provides for better pattern control than screen printing through the use of distinct printing equipment. It is considered as one of the versatile method due its superior pattering ability. Pre-deposition templates and post-deposition processes are not required as they are needed in photolithography. Furthermore, multiple materials can be simultaneously deposited as several ink cartridges are used in inkjet printing [2, 7, 28].

Another printing approach besides screen printing and inkjet printing, which do not necessitate the appropriate substrate or make use of conductive inks, is three-dimensional (3D) printing. It is a simple, rapid, low-cost, and versatile printing technology that is gaining attraction as a production method in several sectors of chemistry. This technology allows for the creation of sensors with intricate designs and customizable features, making them ideal for applications requiring adaptability and conformability to irregular surfaces. This technology allows for the creation of sensors with intricate designs and customizable features, making them ideal for applications requiring adaptability and conformability to irregular surfaces. The design freedom provided by 3D printing enables the integration of multiple sensing elements, microfluidic channels, and electronics into a single, compact device, expanding the sensor's functionality and potential applications. Moreover, the ability to rapidly prototype and customize sensor designs facilitates swift iteration and optimization, propelling research and development in sensor technologies. With a wide range of materials at their disposal, researchers can tailor the sensors' properties to meet specific requirements, ranging from biocompatibility for medical applications to durability for wearable devices. Although challenges exist in achieving high-resolution printing of conductive materials and ensuring consistent performance, the promise of 3D-printed, flexible electrochemical sensors is immense, heralding a future of advanced, customizable, and highly efficient sensing solutions for diverse industries and scientific endeavors [17].

Photolithography is a widely employed fabrication technique used for fabricating flexible electrochemical sensors with high precision and resolution. This process utilizes light to transfer a predefined pattern from a photomask onto a light-sensitive material coated on the flexible substrate. The masked areas shield the material from light exposure, and the unmasked regions are exposed and undergo a chemical change, allowing selective etching or modification. Photolithography enables the fabrication of intricate microscale features, including the precise patterning of electrodes and other sensing elements on flexible substrates. This level of detail ensures consistent and reliable sensor performance, crucial for accurate electrochemical measurements. While photolithography has traditionally been used on rigid substrates, advancements in materials and techniques have made it compatible with flexible substrates, expanding its application to wearable sensors. The integration of photolithography into the fabrication process of flexible electrochemical sensors opens up new opportunities for developing sophisticated and miniaturized sensor devices with high sensitivity and selectivity, paving the way for advancements in fields such as healthcare, environmental monitoring, and consumer electronics. Photolithography for flexible electrochemical sensors offers several advantages. It enables the fabrication of sensors with high spatial resolution, allowing for precise positioning of electrodes and other components. This level of precision ensures consistent and reliable electrochemical measurements critical for accurate sensor performance. Photolithography is also a very helpful and efficient approach for the fabrication of miniaturized sensors with complex structures, maximizing sensor sensitivity and efficiency [11].

Further, electrospinning is also an innovative and promising fabrication technique used to fabricate flexible electrochemical sensors with exceptional properties. The electrospun nanomaterials are deposited on the flexible substrate to form the sensors. General approach involves the electrostatic deposition of nanofibers onto a flexible substrate, forming a porous and interconnected network of electroactive materials. The high surface area of the nanofibers enhances the sensor's sensitivity by providing more active sites for analyte interactions. When fibers, filaments, and yarns are joined with woven, knitted, or non-woven structures, they form smart textiles, which are textile products that can communicate with their surroundings and users. The confluence of textiles with flexible electronics offers the potential to integrate the best features of both technologies, such as the speed and computing capacity of current electronics and the flexible, wearable, and continuous character of fiber assemblies. Additionally, the flexibility of the nanofiber layer allows the sensor to bend and conform to different shapes and surfaces, making it ideal for wearable and flexible electronics. The porous structure facilitates efficient analyte diffusion, leading to faster response times and improved sensing performance. This technique offers versatility in material selection, enabling the integration of various electroactive materials, conductive polymers, or nanoparticles (NPs), tailored to specific sensing requirements [10].

One of the interesting approaches toward manufacturing of flexible electrochemical sensors is the roll-to-roll (R2R) method. R2R manufacturing is an advanced and efficient fabrication technique used for producing flexible electrochemical sensors on a large scale. In this continuous manufacturing process, a flexible substrate, often in the form of a roll, is continuously fed through various processing stations. The substrate undergoes sequential steps, such as coating, printing, and deposition of functional materials. R2R manufacturing offers significant advantages for flexible electrochemical sensors. Firstly, it enables high-speed and cost-effective production, making it suitable for large-scale deployment and commercialization. The continuous nature of the process allows for seamless integration of multiple layers and sensing elements, resulting in complex and sophisticated sensor designs. Additionally, this technique ensures uniformity across the entire production batch, ensuring consistent sensor performance. The flexibility of R2R also permits the incorporation of different materials, including conductive inks, polymers, and electrodes, tailored to specific sensor requirements [26].

7.3 Flexible electrochemical sensors for biomedical applications

7.3.1 Glucose/lactose detection

In the human body, glucose is one of the vital sources of energy for cellular function. Glucose monitoring is critical for managing diabetes. An excess amount of sugar in blood causes hyperglycemic conditions that can be responsible for organ dysfunction, damage, and failure for various delicate organs such as eyes, heart, kidneys, blood vessels, and nervous system in the long run. Thus, glucose detection in human blood is crucial in the diagnosis of diabetes. However, it is difficult to detect glucose levels directly in blood. Flexible electrochemical sensors for glucose sensing have emerged as a trailblazing technology for diabetes management. These sensors offer a non-invasive and continuous monitoring solution, reducing the need for frequent finger-pricking and providing real-time glucose level information to users [7].

Blood glucose level and interstitial fluid (ISF) glucose concentration are tightly correlated. Academics are very interested in a blood glucose measurement method based on ISF analysis: a versatile, three-electrode electrochemical sensor that can be used to precisely measure the glucose level in diluted ISF. In order to increase the electroactive nature of the working electrode surface and get a more uniform distribution of the electrodes' electrochemical active sites, graphene was inkjet printed onto it. This allowed for the detection of glucose at low concentrations. Au NPs were modified onto the graphene layer to increase the rate of electron transfer between the enzyme's activity center and the electrode in order to increase the sensor's sensitivity. The flexible glucose sensor was then combined with the ISF extraction chip to fabricate a wearable device. The wearable gadget establishes a flexible link to the skin, which has important implications [29, 30].

Similar to this, a flexible three-electrode electrochemical sensor based on graphene as the working electrode modified with AuNPs is reported to accurately detect low glucose levels. The sensor electrodes were built on a polyimide substrate using the flexible printed circuit board method. Inkjet printing, a novel technique for micro-scale manufacturing, was used to alter graphene directly onto the working electrode surface to enable glucose detection at low levels. AuNPs were electrodeposited directly onto the graphene layer to increase sensor sensitivity. The experimental findings demonstrate the suggested sensor's capacity to detect hypoglycemia with a linear range of 0–40 mg dl^{-1} and a detection limit of 0.3 mg dl^{-1} (S/N = 3) [30].

An effective flexible glucose sensor based on multi-walled carbon nanotubes –coated carbonized silk fabric (MWCNTs/CSF) with Pt microsphere ornamentation is fabricated. A good conductivity, stability and flexibility achieved by the MWCNTs/CSF prepared in inert atmosphere by carbonization at 950 °C. Furthermore, after being immersed in a glucose oxidase and Nafion mixed solution, this MWCNTs/CSF demonstrated good sensitivity and selectivity for glucose detection [31]. The safe manufacturing process allows the conducting ink to serve as an enzyme matrix for a highly sensitive detection of glucose. A straightforward process for producing poly(3,4-ethylenedioxythiophene):poly(styrene sulfonate) (PEDOT:PSS) sensors on a completely biodegradable and stretchable silk protein fibroin substrate. A tabletop photolithographic setup is utilized to create high fidelity and high resolution PEDOT:PSS microstructures across a large (cm) surface using only water as the solvent. These materials are electroactive, cytocompatible, biodegradable, and are flexible. These sensors can resist repeated mechanical deformations, but enzymatic action entirely destroys them. The technology described here is scalable and can be utilized to create sensitive, resilient, and low-cost biosensors with regulated biodegradability, which could lead to applications in transitory or implantable bioelectronics and optoelectronics [32]. A completely integrated biosensor system was built on a polyimide platform using a one-step laser-scribed graphene (LSG) scribed technique. For targeted glucose detection, glucose oxidase was immobilized on a surface modified with Pt NPs. These results demonstrate that Pt NPs integrated on LSG have a high surface area and robust electrocatalytic activity, which makes them an excellent biosensing platform when paired with a glucose-specific enzyme [33, 34].

Multi-layered biosensor in the form of a wearable patch fabricated by Lee *et al* was used for the detection of glucose by using sweat as shown in figure 7.2. Flexible substrate can adhere well with a skin made from silicone thin film. Herein, sweat was placed between patch and skin for efficient sweat collection along a waterproof band so that evaporation of sweat was prevented, and the patch was very stable, durable, and could be used multiple times. The patch can also be calibrated according to pH and temperature of skin as it also contains pH and temperature sensors [13].

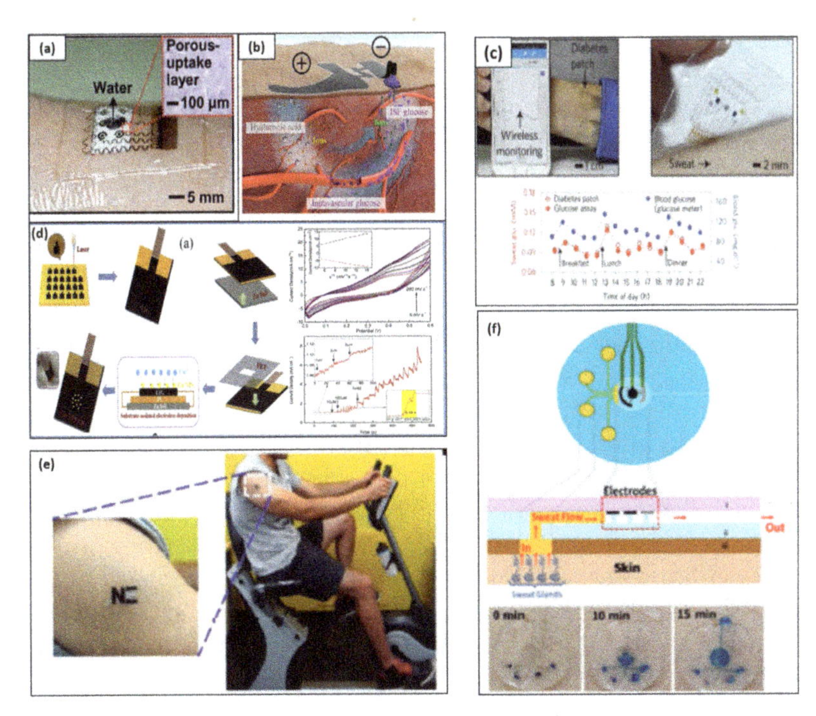

Figure 7.2. Flexible electrochemical sensors for glucose and lactose detection. (a) A stretchable glucose biosensing sweat patch [13], reproduced with permission from the American Chemical Society, Copyright (2021). (b) Iontophoresis module for a sweat glucose patch [13], reproduced with permission from the American Chemical Society, Copyright (2021). (c) Flexible and stretchable diabetes patch based on graphene for glucose monitoring [16], reproduced with permission from *Advanced Functional Materials*, Copyright (2019). (d) Different 2D materials-based glucose sensors. A procedure for fabricating a Cu NPs–laser-induced graphene composite that includes cyclic voltammetry curves and measures the material's amperometric response [7], reproduced with permission from Springer Nature, Copyright (2021). (e) Sweat biosensor for *in situ* continuous lactate monitoring via an epidermal temporary tattoo [5], reproduced with permission from the American Chemical Society, Copyright (2013). (f) Schematic of a lactate detection by microfluidic microchip [25], reproduced with permission from Engineered Regeneration, Copyright (2021).

Further, as shown in figure 7.2(b), the iontophoresis method used by Chen *et al* for increasing accuracy of biosensors by extraction of ISF from subcutaneous tissue to the skin's upper surface. A graphene-hybrid electrochemical device developed for the monitoring of diabetes and its therapy is shown in figure 7.2(c). It contains pH, humidity, and tremor sensors along with a glucose sensor. Fabrication of this device was based on Au-doped graphene synthesized by using a chemical vapor deposition technique. This patch bears a drug release module with a heater along with a temperature sensor and temperature-responsive microneedles. It was attached to electrochemical analyzer for power supply and wireless data transmission. Data related to glucose levels received from this patch match well with commercially monitored glucose levels [16].

Zhang *et al* first used the substrate assisted electrolyte deposition method for the development of Cu NPs–laser-induced graphene, enzyme-free glucose biosensor in which Cu NPs and glucose molecules initiate electron transfer. Excellent efficiency for glucose sensing was facilitated due to acceleration of electron transfer by the Cu/graphene system. Amperometric response of this sensor is shown in figure 7.2(d) [7]. Step-wise current response is a result of continuous addition of sugar. It attends steady state current in just 0.49 s. Rapid glucose molecules diffusion is due to enhanced surface area of the composite.

Furthermore, lactate is the most important anaerobic metabolite. Lactic acidosis develops when lactic acid levels accumulate due to insufficient liver and renal clearance. Lactic acid is thus identified in the blood for clinical hypoxia therapy, lactic acidosis, and other acute heart problems, as well as drug toxicity research. L-lactate dehydrogenase and L-lactate oxidase are the most commonly employed biological recognition factors in the production of L-lactate biosensors. The interfacial synthesis approach is used to create a polyaniline (PANI) film with an ordered structure at the air–water interface. The liquid interface creates a tight region in which the monomers can readily migrate laterally and build a large-area continuous and uniform structure, which facilitates charge transfer. A surfactant-free synthesis process was used due to its low cost, biocompatibility, and environmental friendliness. The PANI film is directly transferred on a flexible substrate with a screen-printed electrode, which facilitates efficient detection of lactate in sweat [35, 36].

Wang *et al* developed a wearable tattoo sensor, as shown in figure 7.2(f), that is capable of real-time extraction of sweat and ethanol tracking with the help of an integrated iontophoresis module [5]. A flexible microfluidic device constructed by Min *et al* consists of a polydimethyl siloxane sensor and a microfluidic layer for enhancing sweat sampling and simple transport, which can enhance sweat collection for continuous monitoring of human health [25].

An Au nanopine needles–programmed flexible sweat sensor has been fabricated for sensitive real-time monitoring of glucose and lactate levels in human perspiration. After the sensor chip production is completed, Au nanopine needles are created by electrochemical deposition on the flexible Au substrate for signal amplification. The appropriate enzymes are immobilized on the chip prior to measuring glucose and lactate using the cross-linker poly(ethylene glycol) diglycidylether. Finally, the developed enzyme sensor is used to measure the concentrations of glucose and lactate in human sweat in real time [36].

The Pt sensor structures were functionalized by crosslinking lactate oxidase with glutaraldehyde and bovine serum albumin, which was followed by polyurethane coating to assure biocompatibility. With a quick response time of 35 s, contact lens sensors can detect lactate in real time without a physical sample. The sensor's average linear range sensitivity was 5 - 25 mM/cm^2. Lactate levels in the body can also be determined using the levels of lactate in perspiration [37].

7.3.2 pH sensor

Miniaturized electrochemical pH sensors have gained popularity in a variety of disciplines, including water-quality monitoring and biomedical applications, due to its high sensitivity, quick response, and low cost of manufacture. The pH value reflects numerous physiological, biological, and medicinal situations [38]. Recently, there has been huge demand for the electrochemical pH sensors in wearable devices for applications such as wound monitoring and sweat- or tears-based health assessments. The accurate and consistent measurement of pH is critical in several domains, including chemical, biological, and environmental study, food science, human healthcare, and illness diagnosis. pH sensing is a fundamental analytical technique used to measure the acidity or alkalinity of a solution. Traditional pH sensors often rely on rigid materials and bulky instrumentation, limiting their application in certain scenarios. The emergence of flexible electrochemical sensors has opened up new possibilities for pH sensing, offering portability, non-invasiveness, and conformability to irregular surfaces. This article explores the recent advancements in flexible electrochemical pH sensors and their diverse applications in various fields [39].

Alginate-based microfibers containing pH-responsive beads have been created for long-term epidermal pH monitoring. After being filled with a pH-sensitive dye, mesoporous polyester beads were then implanted in hydrogel microfibers. The constructed pH-sensors could be put together to form a wearable patch since they were flexible. In addition to creating a biocompatible interface with the wound site, the beads' encapsulation within the hydrogel fibers prevents them from spreading in the wound area [40].

A three-layer electrochemical sensor fabricated by spin coating and drop casting approaches is used for pH sensing. It consists of carbon fiber fabric, MXene/CNTs, and chitosan enzyme layers. During the fabrication by the drop casting method, there is a possibility that the process may develop cracks on the material surface. These cracks can be minimized by introducing pores, which leads to releasing stress on the surface of the ultrathin film. Figure 7.3 represents the flexible sensor fabricated by the same procedure that is used for pH monitoring [3].

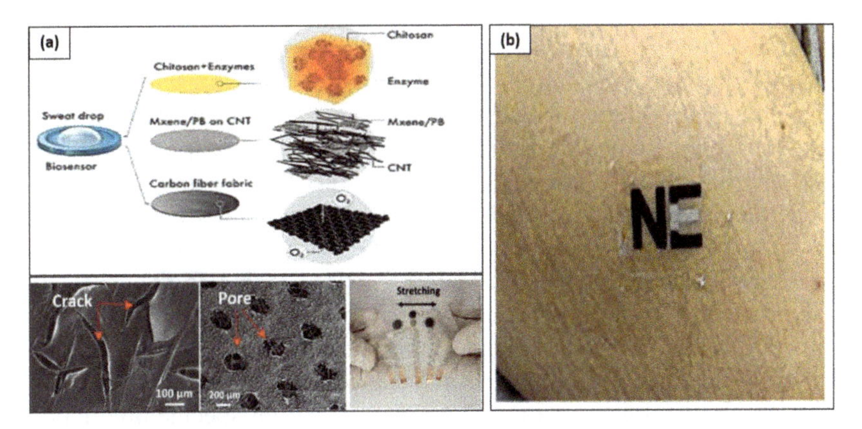

Figure 7.3. Wearable and flexible sweat biosensors. (a) Monitoring of pH concentrations [3], reproduced with permission from The Electrochemical Society, Copyright (2020). (b) Stretchable sensor for sweat pH detection with a serpentine shape on porous polyurethane [5], reproduced with permission from the American Chemical Society, Copyright (2013).

One efficient fabrication technique, i.e. screen printing, was used to implement tattoo on silicone (e.g. Ecoflex, polydimethylsiloxane (PDMS), Solaris) or polymer (e.g. polyethylene terephthalate, polyvinyl alcohol) materials substrates. These platforms have better contact, transpiration, and adhesion due to their human-skin–like properties. Over the tattoo paper sheet, layers of insulators and electrode materials such as carbon and Ag/AgCl were printed with pre-designed designs. Sensing membranes are then added to electrodes [17]. Figure 7.3(b) shows a temporary transfer tattoo flexible sensor with high impact for pH measurement.

Dual-functional Pt-hydroxyethyl cellulose (HEC)/LSG-based flexible electrochemical biosensors were produced for glucose and pH detection using a unique one-step laser fabrication process. HEC thin films and Pt NPs enhanced the electrode's surface area, hydrophilicity, and conductivity using laser-induced, 3D porous graphene. Further alterations to the working electrodes included the addition of glucose oxidase for the detection of glucose and electropolymerization of aniline for the detection of pH. The detection threshold for glucose was determined to be 0.23 M. The pH sensor demonstrated excellent performance with a linear pH 4–8 range and sensitivity of 72.4 mV pH^{-1} [41]. Using a lower-temperature thermal oxidation method, amorphous uniform iridium oxide–film pH-sensing films were fabricated on flexible polyimide substrates using the sol–gel process, which had previously been proven only on rigid substrates. In pH sensing experiments, the flexible iridium oxide film pH sensor demonstrated great reversibility and repeatability, high pH sensitivity, strong potential stability, low potential drift, low ion-interference, and fast time response. The fabrication method for this sensor is easier and could potentially be less expensive. New pH sensor applications, including *in vivo* biomedical, biological, clinical, food monitoring, and lubricant applications, will be made possible by the sensor device architecture based on deformable flexible substrates [42].

7.3.3 Detection of ambient gas molecules

The detection of ambient gas molecules is vital for regulating the healthy surrounding by monitoring air. Traditional gas sensors often lack flexibility, limiting their versatility and adaptability to different environments. However, flexible electrochemical sensors have emerged as innovative solutions, offering portability, sensitivity and the ability to conform to irregular surfaces. Flexible electrochemical sensors operate on the principle of electrochemical reactions between gas molecules and sensing materials. The interaction between the gas molecules and the sensing material causes a change in the electrical properties, leading to the generation of a measurable signal. The sensing materials can be carefully chosen to achieve high selectivity and sensitivity for specific gas molecules. Various sensing materials, such as metal oxides, conducting polymers, and nanomaterials, have been employed in flexible electrochemical gas sensors. These materials offer specific interactions with different gas molecules, enabling the detection of multiple gases with high accuracy [43]. An ionotropic gas-sensing (IGS) sticker reported by Jin *et al* is used for efficient sensing of toxic gases. Figure 7.4(a) shows an image of the IGS sticker attached to a

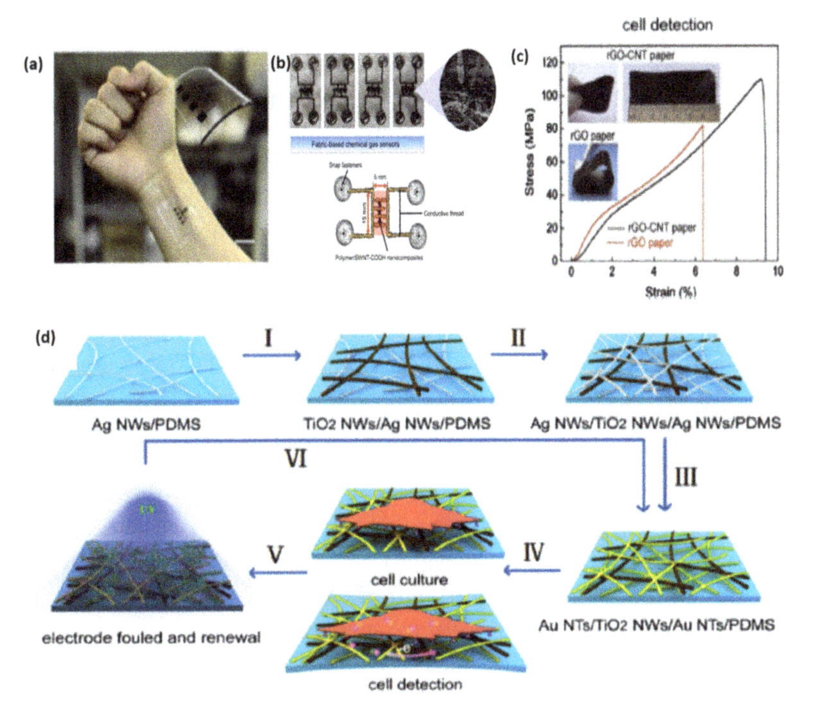

Figure 7.4. (a) Flexible sensors for gas detection, image of skin-attachable sticker for detecting toxic gases [22], reproduced with permission from John Wiley & Sons, Copyright (2019). (b) Gas accumulating smart textiles for medical applications [46], reproduced with permission from Elsevier, Copyright (2022). (c) Stress–strain curves of rGO paper and rGO-CNT paper. Inset images are of rGO-CNTs paper and rGO paper [17], reproduced with permission from Elsevier, Copyright (2022). (d) Schematic representation of fabrication of stretchable Au NTs/TiO_2 NWs/Au NTs/PDMS film [17], reproduced with permission from Elsevier, Copyright (2022).

human arm with very good adhesion. Owing this strong adhesion, IGS exhibits high resistance against compressing, twisting, and stretching deformation of human skin, indicating high stability of IGS [22]. By utilizing homogeneous polypyrrol (PPy) stacking on single-layer graphene (SLG) with polyethylene terephthalate film through *in situ* electrochemically oxidative polymerization, a sub-ppb gas sensor demonstrated for detection of NO_2 and NH_3 gases in ambient circumstances. Due to the 2D atomic layer of sp^2-carbon atoms, the electrically conducting SLG acts as an excellent supporting electrode and substrate with exceptional electron/hole mobility. These SLG characteristics are crucial for the electrochemical procedure that creates a polycrystalline polymer matrix without aggregation. The improved gas sensor film has low detection limits for NH_3. The gas sensor also exhibits exceptional reversible adsorption/desorption for analytes, with response and recovery times for NO_2 and NH_3 gases of only a few seconds without heat treatment or light irradiation [44].

In order to improve the response to NH_3 at room temperature, a specific structure of the PSS-PANI/PANI composite film sensor with more active sites on its surface is being prepared. A simple *in situ* polymerization process is used to fabricate a flexible porous PSS-PANI/polyvinylidene fluoride–based film sensor. The sensor has exceptional flexibility and long-term stability, and it exhibits an excellent linear relationship between the response value and NH_3 concentration (0.1–10 ppm) [42, 45]. Organic molecules that are extremely reactive and classified as major pollutants include volatile organic compounds (VOCs). Compounds having high vapor pressure and limited water solubility are known as VOCs. A variety of compounds are found in VOCs, some of which may have negative long- and short-term health impacts on human health. The most recent and cutting-edge sensing technologies are utilized to detect hybrid sensing materials or a combination of sensitive materials, despite the fact that many previous studies concentrated on establishing a single sensing strategy for PPy-based sensors. Additionally, the detection of sensitive materials or approaches to precisely measure VOCs by fancy hybridization is increasing quickly. PPy has a number of advantageous characteristics, such as a respectable degree of chemical stability, mechanical adaptability that can be turned in to flexibility, and environmental friendliness. Due to these properties, they are interesting candidates for emerging technologies. For applications involving health monitoring, textiles-based gas sensors paired with wearable device gas sensors can be helpful. However, environment and aging can have a significant impact on functioning of sensing devices with electro-conducive textile substrates. However, it has demonstrated the complete capability after including several modalities and accurate sensing response. Figure 7.4(b) shows such an electrochemical sensor fabricated on textile to detect VOCs [46].

Hydrogen peroxide (H_2O_2) gas is a result of several highly selective oxidases that are catalyzed by oxygen metabolism. It performs a crucial physiological role in defending against pathogen invasion. Non-enzyme–based H_2O_2 gas sensors can overcome the problems that arise due to low enzyme stability and interference by electroactive substances. Sun *et al* developed a nanocomposite on reduced graphene oxide (rGO)–CNTs paper modified by using ultra-thin Pt NPs. This nanocomposite can be used to construct an electrochemical sensor that can detect H_2O_2 released by living cells. Introduction of CNTs on graphene paper has improved conductivity, mechanical strength, and surface roughness, providing more nucleation sites for Pt NPs. Hence, flexible Pt/rGO-CNTs paper electrode shows excellent electrochemical activity, active surface area, high flexibility, and stability, which is result of synergistic effect of various components in this nanocomposite [17]. Figure 7.4(c) shows a stress–strain graph of rGO and rGO-CNTs paper. Images of rGO-CNTs and rGO array upon bending are included in the inset of the figure.

Electrochemical sensors based on 2D materials (2DM) have a number of advantages for detecting substances, including convenience and speed, ease of use, low cost, and quick time consumption. They excel in a variety of sectors, including pharmaceutical manufacturing, biological science, environmental monitoring, and food safety. The creation and use of electrochemical sensors based on 2DMs have opened up a brand-new area of study for the detection of biologically active small molecules, heavy metal ions, and pesticide residues. The main focus of this review is to introduce 2DM preparation techniques, structures, and properties, as well as their uses in various electrochemical detection. The test performance of electrochemical sensors is enhanced by fabricating sensing electrodes and designing useful nanomaterials [45, 46].

2D MXenes and molybdenum sulphide (MoS_2) materials can be used to create gas sensors with a large surface area, numerous active surface sites, and high electrical conductivity. A fully functionalized surface and high metallic conductivity could be seen in the Ti_3C_2–MoS_2 combination. The results showed that at ambient temperatures, the Ti_3C_2–MoS_2 composite can detect very low NO_2 concentrations in air. Additionally, the Ti_3C_2–MoS_2 combination picked up harmful chemicals like methane and nitrate [47, 48]. Real-time monitoring of nitric oxide (NO) gas released from mechanically sensitive human umbilical vein endothelial cells have been performed under unstretched and stretched conditions. A photo-catalytically renewable flexible electrochemical sensor fabricated by Wang *et al* based on nano-artwork of Au NTs and TiO_2 nanowires (NWs). It can monitor NO gas released by endothelial cells. External Au NTs were used for electrochemical sensing, and TiO_2 NWs have photocatalytic performance of populant removal. This is the reason that sensors show excellent electrochemical and good mechanical tensile performance as well as high photocatalytic activity [17]. A schematic representation of fabrication of stretchable Au NTs/TiO_2 NWs/Au NTs/PDMS film for NO gas detection is shown is figure 7.4(d).

7.3.4 Detection of bacterial and viral infections

Over time, infections brought on by bacterial over-reproduction in organisms have frequently happened. It is crucial to concurrently identify the presence of antibiotics in the human body due to the widespread overuse of antibiotics, which can cause bacteria to develop resistance. In order to detect meningitis, tuberculosis, or plague, electrochemical biosensors have been developed. Since viruses are the source of numerous human diseases, they pose a significant danger to the world economy. One of the key elements in the development process is a quick and accurate detection of viral genomic DNA/RNA and/or amino acids sequence (epitopes, peptides, proteins), as vaccines have not yet been developed. The ability to accurately identify infected people and restrict the spread of the virus has evolved into a crucial component of effectively managing the current situation. When a person is infected, their immune system responds by producing particular antibodies against the pathogen, which is advantageous when designing electrochemical immunosensors. An electrochemical biosensor based on functionalized TiO_2 NTs has been developed for the rapid detection of the receptor-binding region of the SARS-CoV-2 spike protein. A few examples of novel biosensors used to detect RNA viruses include the paper strip based on CRISPR-Cas9 (clustered regularly interspaced short palindromic repeats associated protein 9) as well as nucleic acid–based, aptamer–based, and antigen–Au/AgNPs–based electrochemical biosensors, optical biosensors, and surface plasmon resonance. Blood samples may currently be examined using lateral flow point-of-care (POC) immunoassays, electrochemical immunosensors, and eye-based devices with detection times less than 10 min after swab introduction. Their main advantage is mobility for on-site testing, effectively without sample treatment [23, 48].

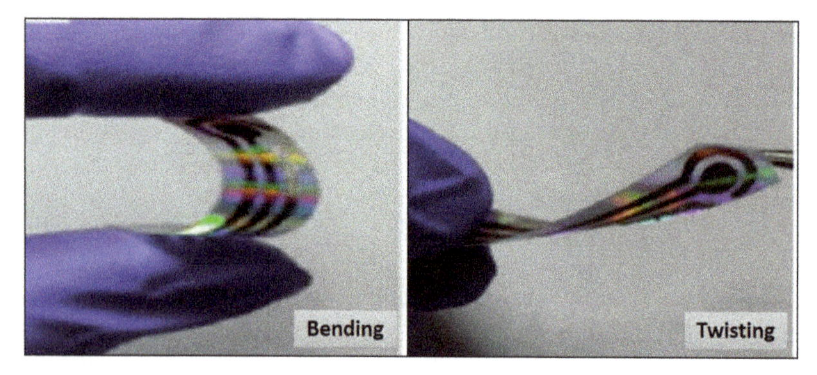

Figure 7.5. Image of bending and twisting of flexible electrochemical sensor fabricated on polyaniline nanoarray having Au and Ti NPs layers for detection of *Escherichia coli* O157:H7 [17], reproduced with permission from Elsevier, Copyright (2022).

In addition to this, food-borne pathogenic bacteria have gained attention of public health surveillance in day-to-day life as the worldwide food industry is growing rapidly. In this family of pathogens, *E. coli* O157:H7 can cause fatal complications to human life. Common medical examination procedures demand complicated operational procedures and costly equipment. Therefore, need for portable, adaptable, and trustworthy detection methods for *E. coli* O157:H7 is essential. Figure 7.5 shows a flexible electrochemical sensor developed by Park *et al* based on a polyaniline nanoarray. Metal evaporation and soft photolithography were used to fabricate highly ordered nanometer electrodes nanoarrays. These arrays were first formed on silicon wafers by soft photolithography, and then vacuum sputtering was implemented to form thin layers of Ti and Au NPs on the array surface [17]. Genetic analysis of amplified genes with high reproducibility and sensitivity from food-borne pathogens derived from real food within 25 s was successfully performed.

7.3.5 Animal health monitoring

Electrochemical biosensors with wireless sensing technologies are used in animal health management, and they represent a emerging industry that has been quickly attracted by global marketing. Development of such platforms is important because the majority of sensing platforms currently in existence are only useful for human applications. But the environment, especially the quality and type of food intake, has a big impact on human health. Today, a wide range of cutting-edge technologies are being considered for the fabrication of electrochemical biosensors for the detection of several animal pathogenic microorganisms, such as viruses and bacteria, stress indicators, biomarkers, or metabolites in biofluids. These technologies include POC diagnostics, hybridization and amplification assays, microfluidics, and others. A large number of original studies are reported on recent advancements in the application of electrochemical biosensors, mainly nanostructured ones, for the detection of drugs and drug residues in food samples. Malachite green, crystal violet, and the medicines chloramphenicol and nitrofurans are among the electrochemical aptasensors used for their detection. To locate antimicrobial medication residues in food derived from animals, a number of detection techniques have been developed. A label-free amperometric immunosensor based on graphene sheet-Nafion-thionine–Pt NPs and an aptasensor for kanamycin are just a few examples of these techniques. Others include MWCNTs, molecularly imprinted polymers, and monoclonal antibodies immobilized in hollow Au nanospheres/chitosan composite [23, 25].

7.3.6 Real-time monitoring of electrochemical sensors

Wireless chemical sensors are hybrid devices that process chemical or biochemical data from an experiment and communicate it to a remote device with or without reduced signal processing interferences using wireless technology, frequently radio-frequency communication. Wireless electrochemical biosensors have a wide range of possible applications, such as environmental monitoring, water-quality testing, and on-body sensing systems. The combination of portable, affordable equipment with electrochemical-sensing devices has opened up new possibilities for real-time monitoring of human healthcare and diagnostics. Routine laboratory tests are traditionally performed by experts, but POC devices and wireless smartphone-based signal registration are increasingly spreading into everyday life. When a sensing platform and a mobile receptor exchange data, generally through Bluetooth, Wi-Fi, or near-field communication, a signal is sent to the analytical equipment. Similar to standard laboratory equipment, smartphone-based systems control every experimental parameter in such systems. It has previously been stated that a number of advanced diagnostic platforms using aptamers as biorecognition components instead of conventional immunoassays have the potential to become established as commercial POC devices. It utilizes 3D printing technology and a modern, smartphone-integrated detection method [23].

Figure 7.6 depicts the layouts on the flexible, soft contact lens where the wireless power transfer circuit, glucose sensor, and display pixel are all fully integrated. The

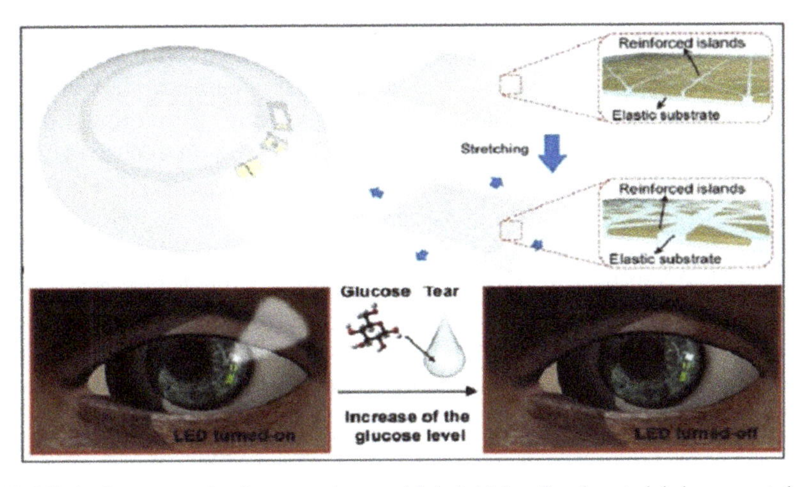

Figure 7.6. Wirelessly connected soft contact lenses with hybrid functional materials incorporated allow for real-time monitoring of glucose in tears [23], reproduced with permission from *Materials Science and Engineering: C*, Copyright (2021).

capacity to remotely monitor the wearer's health (namely, the level of glucose in their tears) through the LED pixel is the key idea behind this soft, smart contact lens device. The transparent and flexible inductive antenna and rectifier circuit of this system were used as the power transfer circuit for wireless operations. The rectifier is made up of Si diodes and a capacitor, and in addition, the elastic portions, which were made of silicone elastomer, a common material for soft contact lenses, are where the flexible antenna and interconnects were constructed. The creation of the soft, smart contact lens was completed by molding the sample produced by the devices into the form of a lens once they have been fully integrated with the glucose sensor and LED pixel [49].

7.4 Conclusion

Flexible electrochemical sensors have shown immense promise for biomedical applications, revolutionizing the field of healthcare and diagnostics. Their unique combination of flexibility, portability, and high sensitivity makes them ideal candidates for non-invasive and continuous monitoring of various biomarkers and analytes. The integration of flexible electrochemical sensors into wearable medical devices has enabled real-time and personalized healthcare, allowing patients to monitor their health parameters in the comfort of their homes. These sensors play a crucial role in providing accurate and frequent glucose monitoring without the need for invasive blood sampling. In addition to glucose sensing, flexible electrochemical sensors have found applications in other areas of healthcare, including lactate sensing for exercise monitoring and wound healing assessment, pH sensing for monitoring body fluids and environmental changes, and detection of specific biomolecules for disease diagnosis and early detection.

The conformable nature of these sensors ensures better skin contact, enhancing accuracy and patient comfort during extended wear. Their wireless connectivity allows for seamless data transmission to healthcare providers, facilitating remote monitoring and timely interventions. While flexible electrochemical sensors have achieved significant advancements in biomedical applications, challenges remain, such as improving long-term stability, enhancing selectivity, and integrating multi-analyte sensing capabilities. Future research efforts should focus on developing novel sensing materials, refining fabrication techniques, and leveraging data analytics and artificial intelligence to optimize sensor performance and analysis.

Flexible electrochemical sensors hold immense potential to transform healthcare and improve patient health. Their ability to provide real-time and continuous monitoring of vital biomarkers offers the promise of personalized medicine and improved disease management, making them invaluable tools in the quest for better health and well-being. As research and technology continue to progress, these sensors are expected to play an increasingly critical role in shaping the future of biomedical applications, paving the way for innovative and patient-centric healthcare solutions.

References

[1] Zhou N, Liu T, Wen B, Gong C, Wei G and Su Z 2020 Recent advances in the construction of flexible sensors for biomedical applications *Biotechnol. J.* **15** 2000094

[2] Zang Y, Zhang F, Di C and Zhu D 2015 Advances of flexible pressure sensors toward artificial intelligence and health care applications *Mater. Horiz.* **2** 140

[3] Jeerapan I and Poorahong S 2020 Review—flexible and stretchable electrochemical sensing systems: materials, energy sources, and integrations *J. Electrochem. Soc.* **167** 037573

[4] Bandodkar A, Jeerapan I and Wang J 2016 Wearable chemical sensors: present challenges and future prospects *J ACS Sens* **1** 464–82

[5] Jia W, Bandodkar A, Valdes-Ramirez G, Windmiller J, Yang Z, Ramirez J, Chan G and Joseph Wang J 2013 Electrochemical tattoo biosensors for real-time noninvasive lactate monitoring in human perspiration *Anal. Chem.* **85** 6553–60

[6] Choi C, Lee Y, Cho K, Koo J and Kim D 2019 Wearable and implantable soft bioelectronics using two-dimensional materials *Acc. Chem. Res.* **52** 73–81

[7] Mathew M, Radhakrishnan S, Vaidyanathan A, Chakraborty B and Rout C 2021 Flexible and wearable electrochemical biosensors based on two-dimensional materials: Recent developments *Anal. Bioanal. Chem.* **413** 727–62

[8] Mahato K and Wang J 2021 Electrochemical sensors: from the bench to the skin *Sensors Actuators* **B 344** 130178

[9] Stetter J, Penrose W and Sheng Y 2003 Sensors, chemical sensors, electrochemical sensors, and ECS *J. Electrochem. Soc.* **150** S11–6

[10] Laschi A and Mascini M 2006 Planar electrochemical sensors for biomedical applications *Med. Eng. Phys.* **28** 934–42

[11] Lin T, Xu Y, Zhao A, He W and Xiao F 2022 Flexible electrochemical sensors integrated with nanomaterials for in situ determination of small molecules in biological samples: a review *Anal. Chim. Acta* **1207** 339461

[12] Maduraiveeran G and Jin W 2020 *Handbook of Nanomaterials in Analytical Chemistry* (Amsterdam: Elsevier)

[13] Yang A and Yan F 2021 Flexible electrochemical biosensors for health monitoring *ACS Appl. Electron. Mater.* **3** 53–67

[14] Lee Y and Wong D T 2009 Saliva: an emerging biofluid for early detection of diseases *Am. J. Dent* **22** 241–8

[15] Dartt D, Hodges R and Zoukhri D 2005 Tears and their secretion *Adv. Organ. Biol.* **10** 21–82

[16] Tu J, Rodriguez R, Wang M and Gao W 2019 The era of digital health: a review of portable and wearable affinity biosensors *Adv. Funct. Mater.* **30** 1906713

[17] Yuan F, Xia Y, Lu Q, Xu Q, Shu Y and Hu X 2022 Recent advances in inorganic functional nanomaterials based flexible electrochemical sensors *Talanta* **244** 123419

[18] Wang B, Huang W, Chi L, Al-Hashimi M, Marks T and Facchetti 2018 High-k gate dielectrics for emerging flexible and stretchable electronics *Chem. Rev.* **118** 5690–754

[19] Wu W and Haick H 2018 Materials and wearable devices for autonomous monitoring of physiological markers *Adv. Mater.* **30** 1705024

[20] Nightingale A, Leong C, Burnish R, Hassan S, Zhang Y, Clough G, Boutelle M, Voegeli D and Niu X 2019 Monitoring biomolecule concentrations in tissue using a wearable droplet microfluidic-based sensor *Nat. Commun.* **10** 2741

[21] Wan H, Yin H, Lu L, Xiangqun Z and Mason A 2018 Miniaturized planar room temperature ionic liquid electrochemical gas sensor for rapid multiple gas pollutants monitoring *Sensors Actuators* B **255** 638–46

[22] Gao Y, Yu L, Yeo J and Lim C 2019 Flexible hybrid sensors for health monitoring: materials and mechanisms to render wearability *Adv. Mater.* **32** 190133

[23] Nemcekova K and Labuda J 2021 Advanced materials-integrated electrochemical sensors as promising medical diagnostics tools: a review *Mater. Sci. Eng.* **120** 111751

[24] Manjakkal L, Shakthivel D and Dahiya R 2018 Flexible printed reference electrodes for electrochemical applications *Adv. Mater. Technol.* **3** 1800252

[25] Lu H, He B and Gao B 2021 Emerging electrochemical sensors for life healthcare *ER* **2** 175–81

[26] Narakathu B, Devadas M, Reddy A, Eshkeiti A, Moorthi A, Fernando I, Miller B, Ramakrishna G, Sinn E and Joyce M 2013 Novel fully screen printed flexible electrochemical sensor for the investigation of electron transfer between thiol functionalized viologen and gold clusters *Sensors Actuators* B **176** 768–74

[27] Metters J, Kadara R and Banks C 2011 New directions in screen printed electroanalytical sensors: an overview of recent developments *Analyst* **136** 1067

[28] Moya A, Gabriel G, Villa R and Campo F 2017 Inkjet-printed electrochemical sensors *Curr. Opin. Electrochem.* **3** 29–39

[29] Kipphan H (ed) 2001 *Handbook of Print Media* (Heidelberg: Springer)

[30] Pu Z, Wang R, Xu K, Li D and Yu H 2015 A flexible electrochemical sensor modified by graphene and AuNPs for continuous glucose monitoring 2015 *IEEE SENSORS (Busan, South Korea)*, pp 1–4

[31] Pu Z, Wang R, Wu J, Yu H, Xu K and Li D 2016 A flexible electrochemical glucose sensor with composite nanostructured surface of the working electrode *Sensors Actuators* B **230** 801–9

[32] Chen C, Ran R, Yang Z, Lv R, Shen W, Kang F and Huang Z 2018 An efficient flexible electrochemical glucose sensor based on carbon nanotubes/carbonized silk fabrics decorated with Pt microspheres *Sensors Actuators* B **256** 63–70

[33] Pal R, Farghaly A, Wang C, Collinson M, Kundu S and Yadavalli V 2016 Conducting polymer-silk biocomposites for flexible and biodegradable electrochemical sensors *Biosens. Bioelectron.* **81** 294–302

[34] Hossain M and Slaughter G 2021 Flexible electrochemical uric acid and glucose biosensor *Biochem* **141** 107870

[35] Khoshroo A, Sadrjavadi K, Taran M and Fattah A 2020 Electrochemical system designed on a copper tape platform as a nonenzymatic glucose sensor *Sensors Actuators* B **325** 128778

[36] Zhu C, Xue H, Zhao H, Fei T, Liu S, Chen Q, Gao B and Zhang T 2022 A dual-functional polyaniline film-based flexible electrochemical sensor for the detection of pH and lactate in sweat of the human body *Talanta* **242** 123289

[37] Yu M, Li Y, Hu Y, Tang L, Yang F, Lv W, Zhang Z and Zhang G 2021 Gold nanostructure-programmed flexible electrochemical biosensor for detection of glucose and lactate in sweat *J. Electroanal. Chem.* **882** 115029

[38] Gao F *et al* 2023 Wearable and flexible electrochemical sensors for sweat analysis: a review *Microsyst. Nanoeng.* **9** 1

[39] Manjakkal L, Sakthivel B, Gopalakrishnan N and Dahiya R 2018 Printed flexible electrochemical pH sensors based on CuO nanorods *Sensors Actuators* B **263** 50–8

[40] Yoon J, Kim S, Park H, Kim Y, Oh D, Cho H, Lee K, Hwang S, Park J and Choi B 2020 Highly self-healable and flexible cable-type pH sensors for real-time monitoring of human fluids *Biosens. Bioelectron.* **150** 111946

[41] Tamayol A *et al* 2016 Flexible pH-sensing hydrogel fibers for epidermal applications *Adv. Healthc. Mater.* **5** 711–9

[42] Wang Y, Guo H, Yuan M, Yu J, Wang Z and Chen X 2023 One-step laser synthesis platinum nanostructured 3D porous graphene: A flexible dual-functional electrochemical biosensor for glucose and pH detection in human perspiration *Talanta* **257** 124362

[43] Huang W, Cao H, Deb S, Chiao M and Chiao J 2011 A flexible pH sensor based on the iridium oxide sensing film *Sensors Actuators* A **169** 1–11

[44] Yoon T, Jun J, Kim D, Pourasad S, Shin T, Yu S, Na W, Jang J and Kim K 2018 An ultra-sensitive, flexible and transparent gas detection film based on well-ordered flat polypyrrole on single-layered graphene *J. Mater. Chem.* A **6** 2257

[45] Lv D, Shen W, Chen W, Tan R, Xu L and Song W S 2021 PSS-PANI/PVDF composite based flexible NH3 sensors with sub-ppm detection at room temperature *Actuators* B **328** 129085

[46] Miah M, Yang M, Khandaker S, Bashar M, Alsukaibi A, Hassan H, Znad H and Awua M 2022 Polypyrrole-based sensors for volatile organic compounds (VOCs) sensing and capturing: A comprehensive review *Sensors Actuators* A **347** 113933

[47] Li T, Shang D, Gao S, Wang B, Kong H, Yang G, Shu W, Xu P and Wei G 2022 Two-dimensional material-based electrochemical sensors/biosensors for food safety and biomolecular detection *Biosensors* **12** 314

[48] Le V, Vasseghian Y, Doan V, Nguyen T, Vo T, Do H, Vu K, Vu Q, Lam T and Tran V 2022 Flexible and high-sensitivity sensor based on Ti3C2–MoS2 MXene composite for the detection of toxic gases *Chemosphere* **291** 133025

[49] Park J *et al* 2018 Soft, smart contact lenses with integrations of wireless circuits, glucose sensors, and displays *Sci. Adv.* **4** eaap9841

Chapter 8

Flexible conformal textile antennas

A Taksala Devapriya and S Robinson

The increasing demand for wearable technology and smart textiles needs effective and undetectable communication capabilities. Textile antennas that are flexible have various benefits over standard rigid antennas, which make them more appropriate for wearable applications. The chapter presents overview of the fundamental principles that govern antenna design, along with the distinctive properties of textile materials employed. Detailed examination of the numerous advantages offered by flexible conformal textile antennas (FCTAs) follows, highlighting aspects such as wearability and durability. Subsequently, the chapter explores a diverse range of design and fabrication processes including the potential applications of 3D-printed flexible antennas, emphasizing their versatility in adapting to distinct parameters. Illustrative examples of various FCTAs are provided, showcasing their applications in various domains, highlighting their applications across different domains, and giving readers a clear understanding of their real-world potential. The design of a microstrip patch antenna for sub-6 GHz 5G applications, which utilizes a Defected Ground Structure (DGS) and is crafted with jeans textile material, is presented. The integration of DGS into the conventional patch design leads to a considerable size reduction and enhanced antenna performance. The chapter also delves into performance assessment and testing techniques for textile antennas, accentuating the influence of factors such as body proximity, bending, and environmental conditions on their efficiency and radiation properties.

8.1 Introduction

In recent years, people are anticipated to possess a wide range of communication devices and medical sensors, allowing them to establish constant connectivity with various systems through wearable devices [1]. In the realm of medical applications, textile antennas have garnered significant interest as of late, particularly in the domains of patient monitoring, body diagnosis, and wireless body area networks, fostering seamless communication between patients and doctors [2]. Hence, the field

doi:10.1088/978-0-7503-5492-9ch8

of wearable and flexible electronics has seen significant growth and innovation. One of the most exciting developments in this field is the emergence of FCTAs. It is a new class of antennas that have been developed to meet the growing demand for wearable and portable communication devices. These antennas offer unique properties that make them ideal for integration into clothing [3] and other wearable devices, providing communication and sensing capabilities that were once thought to be impossible.

According to the latest market analysis report, the revenue generated by flexible electronics was approximately 3 billion USD in 2017. However, this is projected to surge exponentially, surpassing a staggering 300 billion USD by 2028 [4]. Driven by the rising demand for wearable and implantable health-monitoring systems, as well as everyday wireless devices, the markets for flexible wireless devices are experiencing rapid growth.

8.2 Literature review

In this section, the design, fabrication, performance of FCTAs, materials used, antenna geometries, and optimization techniques to achieve the desired resonance frequency and bandwidth is discussed. Also, the different fabrication techniques, such as printing and deposition methods, and how they can be used to create these antennas are discussed. Finally, the application and challenges are discussed.

8.2.1 Existence of FCTAs

The performance of FCTAs is influenced by several factors, including the antenna design, substrate material, and operating frequency. The most important performance parameters for FCTAs are radiation efficiency, bandwidth, and gain. Radiation efficiency refers to the percentage of power delivered to the antenna that is radiated as electromagnetic waves. Bandwidth is the range of frequencies over which the antenna can operate with acceptable performance, and gain refers to the ability of the antenna to concentrate the radiated energy in a particular direction.

Salonen *et al* [4] introduced a dual-band planar inverted-F antenna designed on a flexible substrate. Specifically tailored to be placed on a shirt sleeve, this antenna demonstrated effective operation at 900 MHz and 2.4 GHz frequencies. Notably, the antenna showcased excellent performance at the upper band frequency (2.4 GHz) even in the presence of a human body. However, achieving radiation at the lower band frequency (900 MHz) remained a challenge. Spiral antennas, renowned for their wide bandwidth and compact profile, have gained significant popularity. Extensive research papers have been published focusing on the fabrication processes of wearable spiral antennas [5].

Furthermore, an innovative approach utilizing embroidered nonuniform mesh patch antennas was presented in [6]. This novel technique demonstrated the ability to maintain excellent antenna gain and efficiency while significantly reducing the usage of special conducting threads. Remarkably, the nonuniform mesh patch antennas exhibited comparable current and electric field distribution to that of a solid patch antenna operating in the TM01 mode. Another noteworthy development in wearable antenna technology is highlighted in [7], in which an active wearable antenna was specifically designed for global positioning system (GPS) and satellite phone applications. In a pioneering study conducted by Vallozzi *et al* [8], a dual-polarization patch antenna operating at 2.45 GHz was proposed. This antenna, implemented on a protective foam substrate, offered the unique advantage of polarization diversity using a single, compact, and wearable design.

An exceptional wearable textile substrate integrated waveguide (SIW) antenna system was introduced in [9]. This highly integrated system showcased a compact design that seamlessly incorporated two flexible solar cells, a flexible power management system, and a micro energy cell. Notably, recent research has focused on the fusion of solar energy harvesters with planar textile antennas for the ultra-high frequency (UHF) Radio Frequency Identification (RFID) band. In a notable contribution, Declercq *et al* [10] unveiled an innovative wearable aperture-coupled shorted planar antenna that featured direct integration of solar cells atop the radiating patch. This ingenious approach facilitated the space-efficient combination of two fundamental elements in a wearable textile system: a wearable antenna and an energy harvester. The result was a compact stacked structure that epitomized seamless integration and efficient utilization of available space.

Subsequently, Lemey *et al* [11] leveraged the SIW cavity-backed slot antenna as a foundation for an innovative energy-harvesting platform. This pioneering development involved the compact integration of solar cells and dedicated flexible circuitry onto both the top and back surfaces of the SIW antenna structure. In an earlier investigation [12], a remarkable achievement was made with the design, fabrication, and testing of a flexible, wideband, slotted monopole antenna for millimeter-wave frequencies. Inkjet printing was done using custom-made silver-nanowire ink as conductive material. The antenna exhibited wide bandwidth, ranging from 18 to 44 GHz, while maintaining a high radiation efficiency of 55% and achieving a maximum gain of 1.45 dBi.

In a significant development, Koski *et al* [13] introduced a pioneering 866 MHz patch tag antenna that employed an anisotropic embroidered textile structure as its radiating patch. To assess its performance, the antenna was compared to an equivalent prototype featuring a radiating patch made of conventional copper fabric. Notably, a comprehensive analysis was conducted to evaluate the readability range with the antenna mounted on the arm of a human test subject. The results revealed that the wearable RFID antenna with an embroidered patch showcased comparable link loss and shadowing characteristics when compared to its counterpart with a conventional copper patch. This demonstrated the effectiveness and viability of the embroidering technique in the manufacturing of wearable RFID antennas.

For applications like intrabody telemedicine systems, there is a strong preference for a low profile, lightweight and resilient antenna design. Addressing these requirements, a remarkable proposal was put forth for a flexible antenna utilizing photo paper as its substrate, operating within the frequency range of 2.33–2.53 GHz [14]. In the realm of wearable antennas, the employment of SIW technology offers a distinct advantage. It provides superior human-body shielding capabilities while utilizing a smaller ground plane compared to conventional patch antennas or electromagnetic bandgap (EBG) substrate patches [15]. Wideband multiple-input multiple-output antenna with a fractal shape is introduced for brain and skin implantable applications. The antenna is designed to operate within the 2.4–2.48 GHz band of industrial, scientific, and medical (ISM) standards [16].

Meander-line dual-band implantable antenna suitable for biotelemetry applications [17], operating in the 2.4 GHz (ISM) and 1.4 GHz Wireless Medical Telemetry Service (WMTS) bands. To achieve this, the biocompatible Rogers RT/duroid 6010 material was used as the substrate and incorporated a superstrate in the antenna's design.

This means that SIW antennas can achieve comparable performance with significantly smaller ground plane dimensions than their counterparts.

In a notable recent publication by Mishra *et al* [18], a cutting-edge wearable dual-band patch antenna boasting circular polarization is introduced. The antenna is ingeniously constructed using conductive metalized nylon fabric (Zelt) for both the patch and ground plane, and a durable denim substrate provides structural support. By employing circular polarization, this antenna surpasses linear polarization by maximizing transmit/receive power, thanks to its orientation independence. The antenna design incorporates a modified rectangular slot patch with a distinct L-shaped topology for the feed, surrounded by a coplanar ground plane that is separated from the feed by a peripheral slot. These components are skillfully positioned on top of the substrate, resulting in a high-performance wearable antenna solution.

A remarkable ultra-thin flexible antenna was unveiled in [19], featuring a novel configuration of rectangular fractal patches with a strategically placed stub. This inventive rectangular fractal patch design demonstrated an impressive 30% reduction in size compared to conventional quadrilateral fractal patches. To further enhance antenna performance, a defected ground plane was ingeniously utilized, allowing for precise control over gain, radiation characteristics, and antenna dimensions. Building upon this concept, a compact wearable antenna was meticulously crafted, employing a double flexible substrate, resulting in a highly efficient and compact wearable antenna solution [20].

Pioneering efforts in the realm of wearable textile antennas can be traced back to the seminal work of Vallozzi *et al* [21]. This study marked a significant milestone as it introduced the concept of a wearable GPS antenna designed specifically for operation in the GPS-L1 frequency band [1.563 42, 1.587 42] GHz. Notably, the antenna was ingeniously implemented on a wearable protective-foam substrate, commonly employed in the clothing of rescue workers. The adopted antenna topology featured a patch antenna with a distinctive truncated-corner design. Through meticulous dimensioning, the antenna achieved the crucial requirement of right-hand circular polarization, as mandated by the GPS standard. This remarkable achievement set the stage for further advancements in the development of wearable textile antennas for GPS applications.

Kaivanto *et al* introduced an exceptional breakthrough by presenting a wearable circularly polarized antenna designed for operation in both the GPS and Iridium satellite bands [22]. This remarkable antenna served the dual purpose of enabling positioning and communication functionalities. To achieve coverage in both bands, a wide operational bandwidth was ingeniously attained through the utilization of a polygonal-shaped slot on a square ring patch as the radiating element. The radiating patch, crucial to the antenna's performance, was implemented on a substrate composed of flexible textiles such as Cordura and another ballistic textile, carefully selected for their favorable mechanical properties. Notably, electro textiles made of silver- and copper-plated nylon fabric were skillfully employed to construct the patch and ground plane, ensuring optimal functionality. Through rigorous experimental analysis, it was conclusively demonstrated that the antenna performed admirably in both frequency bands, even under bending conditions. The antenna exhibited exceptional characteristics, maintaining right-hand circular polarization over an impressive 53 MHz band, even in the presence of bending stress. This pioneering work heralded a new era in wearable antenna technology, opening up exciting possibilities for enhanced positioning and communication capabilities in various applications.

Zaric *et al* put forward an impressive innovation in the form of a wearable ultra-wideband (UWB) antenna tailored for operation in the European Commission (EU) UWB band ([6, 8.5] GHz) [23]. This remarkable antenna boasted a remarkably compact profile and a unidirectional radiation pattern and delivered high-fidelity performance. Specifically designed for impulse radio UWB applications, it exhibited outstanding capabilities in accurately localizing human subjects. A miniaturized conformal loop antenna operating at 433 MHz was introduced [24] to transmit real-time video images from inside the body to the outside in capsule endoscope systems. This antenna was designed to provide seamless coverage and efficient communication within the capsule endoscope, enabling effective medical diagnostics and treatments.

A graphene-based antenna sensor was proposed for high strain detection [25]. Operating at 1.63 GHz, this antenna sensor harnessed the remarkable properties of flexible multilayer graphene film with an impressive conductivity of 106 S m^{-1}. Extensive testing under tensile and compressive bending situations showcased the sensor's exceptional performance and reliability. In the pursuit of cost-effective RFID and sensing applications, a meandered line antenna was reported utilizing paper material [26]. This ingenious design not only offered affordability but also demonstrated excellent performance, making it an ideal choice for various low-cost applications requiring reliable wireless communication and sensing capabilities.

Sina *et al* introduced a state-of-the-art wearable antenna tailored specifically for industrial applications, operating at a central frequency of 2.45 GHz [27]. To optimize the antenna's radiation performance while mitigating back radiation and minimizing the specific absorption rate when in close proximity to the human body, the researchers ingeniously incorporated a metal plate that served both as an isolator and a reflector. By utilizing a co-planar waveguide (CPW) feed, the authors achieved remarkable results, including a notable gain of 7.3 dBi at 2.45 GHz and an impressive bandwidth ranging from 2.20 to 2.56 GHz. Furthermore, this innovative design exhibited a substantial improvement in the front-to-back ratio, showcasing an impressive enhancement of 18.2 dB.

The antenna utilized an innovative DGS approach to overcome the limitations commonly associated with traditional antennas, such as restricted gain and bandwidth. A cutting-edge triple-band CPW antenna, incorporating a 2 × 2 electromagnetic bandgap structure, was proposed, resulting in an impressive 95% reduction in specific absorption rate [1, 28]. The study also explored both rectangular and circular microstrip designs, complemented by strategically placed slots, for dual-band applications [1]. The authors strongly recommended the utilization of CPW in combination with a comprehensive ground plane to effectively suppress the propagation of backward waves toward the human body [29, 30].

To enhance the bandwidth, an innovative approach involving a partial ground plane fabricated using foam substrate was introduced [28]. Another remarkable development in antenna design was the creation of a compact, high-gain antenna operating in the ISM band, which featured rectangular and circular slots along with a ground plane adorned with rectangular slots [29]. Additionally, a square microstrip patch with symmetrical etching of four slots was detailed [33], and a planar and flexible UWB antenna utilizing jeans as a textile material, incorporating a hexagonal slot in the patch, was presented [2]. Other wideband applications included the implementation of a slot over both ground and radiating patches [35]. To further enhance the impedance bandwidth and gain of the antenna, a dual-ring antenna backed by an Artificial Magnetic Conductor was developed [1]. Even in the field of optical devices, flexible plastic substrate was used to design white polymer LEDs [36]. Several studies have shown that FCTAs can achieve radiation efficiencies of up to 90% and bandwidths of several hundred megahertz. However, the gain of FCTAs is typically lower than that of conventional antennas due to their conformal nature and reduced size.

8.3 Design of FCTAs

FCTAs have good opinion for their ground-breaking design, revolutionizing the wearable technology landscape. These antennas seamlessly integrate into flexible fabrics, providing a comfortable user experience without compromising performance. Users appreciate the soft and lightweight feel, making them ideal for smart garments and accessories. The antennas' durability and reliability ensure they can withstand various environmental conditions, making them well-suited for long-term use. Moreover, their ability to conform to different shapes optimizes signal reception and radiation efficiency. This adaptability opens up a wide range of applications in industries such as healthcare, sports, military, communication, and the Internet of Things (IoT). One of its key advantages is the aesthetic appeal and design flexibility they offer, allowing customization to suit individual preferences. This feature, combined with potential low-cost manufacturing, makes them appealing to designers and manufacturers alike. Although challenges exist in achieving performance comparable to traditional rigid antennas, ongoing research and development promise continuous improvements. As a result, the future of FCTAs looks promising, with the potential to further revolutionize wearable technology and enhance our daily lives.

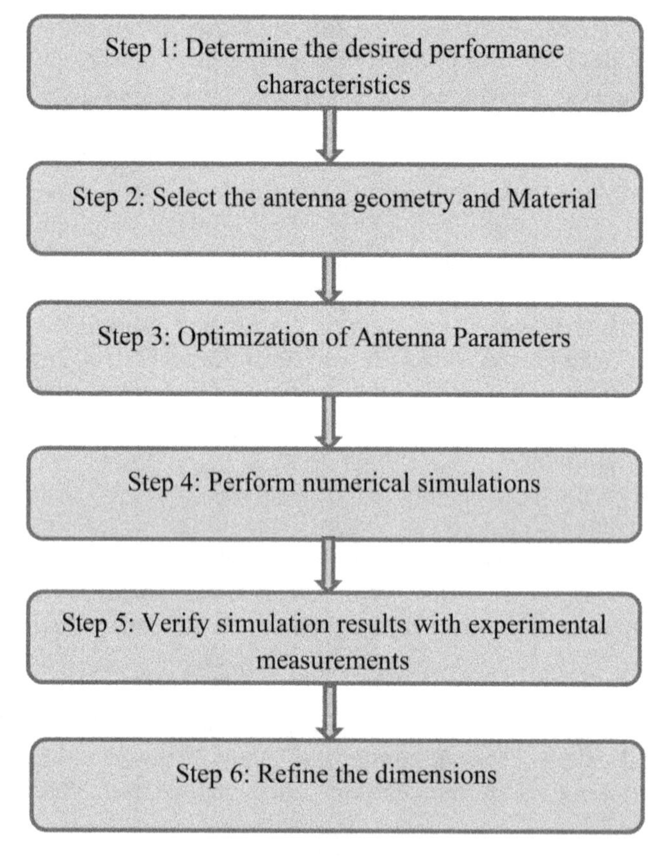

Figure 8.1. Design of FCTAs.

The design of FCTAs is shown in figure 8.1. It involves selecting the appropriate antenna geometry, optimizing its dimensions for the desired resonance frequency and bandwidth, and selecting suitable materials for the antenna and substrate.

8.3.1 Determine the desired performance characteristics

The first step is to determine the desired performance characteristics of the antenna, such as frequency band, gain, radiation pattern, and impedance matching. These characteristics will guide the selection of antenna geometry and material, and ultimately the dimensions of the antenna.

8.3.2 Selection of antenna geometry and materials

Antenna geometry is an important aspect of the design of FCTAs. It is crucial for optimizing performance, ensuring customizability, and achieving durability in antennas. It allows for miniaturization and integration in compact devices while enabling adaptability to various frequency ranges. Cost-effectiveness and improved bandwidth efficiency are achieved through careful choices. Moreover, this process fosters innovation and interdisciplinary collaboration, driving advancements in wireless communication technologies. Thoughtful selection aligns with environmental considerations, promoting eco-conscious designs and reducing electronic waste. The choice of antenna geometry depends on the application requirements, such as the operating frequency, bandwidth, radiation pattern, and impedance matching. FCTAs can be designed in various shapes such as:

- Patch antennas [8, 10, 13, 37, 38]
- Dipole antennas [24]
- Loop antennas
- Meander antennas [24]
- Fractal antennas

Each geometry has unique advantages and disadvantages in terms of their performance as conformal antennas. Patch antennas are simple to design and fabricate, and they have a low profile and wide bandwidth. They are well-suited for high data rate communication applications, but they have relatively low radiation efficiency compared to other types of antennas. Loop antennas are compact and have high radiation efficiency, making them ideal for sensing and medical devices where a longer range is required. They are also compatible with a wide range of substrate materials, and they can be made to be very low-profile. However, loop antennas have a limited bandwidth compared to patch antennas. Dipole antennas are simple to design and have a wide bandwidth. They are well-suited for conformal applications where a broad frequency range is needed, such as military and aviation communication. However, dipole antennas can have a larger footprint than patch and loop antennas, making them less suitable for applications with limited space. Meander antennas consist of a series of repeating elements and are popular for their broadband performance and reduced mutual coupling between elements. They are well-suited for array configurations in conformal applications. However, meander antennas can be more difficult to design and optimize than patch or dipole antennas due to their more complex structure. Fractal antennas are complex structures that can be designed to have a compact size, wide bandwidth, and high radiation efficiency. They are well-suited for conformal applications in which a wide frequency range is required, such as in RFID tags and medical devices. However, fractal antennas can be difficult to fabricate, and their performance can be affected by their shape and size.

8.3.2.1 Conductive materials

Conductive materials are used to create the radiating elements, feed lines, and ground plane of the flexible antenna. These materials provide electrical conductivity and allow for the transmission and reception of electromagnetic waves. Classification of conductive materials is shown in figure 8.2.

Highly conductive materials play a pivotal role in achieving superior gain, efficiency, and bandwidth in various applications. For instance, silver, with its

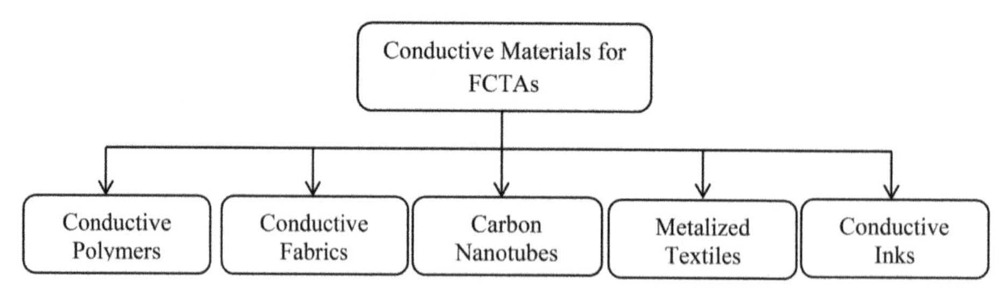

Figure 8.2. Classification of conductive materials.

remarkable conductivity of 6.173 × 107 (S m^{-1}), stands out as an excellent example of such a material [35]. Common conductive materials for flexible antennas include:

- Copper (in the form of thin foil or conductive ink)
- Conductive polymers (polyaniline, polypyrrole) [39]
- Carbon-based materials (carbon nanotubes, graphene, conductive ink) [40]
- Silver (in the form of thin foil or conductive ink) [12, 33, 41]

Conductive polymers, exemplified by polypyrrole [39], exhibits remarkable electrical conductivity achieved through doping or electrochemical synthesis methods. It can be coated onto textiles to create conductive surfaces. Conductive fabrics such as copper-coated nylon, silver-plated polyester, and stainless steel fibers have been made conductive by coating them with a conductive material.

These can be used as the antenna substrate or as a conductor in the antenna design. Metalized textiles, where a thin layer of metal is deposited onto the textile by vacuum deposition. Examples of metals that can be used include copper, silver, and aluminum. Metalized textiles present an excellent option for achieving superior antenna performance while seamlessly integrating into textile structures. Additionally, the utilization of carbon nanotubes, cylindrical carbon molecules, enables the creation of conductive coatings or fibers. By depositing carbon nanotubes onto textiles, the conductivity of surfaces can be significantly enhanced [40]. Studies have reported that carbon nanotubes can provide good antenna performance, especially at higher frequencies. Conductive inks contain conductive particles, typically silver or copper. It can be printed onto textiles to create conductive traces or antennas. These inks are relatively easy to use, and their conductivity is stable over time. However, their conductivity can be lower than other materials, and their performance can be affected by environmental conditions.

8.3.2.2 Substrate and dielectric materials

The choice of material for the antenna and substrate is critical for achieving the desired performance of FCTAs. The dielectric constant of the material affects the impedance matching and radiation efficiency of the antenna. It is important to choose a material with a dielectric constant that matches the desired operating frequency of the antenna. The dielectric constant and dielectric loss of various dielectric substrates is listed in table 8.1. The material should be flexible and conformal, allowing it to conform to the shape of the

Table 8.1. Properties of various dielectric substrate.

Dielectric material	Dielectric constant	Dielectric loss
LCP [42]	2.9	0.0025
PET [43]	3	0.008
PI [54]	2.91	0.005
PDMS [43]	2.65	0.02

underlying surface without affecting the antenna's performance. The chosen material must possess exceptional durability, capable of withstanding the anticipated usage and a range of environmental conditions, including moisture, heat, and mechanical stress. Similarly, a flexible substrate must exhibit remarkable deformability and mechanical robustness, while maintaining superior levels of bending repeatability. The chosen material should exhibit excellent electrical conductivity or be compatible with conductive coatings or metallization techniques. Moreover, it should be cost-effective, ensuring that it does not substantially impact the overall antenna cost. When designing flexible antennas, careful consideration must also be given to the selection of the substrate itself.

Common substrate materials for flexible antennas include:

- Liquid crystal polymer (LCP) [42]
- Polyethylene terephthalate (PET) [43]
- Polydimethylsiloxane (PDMS) [43–47]
- P-Clay [48] [49]
- Tetra ethoxy orthosilicate [44]
- Polyimide (PI) [54]
- Polyethylene naphthalate
- PI/fluoropolymer composite

Jeans [39], felt fabric [50] and PET [37], among various other substrates, are favored options due to their low dielectric constant properties.

PDMS polymer has emerged as a promising substrate option due to its remarkable flexibility and conformality, as indicated by its low Young's modulus of less than 3 MPa [43]. Due to their cost-effectiveness and ease of manufacturing, paper substrates have become increasingly favored for flexible antennas. Ullah *et al* [14] have showcased an innovative design of a flexible antenna for 2.4 GHz ISM bands utilizing a paper substrate tailored for intrabody telemedicine systems. Meanwhile, LCP, recognized for its resemblance to flexible printed circuitry, has emerged as an appealing choice for high-frequency flexible antennas. This is primarily attributed to its low dielectric loss of 0.008 and dielectric constant of 3, minimal moisture absorption, chemical resistance, and impressive ability to withstand temperatures up to 300 °C [42] (table 8.1).

While a low dielectric constant minimizes impedance mismatch, maximizes radiation efficiency, and enhances the bandwidth by providing electrical insulation, impedance matching, and radiation pattern control. They are typically placed between conductive layers or as a coating on the antenna structure. Common dielectric materials for flexible antennas include:

- Air (as a low permittivity medium)
- PI films (as dielectric layers)
- Polytetrafluoroethylene films
- Ceramic-filled polymer composites

It's worth noting that the selection of materials depends on various factors such as frequency range, flexibility requirements, manufacturing process, cost, and performance specifications of the flexible antenna. Different combinations and variations of these materials can be used to achieve the desired electrical, mechanical, and performance characteristics.

The mechanical properties of textiles have some important effect in the performance of FCTAs as listed in table 8.2. Flexibility and conformability allow the antenna to bend and adapt to the shape of the wearer's body or other irregular surfaces, and a low dielectric constant minimizes impedance mismatch, maximizes radiation efficiency, and enhances the bandwidth. Moreover, a substrate with low permittivity increases the resonant frequency of the antenna, and a higher substrate thickness decreases the resonant frequency, broadens the bandwidth, and increases the antenna's weight. However, high moisture content reduces flexibility and may cause corrosion, and high temperatures can impact the mechanical stability of the textile material and antenna structure while increasing the effective electrical length of the antenna.

Table 8.2. Impact of mechanical properties of textile materials.

Mechanical property of textile material	Impact in antenna performance
Flexibility and conformability	Highly flexible and conformable allows the antenna to bend and adapt to the shape of the wearer's body or other irregular surfaces
Dielectric constant	Low dielectric constant minimizes impedance mismatch, maximizes radiation efficiency, and enhances the bandwidth
Permittivity of substrate	Low permittivity increases the resonant frequency of antenna
Thickness of substrate	High substrate thickness decreases the resonant frequency of antenna, broadens bandwidth, and increases weight of the antenna
Moisture content	High moisture content reduces flexibility and causes corrosion
Temperature	High temperature affects mechanical stability of the textile material and the antenna structure and increases effective electrical length of the antenna

8.3.3 Optimization of dimensions

Once the antenna geometry is selected, its dimensions must be optimized for the desired resonance frequency and bandwidth. The resonant frequency of the antenna depends on its physical dimensions, such as its length, width, and height. The bandwidth of the antenna narrows when increasing its quality factor (Q factor). Therefore, the dimensions of the antenna must be optimized to achieve a balance between the resonant frequency and bandwidth.

For the rectangular patch, the width (W) and length (L) of the patch are calculated using equations (8.1)–(8.3) [2].

$$W = \frac{1}{2f_r\sqrt{\mu_0\varepsilon_0}}\sqrt{\frac{2}{1+\varepsilon_r}}, \tag{8.1}$$

$$L = \frac{1}{2f_r\sqrt{\varepsilon_{\text{eff}}}\sqrt{\mu_0\varepsilon_0}} - 0.824\frac{\left(\frac{W}{h}+0.264\right)\left(\varepsilon_{\text{eff}}+0.3\right)}{\left(\frac{W}{h}+0.8\right)\left(\varepsilon_{\text{eff}}-0.258\right)}h, \tag{8.2}$$

$$\varepsilon_{\text{eff}} = \frac{\varepsilon_r+1}{2} + \frac{\varepsilon_r-1}{2}\left[1+12\frac{h}{W}\right]^{-\frac{1}{2}}, \tag{8.3}$$

where f_r is resonance frequency, ε_r is relative permittivity, ε_{eff} is effective permittivity, ε_0 and μ_0 are free space permittivity and permeability, and h is the thickness of the substrate.

8.3.4 Perform numerical simulations

Once the antenna geometry and material are selected, numerical simulations using electromagnetic simulation software such as CST Microwave Studio, HFSS, or FEKO can be performed to model the antenna's behavior and analyze its performance under different conditions. The key parameters and metrics used to evaluate the performance of FCTAs are radiation efficiency, bandwidth, gain, resonant frequency, polarization, impedance matching, directivity, radar cross section, mechanical flexibility and durability, and environment and human interaction.

Vary the dimensions of the antenna and analyze the following performance:

(i) Gain: the measure of how much power the antenna can radiate in a particular direction compared to an ideal isotropic radiator.

(ii) Directivity: the measure of the concentration of radiated power in a particular direction compared to an isotropic radiator.

(iii) Radiation pattern: the graphical representation of how the antenna radiates or receives electromagnetic waves in different directions.

(iv) Return loss: the amount of power reflected back to the source due to impedance mismatch between the antenna and the transmission line.

(v) Bandwidth: the range of frequencies within which the conformal antenna can operate with acceptable performance, usually measured as the frequency range in which the return loss is below a specified threshold.

(vi) Polarization: the orientation of the electric field vector of the radiated wave, which can be linear, circular, or elliptical.

(vii) Conformal shape: the 3D shape of the antenna, conforming to the surface or structure on which it is mounted, enabling integration into curved or irregular surfaces.

(viii) Flexibility and bending performance: the ability of the antenna to withstand bending or flexing without significant degradation in performance.

(ix) Size and weight: the physical dimensions and mass of the conformal antenna, which are important factors for certain applications for which space and weight limitations exist.

(x) Environment and human interaction: the antenna's performance should be evaluated under various environmental conditions, such as moisture, temperature, and electromagnetic interference. The impact of human body interaction should also be considered, especially for wearable applications.

8.3.5 Verify simulation results with experimental measurements

After the simulation results are obtained, the antenna can be fabricated and tested to verify the simulation results. The measurements can be performed using a vector network analyzer or other measurement equipment to analyze the antenna's performance under various conditions. By measuring the antenna's return loss, radiation pattern, and other performance characteristics, designers can determine the optimal dimensions that will result in the desired performance.

8.3.6 Refine the dimensions

Based on the simulation and measurement results, the dimensions of the antenna can be refined to achieve the desired performance characteristics. This may involve iterating through steps 3 and 4 several times until the optimal dimensions are achieved.

8.4 Fabrication of FCTAs

The fabrication of FCTAs involves several steps, including textile preparation, conductive element deposition, and antenna assembly. Antenna assembly involves the integration of the conductive textile material with the antenna feedline and other components such as the ground plane. The assembly process must ensure that the antenna maintains its shape and conformability while providing reliable electrical connections between the various components.

Some of the key techniques used in this process include:

(i) Inkjet printing: This method involves depositing conductive ink directly onto textile substrates in a controlled pattern to create antenna elements. It offers precision and customization.

(ii) Screen printing: a traditional technique that involves pushing conductive ink through a mesh screen onto the textile substrate. It is suitable for larger patterns and thicker conductive layers.

(iii) Conductive thread embroidery: Conductive threads are stitched directly onto the textile substrate to form the antenna pattern. This technique is flexible and can be used for complex designs.
(iv) Conductive yarn weaving/knitting: Conductive yarns are incorporated into the textile structure during the weaving or knitting process. This approach allows antennas to be seamlessly integrated into the fabric.
(v) Conductive fabric lamination: Conductive materials, such as copper or silver films, are bonded onto the textile surface using adhesives or heat-activated lamination techniques.
(vi) Laser ablation: This technique involves using a laser to selectively remove materials from a conductive film or coating, creating the desired antenna pattern on the textile.
(vii) Photolithography: Similar to the semiconductor manufacturing processes, photolithography uses light to transfer a pattern from a photomask onto a photosensitive conductive material. The exposed areas are then etched away, leaving the antenna pattern.
(viii) 3D printing: Additive manufacturing techniques, like 3D printing, can be used to create custom antenna structures directly on the textile substrate. Conductive materials are extruded layer by layer to form the antenna shape.
(ix) Spray coating: Conductive materials in the form of aerosol or liquid sprays are applied directly to the textile to create the antenna pattern. This method is suitable for quick prototyping.
(x) Transfer printing: In transfer printing, the conductive pattern is first fabricated on a separate substrate and then transferred onto the textile using heat or pressure.

These fabrication techniques allow textile antennas to be lightweight, flexible, and conformal to various shapes and surfaces. The choice of method depends on factors such as antenna design complexity, the type of textile substrate, production scale, and the required performance characteristics of the antenna. Among them, conductive element deposition was done widely by using techniques such as embroidery, screen printing, and inkjet printing as shown in figure 8.3.

8.4.1 Embroidery

Patch antennas can be printed directly onto a flexible substrate, such as a polymer film or a textile material. It can also be integrated into a textile structure, such as a garment or a wearable device. The conductive patch can be skillfully sewn or intricately embroidered onto the textile material, allowing the textile to be effortlessly shaped or stretched to achieve the desired form. Embroidered flexible materials necessitate the perfect amalgamation of high conductivity and mechanical strength [30], ensuring that the attire remains both comfortable and visually appealing. Notably, conductive textiles can be categorized into four distinct groups: hybrid metal–textile fabrics, conductive yarn-based fabrics, embroidered conductive

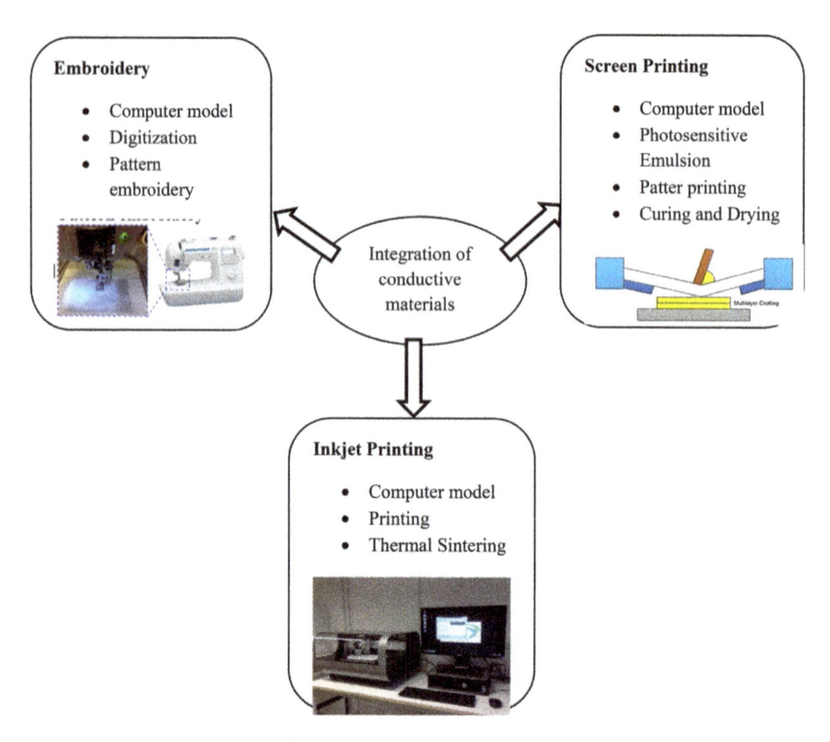

Figure 8.3. Various types of conductive element deposition.

fabrics, and fabrics printed with conductive ink. A range of conductive textiles, including notable examples such as commercial conductive textiles [38] and conductive carbon fiber–braided fabrics [51], showcase remarkable mechanical flexibility and conformality. However, they do face challenges with high conductor loss at radio frequencies (RF). Conversely, metal-coated polymer fibers (e-fibers) [31] exhibit an optimal combination of RF performance, mechanical strength, and load-bearing capability. To integrate the aforementioned embroidered textiles, a low-loss and highly flexible polymer substrate was utilized. In our study, we selected PDMS as the substrate due to its exceptional mechanical flexibility, stretchability, inherent chemical stability, and water resistance [32].

8.4.2 Screen printing

Screen printing has emerged as a highly efficient, rapid, cost-effective, and practical method for fabricating flexible electronics. During this process, the ink is ejected into both the exposed areas of the screen and onto the substrate, resulting in the desired pattern formation. Numerous successful prototypes of RFIDs and transparent flexible antennas have been accomplished using screen printing techniques [52].

This woven screen-based technique allows for variations in thickness and thread densities. In a previous publication, a dual-polarization 2.45 GHz antenna and rectenna were demonstrated on polycotton fabric using screen printing, showcasing the potential for RF power transfer and harvesting [53]. The rectenna underwent rigorous testing and was compared with a similar FR4 rectenna, yielding a performance that was one-third of the standard FR4 rectenna. Screen printing proves to be a cost-effective alternative when compared to other flexible antenna fabrication technologies. However, there are certain limitations inherent to this technique. These include the restricted control over thickness, the number of passes, and the resolution of printed patterns. Ensuring layer consistency poses a challenge due to variations in ink viscosity and changes in substrate surface energy caused by artifacts from thermal curing of solvent-based inks [52].

8.4.3 Inkjet printing

In recent years, the utilization of silver and gold nanoparticles–based conductive inks for inkjet printing of antennas and RF circuits has gained immense popularity. This approach involves the precise deposition of ink droplets, as small as a few picoliters, by inkjet material printers. This remarkable capability allows for the creation of patterns with exceptionally high resolution, making inkjet printing an increasingly preferred method in the field. Inkjet printing technology has emerged as a compelling alternative to traditional fabrication methods like etching and milling. As an additive process, it allows for the direct transfer of the design onto the substrate without the need for masks, resulting in minimal material wastage [3]. This fabrication technique has gained significant popularity, particularly for polymeric substrates such as PI, PET, and paper, owing to its ability to rapidly and accurately prototype designs [54]. The quality of the printed output is primarily determined by two factors: resolution, measured in dots per inch, and the minimum feature size that can be accurately fabricated, typically measured in micrometers.

A remarkable high-gain 4×4 microstrip patch array antenna on a PET substrate and inkjet printing of silver nanoparticle ink was accomplished through the Epson Stylus C88 series printer [41]. It is important to note, however, that inkjet-printed conductive textiles may exhibit relatively lower mechanical strength. The printing resolution of inkjet-printed flexible antennas is contingent upon the surface roughness of the substrate. For smooth substrates like PI, PET, polyethylene naphthalate, LCP, photo paper, etc, an excellent pattern resolution is achievable. For wearable, flexible substrates like e-textiles having the weaving of warp and weft yarns typically have an uneven surface.

8.5 Applications and challenges of FCTAs

FCTAs offer several advantages, such as conforming to irregular shapes, being lightweight, and providing ease of integration into wearable devices or smart textiles. However, they also come with certain challenges and limitations that need to be addressed for optimal performance.

8.5.1 Challenges

Flexibility can lead to deformation or wrinkling, affecting the antenna's shape and performance. Integration into textiles can alter impedance, radiation efficiency, and resonant frequency. Textile materials may have lower conductivity, limiting the

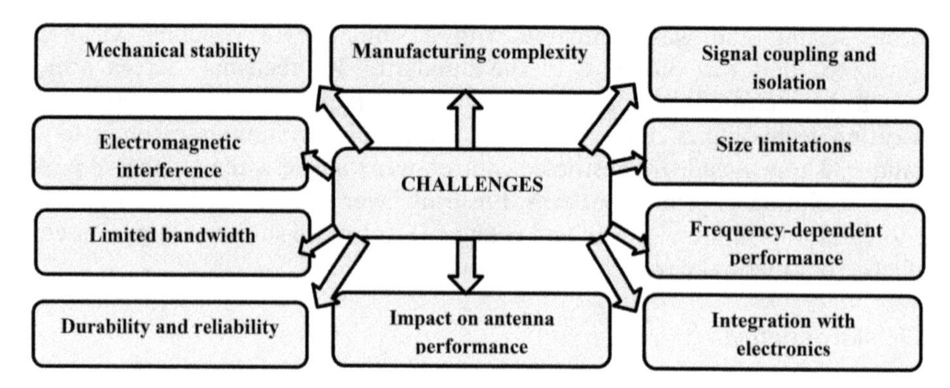

Figure 8.4. Challenges in FCTAs.

antenna's frequency range. Exposure to environmental factors may affect long-term performance. Specialized techniques are required for integration into textiles. Proximity to the body or other objects can cause interference. Integration of multiple textile antennas may lead to coupling issues. Compact form factor can limit radiation efficiency, especially for low frequencies. Dielectric properties may vary at different frequencies. The major challenges are shown in figure 8.4.

8.5.2 Applications

Flexible textile antennas have diverse applications in various industries as shown in figure 8.5. They excel in wearable technology, enabling seamless integration into clothing for communication in smartwatches, fitness trackers, and health monitoring devices. In the IoT, they connect everyday objects, enhancing smart homes and cities

Their use extends to communication systems like military, satellite, and wireless communication due to their conformal nature. Medical applications include patient monitoring and remote health tracking. In transportation, they provide communication in vehicles. Sports, defense, fashion, entertainment, gaming, and industrial sectors also benefit from the versatility of textile antennas. Their lightweight and adaptable design offers solutions for numerous connectivity needs.

8.6 Design of textile antenna

8.6.1 Design process

A novel approach in antenna design involves utilizing jeans material with a permittivity of 1.67 and a thickness of $h = 1.6$ mm for a conventional patch antenna with an inset feed. The proposed antenna, depicted in figure 8.6, has dimensions of $35.6 \times 42.6 \times 1.6$ mm and exhibits a resonant frequency of 3.5 GHz. To excite the antenna, the inset feeding technique is utilized in the design, incorporating four strategically placed slots of varying dimensions within the patch. These design parameters of the proposed antenna can be found in table 8.3.

When slots are introduced into the conventional patch, the permittivity and permeability are altered, leading to changes in the antenna's resistance, capacitance, and inductance depending on the dimensions of the slots. The return loss of the

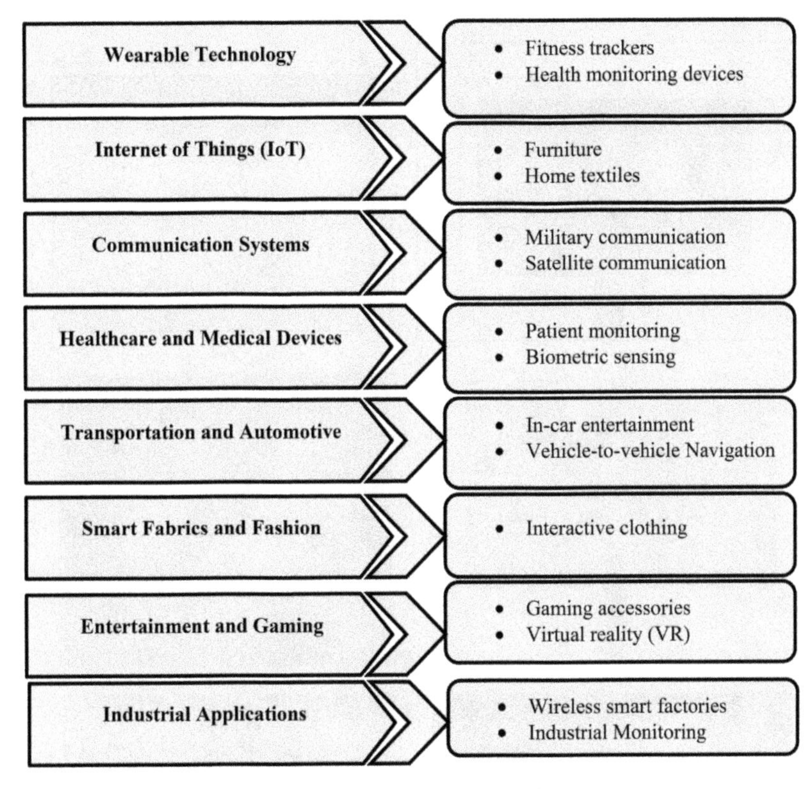

Figure 8.5. Applications of flexible textile antenna.

conventional antenna and proposed antenna with slots is recorded as −19.03 and −22.49 dB, respectively. The antenna's performance is thoroughly assessed across a wide frequency range spanning from 1 GHz to 5 GHz after integrating DGS. Figure 8.7 visually presents the results: slot antennas showcase remarkable return losses of −24.84 and −35.3 dB, respectively. The return loss serves as a crucial metric for gauging power reception efficiency. In comparison to slot antennas lacking DGS, the return loss experiences a notable 10% improvement. The performance parameters of conventional, slot antennas with DGS and without DGS are listed in table 8.4.

Table 8.5 presents a comprehensive comparison of the proposed antenna's parameters with previously documented works. Notably, the coplanar wearable antenna described in [58] achieves an impressive 95% reduction in specific absorption rate and operates across three bands. The E-shaped antenna utilizing denim material, as reported in [31], shares the same bandwidth of 250 MHz as the proposed antenna. However, it exhibits a lower gain of 1.83 dB.

The proposed antenna's performance on various wearable materials is analyzed, and the return loss comparison is illustrated in figure 8.6. It appears that as the relative permittivity of the substrate increases, the resonant frequency decreases, and the bandwidth expands due to the reduced space charge polarization effect.

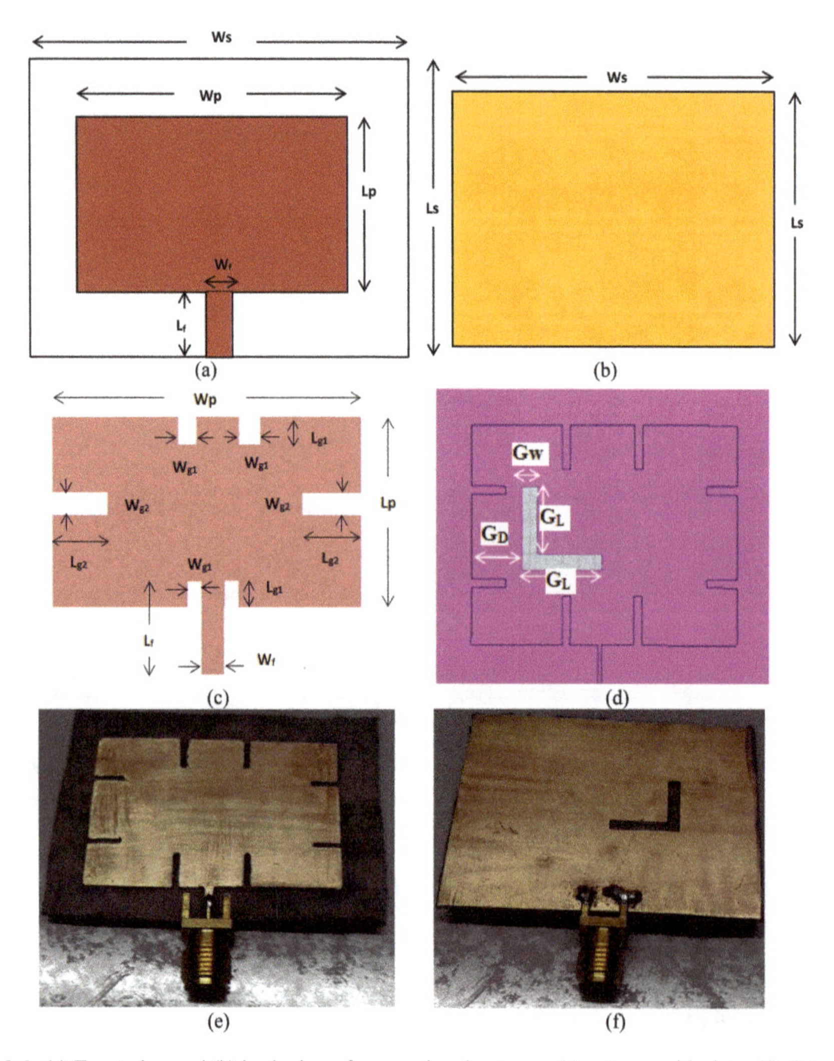

Figure 8.6. (a) Front view and (b) back view of conventional antenna, (c) antenna with slots, (d) slot antenna with DGS, and (e) front view and (f) back view of fabricated antenna.

The relative permittivities of leather, cotton, and silk are 1.8, 2.1, and 4.37, respectively. By selecting wearable jeans material, the proposed antenna offers a lightweight and flexible solution, ideal for IoT applications. Notably, compared to [2, 22, 27], the proposed work exhibits improved gain. Additionally, while similar designs featuring slots and DGS have been documented in [1, 2, 29, 34], the proposed design stands out for its compact size. The proposed antenna offers notable advantages, including a peak gain of 11.8 dB and a peak directivity of 11.6 dB. Furthermore, the DGS-loaded antenna demonstrates a 17% improvement in bandwidth compared to conventional patch antennas.

Figures 8.8(a) and (b) display the radiation patterns generated after incorporating the DGS into the ground plane. Both antennas produce omnidirectional radiation.

Table 8.3. Dimension of proposed antenna (all dimensions in millimeters).

Parameter	Conventional patch	Antenna with slots
Size of substrate (Ls × Ws × Hs)	36.6 × 42.6 × 1.6	
Size of patch (Lp × Wp)	26 × 33	
Length of the feed (Lf)	6.1	
Width of the feed (Wf)	1	0.4
Length of the slot1 (Lg1)	–	6
Width of the slot 1 (Wg1)	–	1
Length of the slot2 (Lg2)	–	5
Width of the slot 2 (Wg2)	–	1

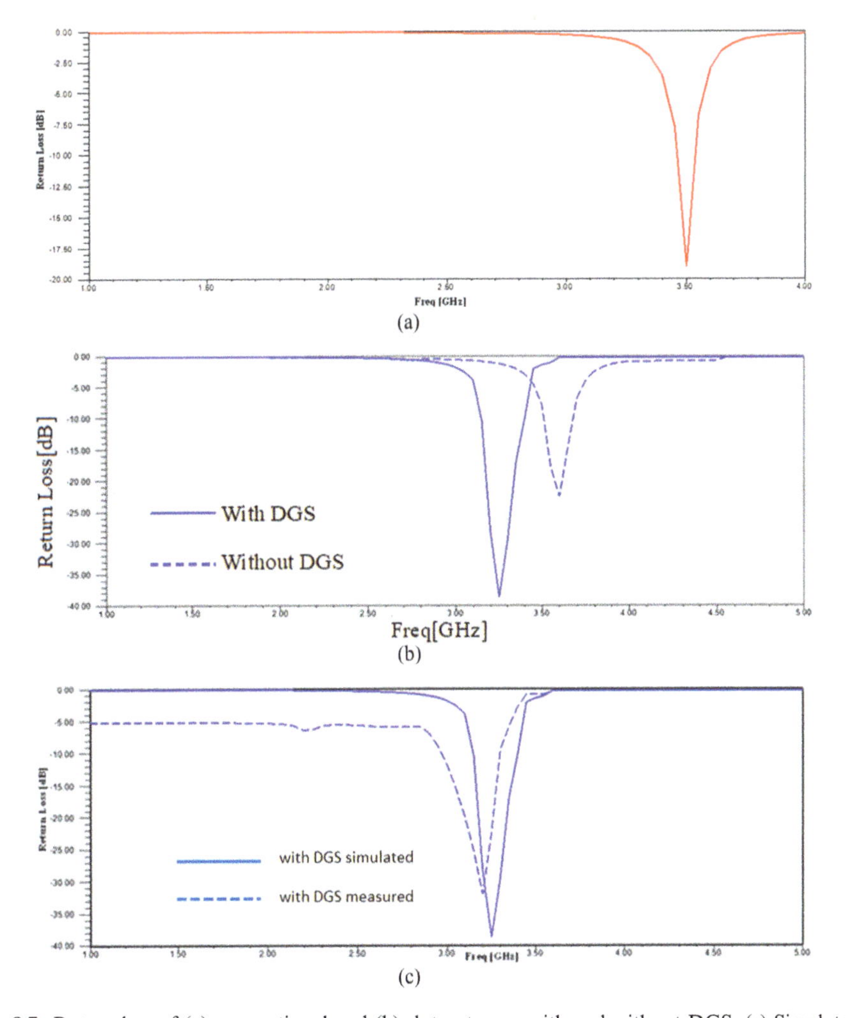

Figure 8.7. Return loss of (a) conventional and (b) slot antennas with and without DGS. (c) Simulated and measured return loss of slot antenna with DGS.

Table 8.4. Performance parameters of conventional and slot antennas with DGS and without DGS.

Parameters	Conventional patch	Simulated results of slot antenna		Measured results
		Without DGS	With DGS	Slot antenna with DGS
Resonance frequency (GHz)	3.5	3.6	3.7	3.65
Return loss (dB)	−19.03	−22.49	−35.30	−34.87
VSWR	1.25	1.18	1.07	1.09
Directivity (dB)	1.60	10.03	12.68	–
Gain (dB)	2.75	7.45	11.8	–
Bandwidth (MHz)	0.08	0.20	250	250

This paper introduces an innovative approach to designing a compact slot antenna for IoT applications, utilizing wearable jeans material. By integrating slots and DGS, the antenna's performance is significantly enhanced. With a sleek and low-profile design measuring $35.6 \times 42.6 \times 1.6\ mm^3$, the proposed antenna operates at a frequency of 3.5 GHz. It sets itself apart with an impressive gain of 11.8 dB and a wide bandwidth of 250 MHz.

8.6.2 Parametric analysis

The proposed antenna's performance on various wearable materials was analyzed, and the effects of different substrate materials with varying relative permittivities on the resonant frequency and bandwidth were investigated. The results of this analysis are presented in table 8.6, which illustrates the comparison of return loss for different wearable materials. A notable observation from table 8.7 is that an increase in the relative permittivity of the substrate leads to a decrease in the resonant frequency of the antenna. This phenomenon can be attributed to the interaction between the electromagnetic waves and the substrate material. With higher relative permittivity, the electric field interactions become stronger, resulting in a lower resonant frequency for the antenna. When the relative permittivity increases, the bandwidth of the antenna expands. This expansion is primarily due to the reduced space charge polarization effect in substrates with higher relative permittivities. Space charge polarization occurs when charge carriers within the material respond to the electric field, causing a delay in their response and affecting the resonant behavior of the antenna. As the relative permittivity increases, the space charge polarization effect diminishes, leading to a broader bandwidth for the antenna. The specific values for the resonant frequencies and bandwidths for each wearable material are available in the study's detailed results and analysis section.

L-shaped defects, with $G_L = 10$ mm, $G_W = 2$ mm, and $G_D = 6$ mm from the patch, are created on the ground of proposed antenna, as depicted in figure 8.6(d).

Table 8.5. Comparison of proposed antenna with other reported work.

References	Size (mm^3)	Material	Resonance frequency (GHz)	Return loss (dB)	Gain (dB)	Band-width (MHz)
[1]	80 × 80 × 2	Wash cotton	2.4	17.29	5.14	84.4
[2]	42 × 40 × 2	Jeans	3.9/10	30/35	–	10 000
[27]	35 × 35 × 0.508	Rogers 4003C	2.45	25	7.3	360
[28]	69 × 69 × 5	Wearable	2.4/3.5/5.8	–	5.11/6.43/ 7.41	10/16/7
[29]	32 × 30 × 2	Denim	2.45	35	1.83	250
[33]	36.4 × 36.4 × 1.6	FR – 4	2.4	30	2.45	220
[34]	85 × 35 × 6	Conductive textile	2.5	1/21, 2/20.5, 3/22, 5/30	–	128%
[55]	54 × 38 × 1	FR – 4	3.3	−22.5	6	200
[56]	215.5 × 115 × 1.15	Leather	0.63/1.47/2.41/ 3.35	12.87/33.07/ 29.42/ 20.51	1.36/1.87/ 3/3.38	70.9/139/ 122/ 125.4
		Silk	0.62/1.49/2.4/ 3.35	12.86/38.95/ 41.76/ 17.96	1.77/1.82/ 2.64/ 2.72	75.4/137.1/ 130.2/ 130.3
	215.5 × 115 × 0.8	Nylon	0.402/3.74	24.02/30.14	0.43/4.96	67.5/368.5
			5.8	15.40	7.14	137.5
[57]	100 × 113 × 2	Rubber	1	18.35	2.52	1690
[58]	40.04 × 29.12 × 0.18	Rubber	0.868/2.44/5.80	25/49/20	−4.5	5500
Proposed work	35.6 × 42.6 × 1.6	Jeans	3.5	34.87	11.8	250

8.7 Advancements in the field of FCTAs

Over the years, significant progress has been made in the field of FCTAs, driven by various advancements in materials, fabrication techniques, and design principles. These developments have led to improved performance, durability, and versatility of textile antennas, making them increasingly relevant in a wide range of applications. In this context, several key trends and innovations stand out, showcasing the potential of this technology for the future. Some of these notable advancements in flexible textile antennas include:

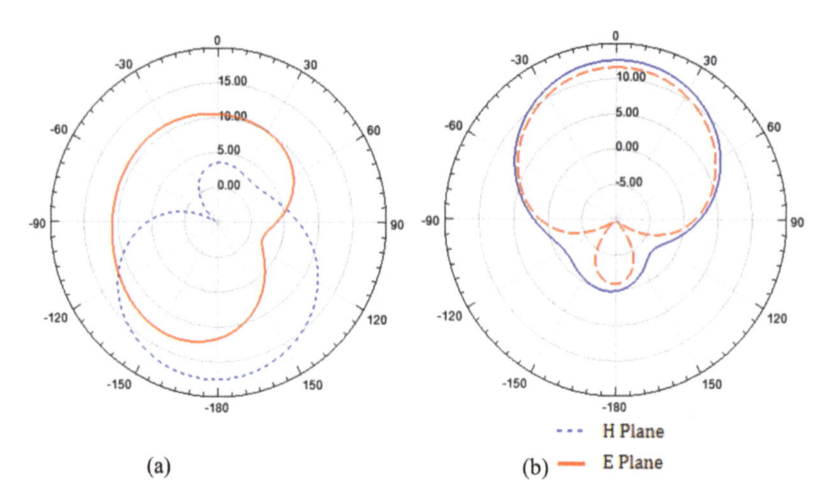

Figure 8.8. Radiation patterns of (a) slot antenna without DGS and (b) with DGS.

Table 8.6. Comparison of proposed conventional antenna parameters for various wearable substrates.

Wearable material	Relative permittivity	Resonance frequency (GHz)	Return loss (dB)	Bandwidth (MHz)
Jeans	1.67	3.5	−19.03	0.0851
Leather	1.8	3.4	−15.64	110
Cotton	2.1	3.25	−23.62	110
Silk	4.37	3	−18.49	70

Table 8.7. Variation in return loss due to the effect of various DGS parameters.

Parameter-G_L (mm)	Return loss (dB)	Parameter-G_W (mm)	Return loss (dB)	Parameter-G_D (mm)	Return loss (dB)
8	−26.33	1	−18.25	2	−13.35
9	−20.61	2	−20.63	4	−15.44
10	−28.70	3	−19.42	6	−18.85
11	−28.04	4	−18.47	8	−18.34
12	−27.45	5	−20.32	10	−17.69

(i) Material advancements: Researchers have been exploring new materials and composites with improved electrical conductivity and mechanical properties. These materials enable higher performance and durability of textile antennas.

(ii) Printed fabrication techniques: Additive manufacturing techniques, such as inkjet printing, screen printing, and 3D printing, have been increasingly used to fabricate flexible textile antennas. These methods allow for rapid and cost-effective production of antennas on various fabrics.

(iii) Miniaturization: Advancements in design and fabrication techniques have led to miniaturized textile antennas, making them suitable for a wider range of wearable applications, including smartwatches, fitness bands, and medical devices.

(iv) Multiband and wideband antennas: Innovations in antenna design have resulted in flexible textile antennas that can operate over multiple frequency bands or exhibit wideband characteristics. This versatility enables compatibility with various wireless communication standards.

(v) Frequency reconfigurability: Researchers have developed flexible textile antennas with frequency reconfigurability, allowing them to adapt to different frequency bands or communication protocols dynamically.

(vi) Integration of electronics: Antennas are increasingly being integrated with electronic components, such as amplifiers, filters, and impedance tuners, to enhance overall system performance and overcome challenges like impedance matching.

(vii) Biocompatible antennas: Specialized textile antennas have been developed for biomedical applications, ensuring biocompatibility and compatibility with the human body for health monitoring and medical devices.

(viii) Stealth technology: Textile antennas with radar-absorbing and stealth properties have been explored for military and defense applications, reducing radar cross-section and enhancing stealth capabilities.

(ix) Energy harvesting: Researchers have integrated energy-harvesting technologies into textile antennas, allowing them to scavenge and convert ambient electromagnetic energy into electrical power for self-sustainability.

(x) Flexible wearable platforms: Advancements in wearable platforms and e-textile manufacturing have facilitated seamless integration of FCTAs into clothing and wearable devices without compromising comfort or aesthetics.

(xi) Machine learning and optimization techniques: Advanced machine learning algorithms and optimization techniques are being applied to design and tune flexible textile antennas for optimal performance and efficiency.

8.8 Conclusion

FCTAs offer unique properties like flexibility, comfort, and conformability, making them ideal for integration into wearable devices. Ongoing research focuses on improving their performance and reliability through new materials and fabrication techniques. The proposed antenna features a low profile, measuring $35.6 \times 42.6 \times 1.6\ mm^3$ for the frequency of 3.2 GHz. It boasts the highest gain of 11.8 dB and a bandwidth of 250 MHz. These antennas have the potential to revolutionize wearables, enabling smart clothing with communication capabilities and unobtrusive health monitoring devices. Biomedical applications benefit from specialized biocompatible textile antennas for continuous health tracking and remote patient monitoring. However, the field is still evolving, with potential for further advancements in design and manufacturing. As interdisciplinary collaboration continues, FCTAs will enhance wearable technology, communication systems, and healthcare, transforming how we interact with technology and improving convenience and connectivity in our daily lives.

References

[1] Sreemathy R, Hake S, Gaikwad S, Saw S K and Behera S 2022 Design, analysis and fabrication of dual frequency distinct bandwidth slot loaded wash cotton flexible textile antenna for ISM band applications *Prog. Electromagn. Res.* M **109** 191–203

[2] Parameswari S and Chitra C 2021 Compact textile UWB antenna with hexagonal for biomedical communication *J Ambient Intell. Human. Comput.* **28** 1–8

[3] Hu J 2010 Overview of flexible electronics from ITRI's viewpoint *28th VLSI Test Symposium (VTS)* pp. 84

[4] Salonen P, Sydanheimo L and Keskilammi M 1999 A small planar inverted-F antenna for wearable applications *Digest of Papers The 3rd Int. Symp. on Wearable Computers*

[5] Wang Z, Lee L Z and Volakis J L 2013 A 10:1 bandwidth textile-based conformal spiral antenna with integrated planar balun *IEEE Antennas and Propagation Society Int. Symp. (APSURSI)* pp. 220–221

[6] Zhang S, Whittow W, Seager R, Chauraya A and Vardaxoglou J(Y) C 2017 Guest editorial: microwave components and antennas based on advanced manufacturing techniques *IET Microw. Antennas Propag.* **11** 1919–2108

[7] Dierck A, Rogier H and Declercq F 2013 A wearable active antenna for global positioning system and satellite phone *IEEE Trans. Antennas Propag.* **61** 532–8

[8] Vallozzi L, Rogier H, Hertleer C and Langenhove L V 2008 Dual Polarized Textile Patch Antenna for Integration into Protective Garments *IEEE Antennas Wirel. Propag. Lett.* **7** 440–3

[9] Lemey S and Rogier H 2015 Substrate integrated waveguide textile antennas as energy harvesting platforms *Int. Workshop on Antenna Technology (iWAT)* pp 23–6

[10] Declercq F, Georgiadis A and Rogier H 2011 Wearable aperture-coupled shorted solar patch antenna for remote tracking and monitoring applications *5th European Conf. on Antennas and Propagation (EUCAP)*

[11] Lemey S, Declercq F and Rogier H 2014 Textile antennas as hybrid energy-harvesting platforms *Proc. IEEE* **102** 1833–57

[12] Tiercelin N, Coquet P, Sauleau R and Senez V 2006 Polydimethylsiloxane membranes for millimeter-wave planar ultra flexible antennas *J Micromechan. Microeng.* **16** 2389

[13] Koski K, Lohan E, Sydanheimo L, Ukkonen L and Rahmat-Samii Y 2014 Electro-textile UHF RFID patch antennas for positioning and localization applications *IEEE RFID Technology and Applications Conf. (RFID-TA)*

[14] Ullah M, Islam M, Alam T and Ashraf F 2018 Paper-based flexible antenna for wearable telemedicine applications at 2.4 GHz ISM band *Sensors* **18** 4214

[15] Zhu S and Langley R 2009 Dual-band wearable textile antenna on an EBG substrate *IEEE Trans. Antennas Propag.* **57** 926–35

[16] Kangeyan R and Karthikeyan M 2023 A novel wideband fractal-shaped MIMO antenna for brain and skin implantable biomedical applications *Int. J. Commun. Syst.* **36** e5509

[17] Kangeyan R and Karthikeyan M 2023 Miniaturized meander-line dual-band implantable antenna for biotelemetry applications *ETRI J.* 1–8

[18] Mishra S, Mishra V and Purohit N 2015 Design of wide band circularly polarized textile antenna for ISM bands at 2.4 and 5.8 GHz *IEEE Int. Conf. on Signal Processing, Informatics, Communication and Energy Systems (SPICES)*

[19] Waterhouse R B, Targonski S D and Kokotoff D M 1998 Design and performance of small printed antennas *IEEE Trans. Antennas Propag.* **46** 1629–33

[20] Iwasaki H A 1996 A circularly polarized small-size microstrip antenna with a cross slot *IEEE Trans. Antennas Propag.* **44** 1399–401

[21] Vallozzi L, Vandendriessche W, Rogier H, Hertleer C and Scarpello M 2009 Design of a protective garment GPS antenna *Microwave Opt. Technol. Lett.* **15** 1504–8

[22] Kaivanto E, Berg M, Salonen E and de Maagt P 2011 Wearable circularly polarized antenna for personal satellite communication and navigation *IEEE Trans. Antennas Propag.* **59** 4490–3

[23] Zaric A, Costa J and Fernandes C 2014 Design and ranging performance of a low-profile UWB antenna for WBAN localization applications *IEEE Trans. Antennas Propag.* **62** 6420–7

[24] 1998 International Commission on Non-Ionizing Radiation Protection Guidelines for limiting exposure to time-varying electric, magnetic and electromagnetic fields (up to 300 GHz) *Health Phys.* **74** 494–522

[25] Tang D, Xu L, Li Y and Ye T T 2018 Highly sensitive wearable sensor based on a flexible multi-layer graphene film antenna *Sci. Bull.* **63** 574–9

[26] Leng T, Huang X, Chang K, Chen J, Abdalla M A and Hu Z 2016 Graphene nanoflakes printed flexible meandered-line dipole antenna on paper substrate for low-cost RFID and sensing applications *IEEE Antennas Wirel. Propag. Lett.* **15** 1565–8

[27] Kiani S, Rezaei P and Fakhr M 2020 A CPW-fed wearable antenna at ISM band for biomedical and WBAN applications *Wirel. Netw.* **27** 735–45

[28] El May W, Sfar I, Ribero J M and Osman L 2021 Design of low-profile and safe low SAR tri-band textile EBG-based antenna for IoT applications *Prog. Electromagn. Res. Lett.* **98** 85–94

[29] Balaji P and Narmadha R 2021 Wearable E-shaped textile antenna for biomedical telemetry *Int. Conf. on Advances in Electrical, Computing, Communication and Sustainable Technologies (ICAECT)* pp 1–5

[30] Wang Z, Zhang L, Bayram Y and Volakis J L 2012 Embroidered conductive fibers on polymer composite for conformal antennas *IEEE Trans. Antennas Prop.* **60** 4141–7

[31] Toyobo Co, Ltd 2005 *PBO Fibers Zylon Technical Information* www.toyobo-global.com/seihin/kc/pbo

[32] Koulouridis S, Kizitas G, Zhou Y, Hansford D J and Volakis J L 2006 Polymer–ceramic composites for microwave applications: fabrication and performance assessment *IEEE Trans. Microwave Theory Techn.* **54** 4202–8

[33] Abdulkawi W M, Sheta A F A, Elshafiey I and Alkanhal M A 2021 Design of low-profile single- and dual-band antennas for IoT applications *J. Electronics* **10** 2766

[34] Olawoyel T and Kumar P 2022 A high gain antenna with DGS for sub-6 GHz 5G communications *Adv. Electromagn.* **11** 41–50

[35] AL-Haddad M A S M, Jamel N and Nordin A N 2021 Flexible antenna: a review of design, materials, fabrication, and applications *J. Phys: Conf. Ser.* **1878** 012068

[36] Shanmuga Sundar D and Sivanantharaja A 2012 High efficient plastic substrate polymer white light emitting diode *Opt. Quantum Electron.* **45** 79–85

[37] Purohit S and Raval F 2014 Wearable-Textile Patch Antenna using Jeans as Substrate at 2.45 GHz *Int. J. Eng. Res. Technol.* **3** 1–5

[38] Hasliza A, Fareq A M M, Ismahayati A, Sahadah A, Baya M H N and Hall P 2012 Design and simulation of a wearable textile monopole antenna for Body Centric Wireless Communications *Prog. Electromagn. Res. Symp. Proc.*

[39] Mo L, Guo Z, Wang Z, Yang L, Fang Y, Xin Z, Li X, Chen Y, Cao M and Zhang Q *et al* 2019 Nano-silver ink of high conductivity and low sintering temperate for paper electronics *Nanoscale Res. Lett.* **14** 197

[40] Guerchouche K, Herth E, Calvet L E, Roland N and Loyez C 2017 Conductive polymer based antenna for wireless green sensors applications *Microelectron. Eng.* **182** 46–52

[41] Khaleel H 2015 *Innovation in Wearable and Flexible Antennas* (Southampton: Wit Press)

[42] DeJean G, Bairavasubramanian R, Thompson D, Ponchak G E, Tentzeris M M and Papapolymerou J 2005 Liquid Crystal polymer (LCP): a new organic material for the development of multilayer dual-frequency/dual-polarization flexible antenna arrays *IEEE Antennas Wireless Prop. Lett.* **4** 1663–9

[43] Vallozzi L, Rogier H and Hertleer C 2009 A textile patch antenna with dual polarization for rescue workers' garments *Eur. Conf. Antennas Prop. (Berlin)* 1018–21

[44] Sundar D S, Sivanantha Raja A, Sanjeeviraja C and Jeyakumar D 2016 High Temperature Processable Flexible Polymer Films *Int. J. Nanosci.* **15** 1660038

[45] Sundar D S, Sivanantha Raja A, Sanjeeviraja C and Jeyakumar D 2016 Highly transparent flexible polydimethylsiloxane films – a promising candidate for optoelectronic devices *Soc. Chem. Ind., Polym. Int.* **65** 535–43

[46] Dhanabalan S S, Sitharthan R, Madurakavi K, Thirumurugan A, Rajesh M, Avaninathan S R and Carrasco M F 2022 Flexible compact system for wearable health monitoring applications *Comput. Electr. Eng.* **102** 108130

[47] Dhanabalan S S, Thirumurugan A, Periyasamy G, Dineshbabuy N, Chidhambaram N, Avaninathan S R and Carrasco M F 2022 Surface engineering of high-temperature PDMS substrate for flexible optoelectronic applications *Chem. Phys. Lett.* **800** 139692

[48] Sundar D S, Sivanantharaja A, Sanjeeviraja C and Jeyakumar D 2015 Synthesis and characterization of transparent and flexible polymer clay substrate for OLEDs *Conf. Recent Adv. Nano Sci. Technol.*

[49] Hassan A, Ali S, Bae J and Lee C H 2016 All printed antenna based on silver nanoparticles for 1.8 GHz applications *Appl. Phys.* A **122** 768

[50] Tao Y, Tao Y, Wang L, Wang B, Yang Z and Tai Y 2013 Transport properties of two finite armchair graphene nanoribbons *Nanoscale Res. Lett.* **8** 1

[51] Mehdipour A, Sebak A R, Trueman C W, Rosca I D and Hoa S V 2010 Full-composite fractal antenna using carbon nanotubes for multiband wireless applications *IEEE Antennas Wirel. Propag. Lett.* **9** 891–4

[52] PCI 2001 Dynamic surface tension and surface energy in ink formulations and substrates, https://www.pcimag.com/articles/85879-dynamic-surface-tension-and-surface-energy-in-ink-formulations-and-substrates

[53] Chen K, Gao W, Emaminejad S, Kiriya D, Ota H, Nyein H Y Y, Takei K and Javey A 2016 Printed carbon nanotube electronics and sensor systems *Adv. Mater. Technol.* **28** 4397–414

[54] Guo X, Hang Y, Xie Z, Wu C, Gao L and Liu C 2017 Flexible and wearable 2.45 GHz CPW-fed antenna using inkjet-printing of silver nanoparticles on pet substrate *Microwave Opt. Technol. Lett.* **59** 204–8

[55] Raviteja V, Ashok Kumar S and Shanmuganantham T 2019 CPW-Fed Inverted Six Shaped Antenna Design for Internet of Things (IoT) Applications *TEQIP III Sponsored Int, Conf. on Microwave Integrated Circuits, Photonics and Wireless Networks (IMICPW)* pp 1–11

[56] Jayabharathy K and Shanmuganantham T 2019 Design and development of textile antenna for multiband applications *Int. Conf. Commun. Signal Process.* 0348

[57] Al-Sehemi A G, Al-Ghamdi A A, Dishovsky N T, Atanasov N T and Atanasova G L 2021 A flexible planar antenna on multilayer rubber composite for wearable devices *J. Electromagn. Waves Appl.* **75** 31–42

[58] Al-Sehemi A, Al-Ghamdi A, Dishovsky N, Atanasova G and Atanasov N 2020 A flexible multiband antenna for biomedical telemetry *IETE J. Res.* **63** 189–202

IOP Publishing

Advances in Flexible and Printed Electronics
Materials, fabrication, and applications
Shanmuga Sundar Dhanabalan and Arun Thirumurugan

Chapter 9

Recent developments in flexible and printed reconfigurable antennas for medical and Internet of things applications

Madurakavi Karthikeyan, J Pradeep, M Harikrishnan, R Sitharthan, Naveen Mishra, M Rajesh and Thamizharasan Sivanesan

There are numerous possibilities to achieve complex and life-saving applications thanks to high data rates and developing wireless communication standards. The comfort and ease with which humans live have been enhanced by modern electronic communication devices that take advantage of the Internet of things (IoT). IoT helps to solve various patient monitoring issues together with updates to implantable medical devices. The antenna is the main part of IoT and implantable devices. Due to shifting standards, novel antenna design is always needed for these applications. The innovative guidelines place strict criteria on antenna design elements such as operating frequency, gain, and size. Additionally, the majority of IoT and medical applications require that antenna performance be unaffected by antenna bending. Recently, due to emerging flexible substrates, a lot of sophisticated applications in the field of IoT and medical areas have been reported.

The optimal substrate for the flexible antenna has a low dielectric loss, a small coefficient of thermal expansion, low relative primitivity, and high thermal conductivity. The substrate's dielectric constant must be high to reduce the antenna's size. The trade-off between efficiency and antenna design size is that increasing antenna efficiency will cost more with larger antenna sizes.

To enable wireless communication, which is widely desired by today's information-oriented culture, flexible antennas operating in specified frequency bands must be integrated into flexible electronic systems. For instance, industrial, scientific, and medical (ISM) frequency bands are used for biomedical antennas. The flexible antenna design in this frequency band is attracting a lot of researchers in the industry as well as academics.

doi:10.1088/978-0-7503-5492-9ch9

This chapter addresses progress in flexible antenna design, simulation, various novel flexible substrates, measurement, and suitability for various applications. Furthermore, various challenges that need to be addressed by the researchers have been presented.

9.1 Introduction

Flexible and printed reconfigurable antennas have gained significant attention in recent years due to their ability to meet the requirements of medical and IoT applications. Furthermore, they are a new class of antennas that provides improved flexibility and conformability compared to traditional antennas [1]. With the advent of advanced materials and manufacturing techniques, these antennas have become cost-effective and high performing, providing improved conformability, portability, and a low profile [2–5]. The ability to integrate them into wearable devices, implantable devices, and other medical and IoT applications has significantly increased their value in the market. Flexible and printed reconfigurable antennas are made using flexible substrates, which allow them to be bent or shaped to conform to various surfaces or objects [6]. In addition, they can be printed using a variety of techniques such as inkjet printing, screen printing, and laser ablation, making them cost-effective and scalable [7]. Flexible and printed antennas offer several advantages over traditional antennas, including improved portability, a low profile, and the ability to integrate them into various devices [8], hence making them comfortable to be used in a wide range of applications such as wearable devices, implantable devices, and IoT devices. Reconfigurable antennas are a type of antenna that can change their radiation properties in response to different operating conditions [9]. These antennas can be reconfigured to adjust their operating frequency, polarization, and radiation pattern [10, 11]. The ability to reconfigure these antennas provides greater flexibility in wireless communication and sensing applications [12]. Reconfigurable antennas can be divided into two categories: analog and digital. Analog reconfigurable antennas use tunable components, such as varactors or micro-electromechanical systems, to change the antenna's radiation properties. Digital reconfigurable antennas, on the other hand, use switches to connect or disconnect different parts of the antenna, thereby changing its radiation properties [13–15].

Flexible antennas offer several qualities, as depicted in figure 9.1, that make them useful for a range of uses. The following are some crucial traits of flexible antennas:

- Flexibility: Flexible antennas can stretch and adapt to many sizes and shapes, as their name suggests. They are therefore perfect for usage in situations where conventional stiff antennas cannot be used.
- Light weight: Since flexible antennas are often lightweight, they can be used in wearables and portable electronics.
- Low profile: Flexible antennas are perfect for use in devices with limited space because they may be made to have a low profile.
- Durability: Flexible antennas are strong and long-lasting because they are made to endure bending and twisting without breaking.

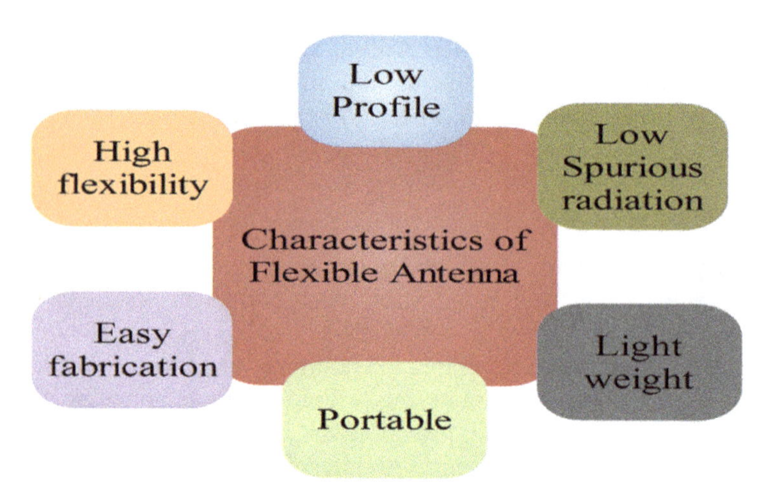

Figure 9.1. Several characteristics the flexible antenna should possess.

- Ease of integration: Flexible antennas can be made to function with several wireless protocols, including Bluetooth, Wi-Fi, and cellular networks, and are simple to integrate into a wide range of gadgets.
- High gain: Flexible antennas with high gain can be made to transmit and receive signals over long distances.
- Wide frequency range: Flexible antennas are appropriate for use in a variety of applications because they can be made to work over a broad frequency range.

Overall, flexible antennas are a preferable choice for a range of applications, including wearables, IoT devices, and medical devices. The overview of the application of flexible antennas is depicted in figure 9.2. Due to their flexibility and capacity to adapt to a variety of shapes and sizes, flexible antennas are useful in a wide range of applications. The following are a few typical uses for flexible antennas:

- Wearable devices: Flexible antennas are frequently employed in wearable technology, including smartwatches, fitness trackers, and medical equipment. They can be made to deliver dependable wireless connectivity while molding to the shape of the body.
- Military and defense applications: These include the employment of flexible antennas in communication systems, surveillance apparatus, and unmanned aerial vehicles. They may be made to tolerate abrasive conditions and high temperatures.
- Automotive industry: In the automotive sector, flexible antennas are utilised for satellite radio, the Global Positioning System, and keyless entry systems. They are capable of being made to fit into small locations and offer dependable connectivity.

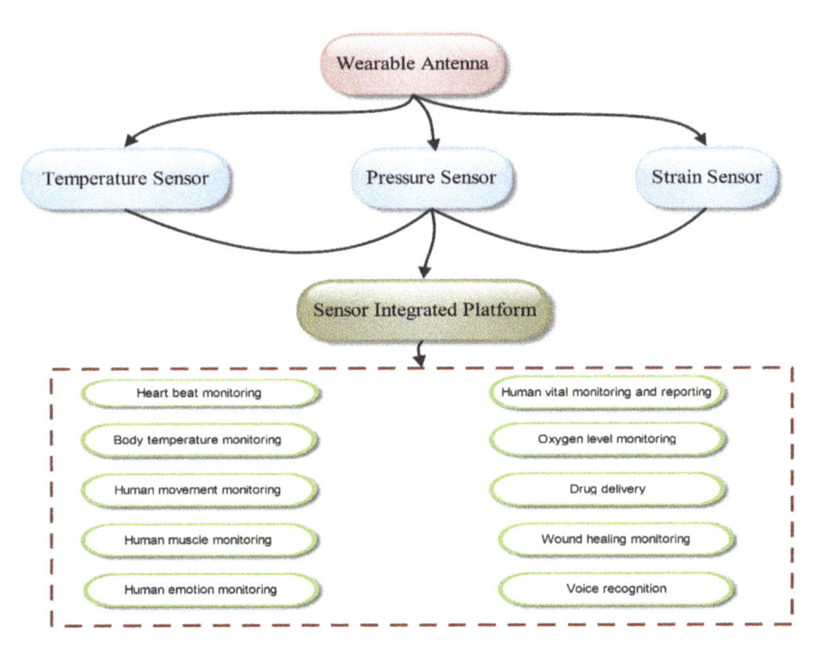

Figure 9.2. Various applications in which flexible antennas are deployed.

- IoT devices: Small, flexible antennas that may be incorporated into a range of devices, including sensors and smart home appliances, are in high demand as a result of the IoT.
- Medical devices: Medical gadgets like pacemakers and implanted medical equipment require flexible antennas. They can be made to be reliable wireless connectors and biocompatible.
- Consumer electronics: Several consumer goods, including laptops, tablets, and smartphones, feature flexible antennas. They can be made to be compact and offer dependable wireless connectivity.

Furthermore, flexible and printed reconfigurable antennas require flexible substrates that are compatible with the manufacturing process and can provide the desired mechanical and electrical properties. Recent developments in flexible substrates have significantly improved the performance of these antennas and enabled new applications [16].

Although flexible antennas have many advantages, they also present several design problems. Some of the key challenges in flexible antenna design include:

- Material selection: The performance of flexible antennas depends heavily on the materials chosen for their design. The materials must have strong conductivity and be able to bear bending and twisting without breaking.
- Fabrication complexity: Fabricating flexible antennas can be challenging, since they frequently call for specialized tools and manufacturing processes.

- Tuning and matching: Flexible antennas can be sensitive to changes in their environment, which can affect their tuning and matching. Designers must carefully tune and match the antenna to ensure optimal performance.
- Performance trade-offs: In comparison to more conventional rigid antennas, flexible antennas frequently have performance trade-offs such as decreased gain or efficiency. To guarantee that the antenna matches the required standards for the intended application, designers must carefully weigh the trade-offs.
- Durability: Flexible antennas must be strong enough to endure frequent bending and twisting without breaking or performing worse.
- Integration: Since the antenna must be made to fit into the limited space and cooperate with the device's wireless protocol, integrating a flexible antenna into a device might be difficult.

In general, careful consideration of the materials utilized, performance trade-offs, fabrication complexity, tuning and matching, integration, and durability are necessary when developing flexible antennas. Notwithstanding these difficulties, flexible antennas provide several advantages that make them a desirable choice for numerous applications.

The substrates play a major role in the development of flexible and printed reconfigurable antennas [17]. The substrates such as polyimide (PI), liquid crystal polymer (LCP), polyethylene terephthalate (PET), and paper-based substrates are common types of substrates used in the development of flexible, printed reconfigurable antennas and biodegradable substrates [1]. Each substrate has its unique feature. PI is a popular choice for flexible substrates due to its high-temperature stability, chemical resistance, and mechanical strength [7]. It is widely used in the aerospace, automotive, and medical industries. PI films can be used as a substrate for inkjet printing or as a carrier for transferring printed antennas onto other substrates. PI can also be coated with conductive materials, such as copper or silver, to create a conductive layer for antenna fabrication [18]. LCP is a thermoplastic material that provides excellent flexibility, low dielectric constant, and low loss [19]. It has excellent mechanical properties and is widely used in high-frequency and high-speed applications. LCP films are capable of being used as a carrier to transfer printed antennas onto other substrates or as a substrate for inkjet printing. To generate a conductive layer for the construction of antennas, LCP may additionally be coated with conductive materials like copper or silver [20]. PET is a widely used flexible substrate due to its low cost, high flexibility, and good thermal stability [21]. PET films have two uses: they can be a carrier for printed antennas to be transferred to other substrates or a substrate for inkjet printing. Additionally, conductive materials like copper or silver can be applied to PET to provide a conductive layer for the construction of antennas [22]. Paper-based substrates are a relatively new class of flexible substrates that have gained significant attention due to their low cost, biodegradability, and ease of disposal [23]. Printed antennas can be transferred to different substrates using paper-based substrates or as a substrate for inkjet printing. A conductive layer for antenna construction can also be created by coating paper-based substrates with conductive substances like copper or silver [24]. Biodegradable substrates are another new class of flexible substrates that have gained significant attention due to their biocompatibility and sustainability [25]. Biodegradable substrates offer versatility in their application, serving as an ideal canvas for inkjet printing or a conveyance medium for transferring printed antennas onto alternative substrates. Additionally, these substrates can undergo coating with conductive materials like silver or copper, resulting in the formation of a conductive layer suitable for the fabrication of antennas [26].

There are several recent developments in flexible and printed reconfigurable antennas for medical and IoT applications. These antennas offer several advantages over traditional rigid antennas, including improved conformability, portability, and low profile [27]. In addition, they can be easily integrated into wearable devices, implantable devices, and other medical and IoT applications. Flexible and printed reconfigurable antennas have been extensively researched and developed for medical and IoT applications. In medical applications, these antennas can be used for wireless body area networks (WBANs) and implantable devices [28]. For example, they can be used to monitor vital signs, such as heart rate and blood pressure, as well as to transmit data from implantable devices to external monitoring systems. Recent research has shown that reconfigurable antennas can significantly improve the performance of existing medical devices, such as magnetic resonance imaging (MRI) machines [29]. These antennas can reduce interference and improve signal quality, leading to more accurate diagnosis and treatment. In IoT applications, flexible and printed reconfigurable antennas can be used to improve wireless communication between devices and networks [30]. They can improve the range, data rate, and reliability of wireless communication, as well as reduce interference and improve signal quality. These antennas can also enable new applications such as low-power IoT devices that can operate for long periods without the need for frequent battery replacement. The use of advanced materials, such as conductive polymers, graphene, and metal alloys, has significantly improved the performance of flexible and printed reconfigurable antennas [31]. These materials provide improved conductivity, flexibility, and durability, making them ideal for medical and IoT applications [32]. Flexible and printed reconfigurable antennas represent a new class of antennas that are gaining significant attention in medical and IoT applications. Recent developments in advanced materials and manufacturing techniques have significantly improved their performance, making them cost-effective and high performing. The ability to integrate them into wearable devices, implantable devices, and other medical and IoT applications has significantly increased their value in the market. As these technologies continue to evolve, they are likely to play an increasingly important role in the future of wireless communication and sensing.

Substrate and conducting materials for flexible printed antennas.

Several conductive materials and substrates are used to create flexible antennas. The dielectric characteristics, mechanical deformation tolerance (including bending, twisting, and wrapping), sensitivity to miniaturization, and environmental durability of the substrate are taken into consideration while selecting it. The choice of conductive material, on the other hand, determines the antenna performance, including radiation efficiency (based on electrical conductivity). In recent days, flexible as well as transparent substrates are used in fabricating antennas for IoT and biomedical applications.

The rest of this chapter is divided up as follows. The discussion of the various materials that have been used to create flexible antennas is covered in detail in section 9.2. The factors that aid in evaluating the performance of the flexible antenna have been described in section 9.3. Section 9.4 discusses the current flexible antenna design as well as some noteworthy recent proposals for three key application areas, including 5G and beyond, IoT applications, and biomedical applications. Section 9.5 then brings this chapter to a close by discussing potential future issues.

9.2 Materials for flexible antenna design

9.2.1 Substrates

The substrate material for flexible antennas must have a low coefficient of thermal expansion, good thermal conductivity, and minimal dielectric loss. Such a restriction is caused by the demand for higher efficiency (in varied settings) at the cost of larger antenna size. An exception to the above-mentioned fact is the need for a large dielectric constant for tiny antennas. In the construction of flexible antennas, substrates made of thin glass, metal foils, and plastics or polymers have all frequently appeared [33, 34]. Additionally, researchers are more likely to choose substrates with the qualities of durability, washability, flexibility, and stretchability in many applications for flexible wearable antennas. Recently, natural rubber, foam, and buffalo leather have been used for fabricating flexible antennas. Metal foils can withstand elevated temperatures and provide an area for the deposition of inorganic compounds, but their uses are limited by their rough surface and expensive material costs [35]. The best plastic or polymer materials for flexible antenna applications are PET and polyethylene naphthalate (PEN), which are thermoplastic semicrystalline polymers; polycarbonate and polyethersulphone, which are thermoplastic noncrystalline polymers; and PI, which has a high glass transition temperature. Among these, PEN has excellent qualities: it is an affordable, transparent material. Moreover, it exhibits a very high heat shrinkage coefficient and good acid and base resistance. Three low-cost flexible material categories for wearable antennas, buffalo leather, natural rubber, and foam, were recently chosen for evaluation [36]. SPEAG's Dielectric Evaluation Kit was used to test these materials' relative permittivity and loss tangent. Two substrates—buffalo leather and natural rubber—were chosen from the tested materials for the design and production of antenna prototypes. Due to its high permittivity, high resilience, low loss, low cost, and water-resistant qualities, natural rubber has been proven to be a suitable flexible material for wearable antennas. Next, fabric has been looked at as a potential substrate material for flexible antennas. These antennas are made by directly sewing conductive threads into the garment and embroidering the radiating parts into the desired shape. Embroidered antennas successfully meet the unobtrusiveness criteria of wearable electronics due to their seamless integration with clothing. Also, they possess a strong physical capacity to withstand repeated deformations brought on by a human body's contour alterations. Due to their porous nature, textiles generally have the desirable property of low permittivity. In [37], a resonance-based approach

Table 9.1. List of substrates used for flexible antennas.

Substrate	Dielectric loss	Dielectric constant	Thickness (mm)
Natural rubber	0.001	6.46	4.1
Buffalo leather	0.058	2.80	3.0
Foam	0.025	1.22	3.1
Wash cotton	—	1.51	3.0
Jeans cotton	—	1.67	2.84
Polycot	—	1.56	3.0
Curtain cotton	—	1.47	3.0
Polyester	—	1.44	2.85
Bed sheet	—	1.46	3
PET	0.008	3	0.140
PEN	0.025	2.9	0.2
PI	0.005	2.91	—
PDMS	0.02	2.65	0.2
LCP	0.004	3	0.05
Kapton PI	0.004	3.9	0.007–0.0125
PTFE	0.025	10.2	0.1

for determining a fabric's dielectric constant has been reported. The dielectric constants of six types of fabric, including polyester, cotton-polyester blends, and denim cotton have been determined in benchmarking with standard material Teflon and are shown in table 9.1.

Flexible polymer-based wearable antennas are an alternative to embroidered antennas made of fabric. Polymers are an excellent choice for higher-frequency applications because they typically have low permittivity and low loss. As a result, the antenna substrate is made of flexible polymers like polydimethylsiloxane (PDMS) and LCP [38]. An extremely popular core flexible material for supporting overlays for soft printed circuit board (PCB) processes, such as PI films [39, 40], is PI, a thin, flexible, and light polymer [41, 42]. Because of its flexible endurance, dissipation factor, low dielectric constant, superior tear resistance, moisture absorption, and coefficient of linear thermal expansion, PIs are frequently utilized in flexible printed circuits. A PI noted for its flexibility and a good balance of physical, chemical, and electrical qualities is the Kapton PI [43]. Kapton is a dependable flexible substrate that is inexpensive, has mechanical toughness, a low loss factor over a wide frequency range, and thermal durability. Various flexible antennas that use Kapton have been reported in the literature. The main advantage of Kapton polymer is its capability to withstand high temperatures leading to better soldering tolerance [44]. Next, polytetrafluoroethylene (PTFE) has been used to fabricate flexible antennas. It is water-resistant, chemically stable, and has high-temperature handling capabilities [44, 45]. Teflon is a PTFE polymer that is readily accessible on the market and has excellent mechanical, electrical, thermal, and anti-friction characteristics. Teflon is a well-known flexible material because of its thermal stability, resistance to change in temperature, resistance to corrosion, and steady dielectric constant over a broad frequency range. Teflon has an extremely high melting point, where its capabilities persist at 260°C. With PTFE Teflon being successfully used in radio frequency identification (RFID) tag antenna designs, its features serve to enable versatility in antenna designs [46]. The comparison of various flexible substrates is given in table 9.1.

9.2.2 Conducting materials for flexible antennas

Conductive polymers, carbon nanotubes, graphene, and metallic solutions of gold, silver, copper, and nickel nanoparticles are some options for conductive material. When utilized in applications that demand the flexing of the antenna from its normal straight shape, these materials must have the ability to bend, crumple, and stretch without impairing the performance of the antenna. They also need to be resistant to material deterioration. Due to their great electrical conductivity, metal-plated textiles or metal nanoparticles, such as nickel-plated, copper-plated, and silver-plated textiles, are widely used to create flexible antennas [46–48]. Recently, due to its outstanding electrical qualities, graphene, a single sheet of carbon atoms organized in a hexagonal lattice, has drawn attention. It can be incorporated into flexible substrates to produce flexible, highly conductive antennas.

Liquid metal alloys based on gallium, like Galinstan, are malleable and have a high electrical conductivity. These alloys can be injected or printed onto appropriate substrates to produce flexible antenna structures.

9.3 Measures to evaluate a flexible Antenna's performance

Apart from the substrate and conductive material selection, the performance of flexible antennas is influenced by so many factors such as reflection coefficient, gain, specific absorption rate (SAR) (for implantable antennas), bending analysis, radiation pattern, and efficiency.

9.3.1 Reflection coefficient

The reflection coefficient is one of the key factors in the construction of a flexible antenna. The reflection coefficient of an antenna is defined as the difference between the amplitudes of the incident and reflected waves. The amount of power reflected from the antenna is shown by the reflection coefficient (S11). If the antenna is designed with a low loss, most of the power applied to it will be radiated because the allowable value of S11 is −10 dB.

The radiation characteristics of the flexible antenna can be described using the resonance bandwidth and resonant frequency of the antenna. The antenna's resonant bandwidth is the range of resonance frequencies with the stated return loss of −10 dB.

As a result, the reflection coefficient is a factor that must be considered when designing a flexible antenna to ensure that any shift in frequency caused by bending, crumpling, temperature sensor differences, or contact with the human body will not have an impact on the flexible antenna's significant return loss.

9.3.2 Gain and efficiency

The ratio of power transmitted from one direction to another with a particular reference point is known as antenna gain. Gain and efficiency are correlated by the formula $G = e \times D$, where D stands for directivity and e for efficiency. The amount of radio frequency (RF) power given to the antenna (via radio) that is transmitted into the air is known as antenna efficiency. If a flexible antenna wants to be more effective for wearable applications, it should have a positive gain.

The gain of the antenna can be improved by utilizing the multiple input multiple output (MIMO) idea and expanding the patch area. However, since it increases the size of the antenna, increasing the antenna size will be a disadvantage in wearable and implantable applications.

In situations where the broadcast signal needs to be directed in a certain direction, such as in satellite communication, radar systems, and wireless networking, high-gain antennas are frequently utilized. High-gain antennas can have narrow beam-width, which means they might not be as good at connecting to objects that are not in the antenna's direct line of sight.

The effectiveness with which a flexible antenna transforms electrical energy into radiated electromagnetic waves is measured by its radiation efficiency. The antenna may be used to send or receive signals into or from the human body in the context of biomedical applications.

Maximizing a flexible antenna's radiation efficiency is crucial for maximizing performance. To do this, the antenna's design can be improved, the right materials can be used, and the antenna must be correctly matched to the transmission line.

The proportion of the antenna's output power to its total input power is known as the radiation efficiency of an antenna. It can be mathematically stated as

$$\text{(Radiated power)} / = \text{Radiation efficiency (Input power)}$$

The radiated power can be indirectly measured in practice. Instead, the radiation efficiency is frequently calculated by measuring the antenna's input impedance and contrasting it with the impedance of an ideal, lossless antenna of the same dimensions. This approach is called the "gain-over-loss" strategy.

Several variables, such as the antenna's size and shape, operating frequency, closeness to the body, and the surrounding tissue's dielectric characteristics, might affect how well a flexible biomedical antenna radiates. By making these adjustments, it is possible to increase the antenna's radiation efficiency and guarantee dependable and efficient communication for biomedical applications.

9.3.3 Bending analysis of flexible antenna

In situations where the antenna must adapt to a curved or irregular surface, such as in wearable technology, medical implants, and aerospace applications, flexible antennas are frequently utilized. A flexible antenna must undergo bending analysis to make sure it will continue to operate as intended even under mechanical stress from bending or twisting.

Analytical modelling and experimental testing are commonly used in the bending study of flexible antennas. In analytical modelling, the deformation of the antenna under various bending situations is predicted using mathematical formulae. Techniques like finite element analysis or analytical modeling of the antenna's mechanical properties can be used for this [49, 50].

A flexible antenna's bending analysis can help to spot any potential design flaws and make sure that the antenna will still work as intended under mechanical strain. Also, it may be beneficial to change the antenna's geometry to better fit the surface it is mounted on or use materials that are more flexible to increase the antenna's performance under bending conditions.

In experimental testing, the antenna is bent under various conditions and its performance is evaluated. This can be accomplished by putting the antenna on a bending fixture and simulating various bending circumstances by applying controlled forces to the fixture. The performance of the antenna can then be assessed using tools like network analyzers, spectrum analyzers, or measures of radiated power.

9.3.4 Specific absorption rate

When exposed to electromagnetic radiation, such as radio waves, the human body absorbs energy at a rate known as the SAR. SAR is a common unit of measurement used to assess the potential health effects of electromagnetic field exposure in watts per kilogram (W kg^{-1}) [51].

The distribution and concentration of electromagnetic fields can be affected by the flexibility of an antenna when it is in close contact to the body, as in wearable technology or body-worn applications. The SAR values that the user experiences as a result may change.

SAR is specifically employed to evaluate the security of wireless equipment, including mobile phones. To guarantee that their products are secure for use by the general population, producers must adhere to SAR limitations set by regulatory agencies all around the world.

If flexible antennas are used in medical applications, it is necessary to ensure that the SAR value is by the advised value to ensure its safety. According to US regulations, the maximum SAR for 1 gm of tissue is 1.6 W kg^{-1}, and the International Commission on Non-Ionizing Radiation Protection for Europe has set the maximum SAR for any 10 g of tissue at 2 W kg^{-1} [52].

It is crucial to keep in mind that SAR is only one of many elements that must be taken into account when assessing the potential health impacts of electromagnetic radiation. It is also necessary to consider other elements, such as the frequency and duration of exposure, as well as the person's general health and sensitivity.

9.4 Notable flexible antenna designs

9.4.1 Flexible antenna design for 5G and beyond

An antenna must be small and able to mount on any uneven surface to meet the growing demand for smart gadgets. Additionally, these antennas must continue to function well in terms of several criteria including gain, impedance bandwidth, radiation efficiency, and radiation pattern. Research on various antenna structures and flexible substrates is required to convert an antenna from a nonflexible to a flexible antenna. So, a thorough description of the flexible antenna for the recently available sub-6 GHz 5G band is necessary. This section covers the various design techniques that have been followed in designing flexible antennas for 5G and beyond. Mostly, the single-element antennas have been designed based on any one of the following categories such as defected ground structure based, slot/slit based, metamaterial based, and dielectric resonator antenna (DRA) based.

The authors in [53] propose a wideband and tri-band antenna with separately controlled notch bands that can be used for 5G sub-6 GHz transmission. The octagonal shape of the wideband monopole antenna has truncated edges, which increases the impedance bandwidth of the antenna. A rectangular stub spanning the entire sub-6 GHz band widens the bandwidth. To achieve the band notches to reduce the C-band and Wi-Max band, slots are added to the wideband antenna structure. The tri-band antenna resonates at frequencies of 2.45, 3.5, and 4.7 GHz, whereas the wideband antenna has a frequency range of 2.8–5.35 GHz.

Flexible and bendable IoT devices are becoming more and more common as new 5G and 6G network infrastructures are introduced. The idea of wearable and flexible antennas is a critical problem to solve in this regard [54, 55]. Research focused on developing small, flexible antennas that could be worn to meet modern requirements. Many novel designs have been reported in the literature. In [56], an antenna has been produced using a novel fabrication technique that includes three-layer three-dimensional (3D) inkjet printing and an alignment step on a flexible and biocompatible PEN substrate measuring 250 μm. The final antenna shown in figure 9.3 has been obtained because of optimization through a genetic algorithm.

Next, with a compact dimension of 40 $\times$ 40 mm^2, the antenna wireless sensor is made up of a conductive MXene patch, copper ground plane, and a PDMS dielectric layer [57]. The MXene varying conductivity will affect the resonance frequency when the antenna deforms. By sacrificing the gain of the proposed antenna, compactness in design has been achieved. The antenna exhibits resonance at 4.8 GHz, which is suitable for sub-5G communication. It was also shown that the proposed antenna can work as a strain sensor as shown in figure 9.4.

In [58], a flexible wideband ink-printed dipole array antenna is presented by the authors for burgeoning medical sensing and wearable electronic applications for the human body. The antenna operates between 23 and 30 GHz, with 24.5 and 28 GHz being used for 5G applications. The antenna is a straightforward, layered screen-printed construction that may be quickly prototyped using printing technology. The authors in [59] designed a MIMO antenna on a surface-treated PET film with a dimension of 11 $\times$ 25.4 mm^2.

On the Ultralam 3850 substrate, an antenna is created with a small footprint of 60 $\times$ 40 $\times$ 0.1 mm^3 [60]. This antenna operates in the 1.74–100 GHz range. The reported antenna can operate in a wide range of wireless application domains, including 5G and the IoT, thanks to its extensive frequency coverage. The comparison of recently reported flexible 5G antennas is given in table 9.2.

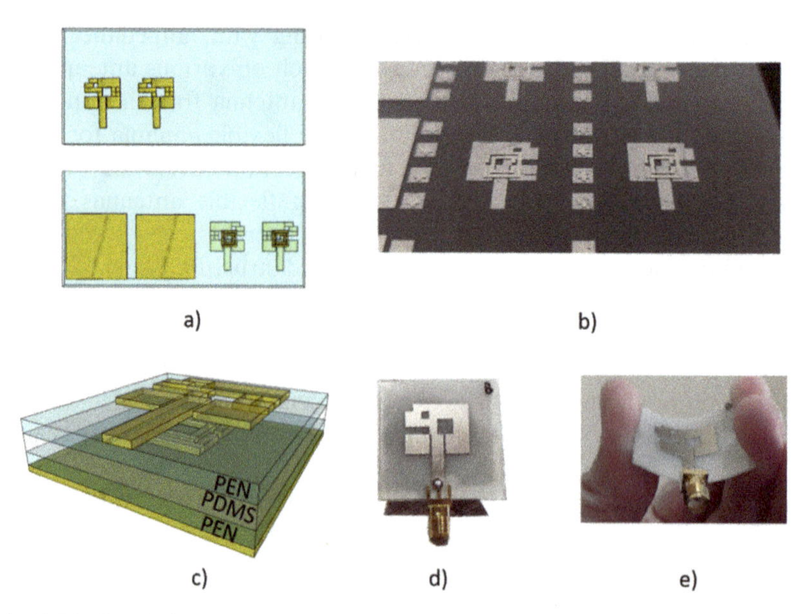

Figure 9.3. (a) The 3D printing process steps, (b) alignment between the patch, split ring resonator (SRR), and printed ground planes, (c) multilayer view, (d) flat and (e) bent fabricated prototype. Reprinted from [56] with permission from Springer Nature, Copyright (2022).

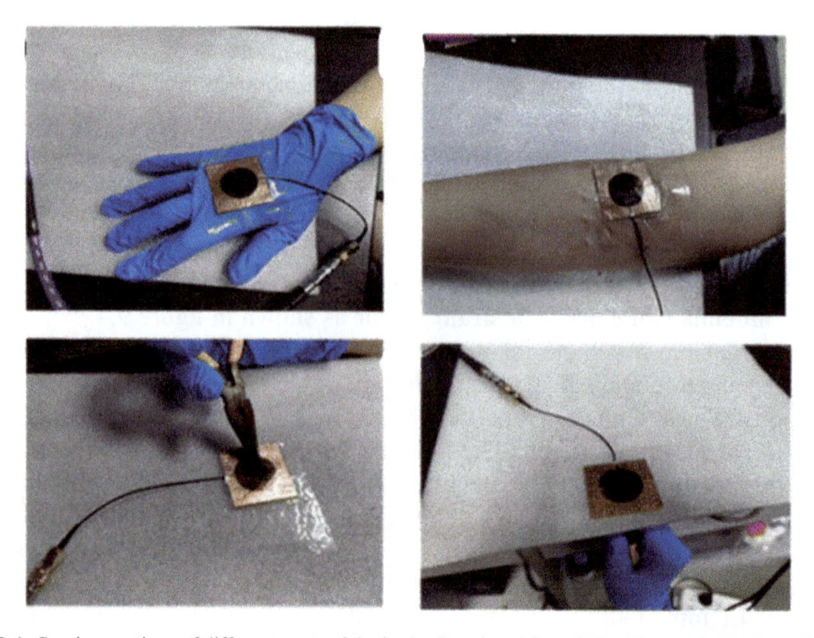

Figure 9.4. Sensing motions of different parts of the body. Reprinted from [57] with permission from the Royal Society of Chemistry.

Table 9.2. Comparison of recently reported flexible antennas for 5G communications.

Reference	Dimension	Substrate	Operating frequency	Gain (dB)	Application	Bandwidth	No. of layers
[56]	16.8 × 13.7 mm^2	PEN	3.98 GHz	5.8	Sub-6 GHz band of the 5G spectrum	15.4 MHz	5
[61]	Diameter: 25 mm	PDMS	4.8 GHZ	—	Sub-6 GHz band of the 5G spectrum	—	1
[53]	32 mm × 25 mm	RO4835T	2.45, 3.45 and 4.7 GHz	3.75		2.5 GHz	1
[58]	25 mm × 12.7 mm	Polyester	24.5 and 28 GHz	4.2		3 GHz	1
[59]	11 mm × 25.4 mm	PET	28, 30, 32, 34 GHz	6.5		6 GHz	1
[60]	60 mm × 40 mm	Ultralam 3850	1.75–100 GHZ	10.35		67GHz	1

The design of flexible 5G MIMO antennas faces numerous difficulties. Here are a few of the main difficulties:

- Size and weight constraints: To be easily incorporated into handheld devices and tiny base stations, 5G MIMO antennas must be compact and lightweight. Little antennas with excellent performance are difficult to design, though.
- Complex radiation patterns: For MIMO systems to achieve high data speeds and minimal interference, antennas with complex radiation patterns are necessary. Designing such antennas can be challenging, especially for 5G systems that use large numbers of antenna elements.
- High-frequency operation: Due to their high propagation losses and high penetration losses, millimeter-wave frequency bands (such as 28 and 39 GHz) where 5G works present substantial architectural issues.
- Beamforming: For 5G MIMO systems to deliver high data speeds and little interference, beamforming plays a significant role. This calls for intricate antenna designs with exact control over each antenna element's phase and amplitude.
- Crosstalk and mutual coupling: Due to the antennas' proximity in MIMO systems, crosstalk and mutual coupling between the antennas may occur. This needs to be carefully managed in the aerial design because it can lead to decreased performance and increased interference.
- Thermal management: The performance of the antennas may be impacted by the heat generated by the 5G systems' high data rates and processing demands. For 5G MIMO systems to operate reliably, proper thermal management must be ensured.

Overall, creating 5G MIMO antennas is a difficult operation that involves carefully taking into account a wide range of variables such as size, radiation patterns, frequency, crosstalk, beamforming, and thermal management.

9.4.2 Flexible antenna proposed for biomedical applications

Due to their versatility and ability to follow the contours of the human body, flexible antennas are becoming more and more common for biomedical applications. They also offer dependable wireless communication in a variety of contexts. The following list includes some prospective study areas for flexible antennas in biomedical applications:

- Wearable sensor networks called WBANs gather and send information about a patient's health. In these networks, flexible antennas could be employed to increase data collection accuracy and deliver dependable wireless connectivity.
- Reliable wireless connectivity is essential for implantable medical devices like pacemakers and defibrillators to transmit and receive commands. These gadgets could integrate flexible antennas to offer this connectivity while also lessening patient discomfort.
- Systems for remotely monitoring patients are used to keep an eye on them at home or in other remote locations. These devices might broadcast data from the patient's gadget to a distant site via flexible antennas, enabling real-time monitoring and better patient outcomes.
- To enhance the precision and dependability of the imaging process, flexible antennas could be employed in MRI machines. These antennas could enhance the quality of the photographs captured by providing dependable wireless communication in the presence of high magnetic fields.
- Fitness trackers and glucose monitors are two examples of wearable health monitoring technology that are gaining popularity. To ensure persistent wireless communication and increase the precision of the data acquired, these devices could utilize flexible antennas.

Even though many novel designs on nonflexible antennas have been proposed [62, 63], flexible antennas have a plethora of possible uses in the biomedical industry. There will certainly be additional opportunities to use flexible antennas in biomedical applications as technology develops. There are many designs reported by the research community. Some of the recent and notable designs are discussed in the following section.

The authors in [64] designed a circular polarised flexible antenna using a novel method: namely, holey superstrate and defective ground construction as shown in figure 9.5. This has led to the creation of an antenna with a small footprint of 2.5 mm × 2.5 mm × 1.28 mm. This antenna was created using the Taconic CER-10, a flexible and biocompatible material. The other benefits of this reported antenna are low

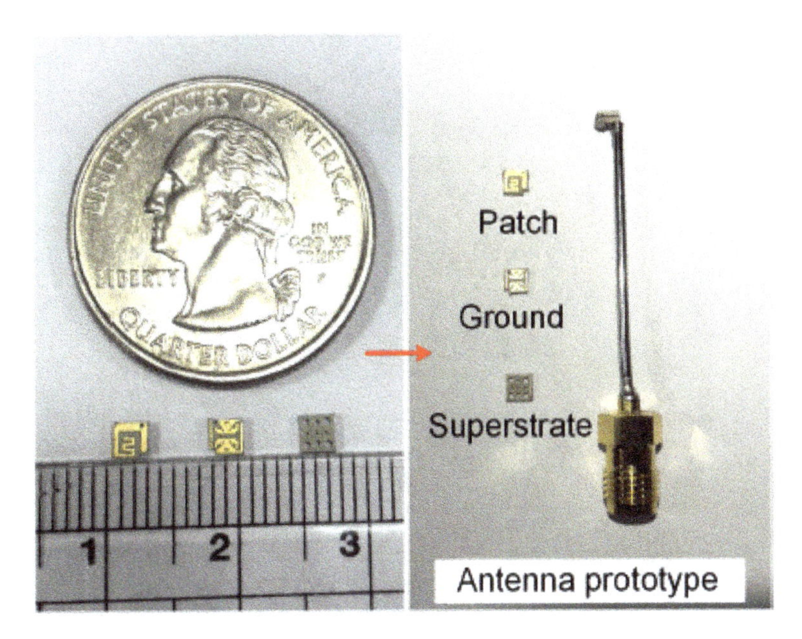

Figure 9.5. Miniaturized flexible antenna. Reprinted from [64] with permission from IEEE, Copyright (2023).

SAR and a high peak gain of −14.3 dBi. The antenna is made to function in the 2.4 GHz ISM band.

The authors in [65] have proposed a spiral-shaped flexible antenna with a dimension of 9.2 × 9.2 × 0.5 mm^3. The antenna operates in two bands: 2.4 and 5.8 GHz. The proposed antenna achieves a high gain of around −15 dBi.

In [66], a dual-band implantable antenna with an oval form was suggested. Using their suggested design, which has a dimension of 12 × 8 mm^2, dual-band operation of 2.4–3.4 GHz (ISM band) and 394–407.61 MHz (medical implant communication service (MICS) band) was made possible. The flexible RO3010 substrate was used in the fabrication of the antenna. Three layers of tissue, including layers of skin, fat, and muscle, were used to imitate it. At 2.4 GHz, the antenna's gain is incredibly poor, at about −30 dB.

For capsule endoscopy, a meandering inverted F-shaped antenna has been proposed [67]. Using two positive intrinsic negative (PIN) diodes, the antenna additionally reported a programmable radiation pattern as shown in figure 9.6. With a volume of 30 × 15 × 0.254 mm^3 and a Rogers 5850 substrate, it attained resonance at 433 MHz. The claimed antenna's maximum achieved gain is −33.64 dBi.

In [68], a new antenna with resonance at 915 MHz was proposed. The 915 MHz is more interference resistant and more power efficient. The antenna was created on an RT Duroid 5850 substrate with a 0.254 mm thickness. The gain for the antenna is −28 dB.

A small, flexible antenna in the shape of a helix has been shown in [69], measuring 6.1 mm in length and 0.352 mm in diameter. The proposed design was made using a PTFE substrate. The 4.7 GHz frequency band is the intended operating range for the antenna. Comparatively, this paper achieves a very high gain of −4.7 dBi.

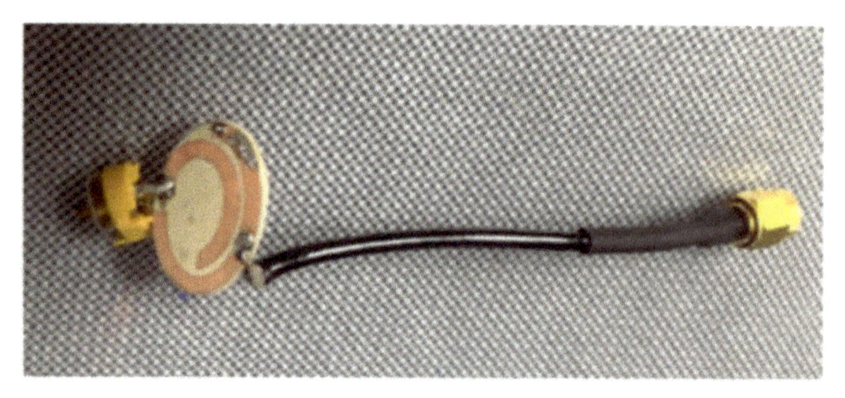

Figure 9.6. Elliptical-shaped antenna. Reprinted from [66]. CC BY 4.0.

Table 9.3. Comparison of recently reported flexible antennas for biomedical applications.

Reference/ year	Frequency (GHz)	Dimension (mm)	Gain (dBi)	SAR		Implantation depth	Polarization
				1-g	10-g		
[64]/2022	2.4	2.5 × 2.5 × 1.28	−14.3	352	48.3	12.5	Circular
[65]/2023	2.4/5.8	9.2 × 9.2 × 0.5	−15				
[66]/2022	2.4	12/8	−31	81.3	35.7	6	Linear
[67]/2022	0.433	30 × 15 × 0.254	−33.64		89.5	—	Linear
[68]/2022	.915	7 mm × 7 mm × 0.254	−28	8.22	—		Linear
[63, 69]/ 2022	4.7	0.610 × 0.352	−4.7	—	—	—	Linear

Table 9.3 provides the comparison of various flexible antennae reported recently for biomedical applications.

9.4.3 Flexible antenna for IoT applications

Due to their adaptability to various forms and sizes, flexible antennas are becoming more and more popular for IoT applications. This makes them perfect for use in small, compact devices. For IoT applications, the following factors should be considered when choosing a flexible antenna:

9.4.3.1 Frequency range

The frequency range needed for your device is the first factor to take into account when choosing a flexible antenna for IoT applications. It is critical to select an antenna that can operate inside the frequency range of your device because different types of antennas are built for various frequency ranges. The common frequency bands for some of the well-known wireless protocols used in IoT applications are

Wi-Fi (2.4 and 5 GHz), Bluetooth (2.4 GHz), Zigbee (2.4 GHz and 900 MHz), which is frequently used for low-power and low data rate applications, LoRa (433, 868, and 915 MHz), which is used for long-range IoT applications, and NB-IoT (800, 900 MHz, and 1.8 GHz bands), which is used for wide-area IoT applications.

9.4.3.2 Radiation pattern

The orientation in which an antenna transmits or receives signals is determined by its radiation pattern. The specific application and the needs of the wireless communication system will determine the required emission pattern for an IoT antenna. Nevertheless, omnidirectional radiation patterns are typically the most preferred for IoT antennas. A 360-degree coverage area is provided by an antenna with an omnidirectional radiation pattern, which broadcasts or receives signals equally in all directions. For IoT applications where the device may be mobile, oriented differently, or require multidirectional communication with other devices, this type of radiation pattern is optimal.

Yet, a directed radiation pattern might be preferable in some IoT applications. To increase the range and dependability of a point-to-point communication system, for instance, a directional antenna may be used to focus the signal in a certain direction.

In the end, the precise needs of the application and the wireless communication technology will determine the radiation pattern selection for an IoT antenna. While choosing the right radiation pattern for an IoT antenna, it is also important to take the surroundings, the distance between devices, and the level of interference into account.

9.4.3.3 Environmental factors

The kind of antenna you select will also depend on your IoT device's operational environment. You might need to choose an antenna that is resistant to these conditions, for instance, if your device will be exposed to severe weather.

All things considered, it is critical to take the frequency range, radiation pattern, efficiency, size, form, and environmental considerations into account when selecting a flexible antenna for IoT applications. You can choose an antenna that is appropriate for your application by taking these aspects into account. Some of the recently reported antennas have been discussed in the following paragraph.

A novel flexible ceramic substrate i.e. ENrG's Thin E-strate has been exploited in [70] for IoT applications. The designed antenna achieves resonance at 2.4 GHz and 5.7 GHz. The unique structure was tested on three different substrates. Fabrication has been done using electro-textile and inkjet printing.

In [71], using a PET substrate, a monopole antenna with a circular-shaped patch has been proposed as shown in figure 9.7. The reported antenna had a dimension of $47 \times 25 \times 0.135$ mm^3 and operated in the range of 3.04–10.70 GHz and 15.18–18 GHz. The reported antenna exhibits an omnidirectional radiation pattern.

With a simple geometry, an ultra-wide band (UWB) antenna using ROGERS 5880 has been reported in [72] as depicted in figure 9.8. The antenna patch has been designed by etching a pair of slots and a loading stub. The authors utilized two triangular slots to convert the narrowband response into an ultra-wideband

Figure 9.7. Flexible antenna for IoT applications designed on terephthalate substrate. Reprinted from [71]. CC BY 4.0.

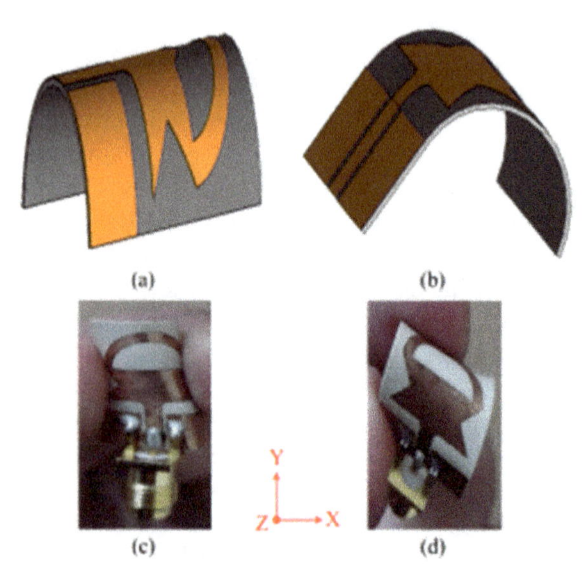

Figure 9.8. Flexible antenna for IoT applications designed on ROGERS 5880 substrate. Reprinted from [72]. CC BY 4.0.

response. The antenna has co-planar waveguide feeding, which makes it easier to integrate with other electronic circuits.

In [73], as a novelty, a graphene-based coplanar waveguide (CPW) antenna has been designed and fabricated for IoT applications as shown in figure 9.9. The substrate material and conductive material used for the proposed antenna are glass and graphene, respectively. The antenna was designed to operate at 2.45 GHz ISM band and 4–6 GHz band. The antenna achieved a peak gain of 0.765 dB.

With a circular-shaped patch and CPW feeding technique on one-sided photo paper, a novel inkjet-printed flexible UWB antenna has been proposed [74]. The antenna was designed to operate from 3.2 to 30 GHz. The antenna achieves a peak radiation gain and efficiency of 4.87 dB and 86.61%, respectively. However, the antenna has a larger dimension of 33.1 mm × 32.7 mm × 0.254 mm.

In [75], using a novel metamaterial superstrate, a reconfigurable antenna working in the range of 4–8 GHz has been proposed. The antenna has reported a high gain of 8.3 dB. The reconfigurability has been achieved by three PIN diodes. Analysis of the proposed antenna has been carried out by loading three different materials such as copper sea water and distilled water. Table 9.4 provides a comparison of various flexible antennas reported recently for IoT applications.

Figure 9.9. Graphene-based flexible antenna. Reprinted from [73]. CC BY 4.0.

Table 9.4. Comparison of recently reported flexible antennas for IoT applications.

References	Dimension (mm^3)	Frequency (GHz)	Peak gain (dBi)
[70]	62 × 60 × 0.812	2.7 and 5.7	5.4
[71]	47 × 25 × 0.135	3.04–10.70	5.7
[72]	15 × 20 × 0.254	2.73–9.68	2.5
[73]	30 × 30 × 2	2.45, 4–6	0.765
[74]	33.1 × 32.7 × 0.254	3.2–30	4.87
[75]	30 × 33 × 0.254	4–8	8.3

9.5 Conclusion and future work

In conclusion, because of its many benefits, including conformability, light weight, and low profile, flexible antennas have emerged as an attractive option for wireless and biomedical applications. These antennas have made it feasible to develop a fresh category of wireless devices that may be integrated into clothing and worn on the body, offering a seamless and unobtrusive user experience. Flexible antennas have shown tremendous potential in biomedical applications for wireless power transfer and communication inside the body, enabling new medical procedures and therapies. The advancements made recently have shown the potential of flexible antennas in a wide range of applications, despite the difficulties involved in their design and fabrication. Future developments in the utilization of flexible antennas for wireless and biological applications are anticipated as this field of study advances. The advanced study should focus on increasing the operating frequency range, improving radiation efficiency, integration of multiple antennas, integration of active components, and exploration of new applications.

Flexible antennas can also be applied in a wide range of other industries such as aerospace, automotive, and military. They can be utilized in the creation of sensors for use in the military as well as the construction of thin, light communication systems for vehicles and aircraft.

Flexible and printable antennas have completely changed how we think about wireless communication systems, especially in the context of IoT and medical applications. These antennas' innate flexibility and conformability allow for seamless integration with a variety of items, gadgets, and even the human body, creating new opportunities for wireless networking in as-yet-undiscovered areas.

Important design factors, fabrication processes, and performance assessment strategies for flexible and printed reconfigurable antennas have been discussed in this chapter. In order to make these antennas with better performance, durability, and affordability, it has also delved into the investigation of novel materials and cutting-edge production techniques.

In conclusion, the chapter on current innovations in flexible and printable reconfigurable antennas for medical and IoT applications offers a thorough overview of the most recent scientific findings and technology developments in this fascinating area. It clarifies the possible effects of these antennas on the IoT and health care sectors, encouraging a world in which wireless connectivity is easily incorporated into daily life and changes how we observe, identify, and communicate with our surroundings.

Future applications of flexible antennas are likely to be significantly more innovative as this field of study progresses. They have the potential to transform the way we think about wireless communication and sensing by adapting to a range of shapes and surfaces, creating new possibilities for technological advancement.

References

[1] Kirtania S G *et al* 2020 Flexible antennas: a review *Micromachines* **11** 847

[2] Khan M U A, Raad R, Tubbal F, Theoharis P I, Liu S and Foroughi J 2021 Bending analysis of polymer-based flexible antennas for wearable, general iot applications: a review *Polymers (Basel)* **13** 1–34

[3] Karthikeyan M, Sitharthan R, Ali T, Pathan S, Anguera J and Shanmuga Sundar D 2022 Stacked T-shaped strips compact antenna for WLAN and WiMAX applications **123** 1523–36
[4] Karthikeyan M, Sitharthan R, Ali T and Roy B 2020 Compact multiband CPW fed monopole antenna with square ring and T-shaped strips *Microw. Opt. Technol. Lett.* **62** 926–32
[5] Priyadharshini A S, Arvind C and Karthikeyan M 2023 Novel ENG metamaterial for gain enhancement of an off-set fed CPW concentric circle shaped patch antenna *Wirel. Pers. Commun.* **130** 2515–30
[6] Suliman Munawar H 2020 An overview of reconfigurable antennas for wireless body area networks and possible future prospects *Int. J. Wirel. Microw. Technol.* **10** 1–8
[7] Mohamadzade B, Hashmi R M, Simorangkir R B V B, Gharaei R, Rehman S U and Abbasi Q H 2019 Recent advances in fabrication methods for flexible antennas in wearable devices: state of the art *Sensors* **19** 2312
[8] Yang L *et al* 2021 Review on wearable antenna design *Lect. Notes Electr. Eng.* **654** LNEE 731–42
[9] Arab Hassani F *et al* 2020 Smart materials for smart healthcare– moving from sensors and actuators to self-sustained nanoenergy nanosystems *Smart Mater. Med.* **1** 92–124
[10] Kamran Shereen M, Khattak M I and Witjaksono G 2019 A brief review of frequency, radiation pattern, polarization, and compound reconfigurable antennas for 5G applications *J. Comput. Electron.* **18** 1065–102
[11] Dhanabalan S S *et al* 2022 Flexible compact system for wearable health monitoring applications *Comput. Electr. Eng.* **102** 108130
[12] Ni C, Chen M S, Zhang Z X and Wu X L 2018 Design of frequency-and polarization-reconfigurable antenna based on the polarization conversion metasurface *IEEE Antennas Wirel. Propag. Lett.* **17** 78–81
[13] Sitharthan R *et al* 2022 Performance enhancement of an economically operated DC microgrid with a neural network–based tri-port converter for rural electrification *Front. Energy Res.* **10** 972
[14] Karthikeyan M, Jayabala P, Ramachandran S, Dhanabalan S S, Sivanesan T and Ponnusamy M 2022 Tunable optimal dual band metamaterial absorber for high sensitivity THz refractive index sensing *Nanomater* **12** 2693
[15] Fu X, Yang F, Liu C, Wu X and Cui T J 2020 Terahertz beam steering technologies: from phased arrays to field-programmable metasurfaces *Adv. Opt. Mater.* **8** 1900628
[16] Shanmuga Sundar D, Sridarshini T, Sitharthan R, Karthikeyan M, Sivanantha Raja A and Carrasco M F 2019 Performance investigation of 16/32-channel DWDM PON and long-reach PON systems using an ASE noise source *Advances in Optoelectronic Technology and Industry Development* ed G Jose and M Ferreira (London: CRC Press) Proceedings of the 12th International Symposium on Photonics and Optoelectronics (SOPO 2019)1st edition 93–9
[17] Kumar P, Ali T and Sharma A 2021 Flexible substrate based printed wearable antennas for wireless body area networks medical applications (review) *Radioelectron. Commun. Syst.* **64** 337–50
[18] Yi C *et al* 2020 High-temperature-resistant and colorless polyimide: preparations, properties, and applications *Sol. Energy* **195** 340–54
[19] Takata K and Pham A V 2007 Electrical properties and practical applications of liquid crystal polymer flex *6th Int. IEEE Conf. Polym. Adhes. Microelectron. Photonics, Polytronic 2007, Proc.* pp 67–72

[20] Vyas R, Rida A, Bhattacharya S and Tentzeris M M Liquid crystal polymer (LCP): the ultimate solution for low-cost RF flexible electronics and antennas
[21] Zardetto V, Brown T M, Reale A and Di Carlo A 2011 Substrates for flexible electronics: a practical investigation on the electrical, film flexibility, optical, temperature, and solvent resistance properties *J. Polym. Sci., Part B: Polym. Phys.* **49** 638–48
[22] Formica N, Sundar Ghosh D, Chen T L, Eickhoff C, Bruder I and Pruneri V 2012 Highly stable Ag–Ni based transparent electrodes on PET substrates for flexible organic solar cells *Sol. Energy Mater. Sol. Cells* **107** 63–8
[23] Lin Y, Gritsenko D, Liu Q, Lu X and Xu J 2016 Recent advancements in functionalized paper-based electronics *ACS Appl. Mater. Interfaces* **8** 20501–15
[24] Solhi E, Hasanzadeh M and Babaie P 2020 Electrochemical paper-based analytical devices (ePADs) toward biosensing: recent advances and challenges in bioanalysis *Anal. Methods* **12** 1398–414
[25] Baldo T A, De Lima L F, Mendes L F, De Araujo W R, Paixão T R L C and Coltro W K T 2021 Wearable and biodegradable sensors for clinical and environmental applications *ACS Appl. Electron. Mater.* **3** 68–100
[26] Li W *et al* 2020 Biodegradable materials and green processing for green electronics *Adv. Mater.* **32**
[27] Koul S K and Bharadwaj R 2021 Flexible and textile antennas for body-centric applications *Lect. Notes Electr. Eng.* **787** 99–124
[28] Cornet B, Fang H, Ngo H, Boyer E W and Wang H 2022 An overview of wireless body area networks for mobile health applications *IEEE Netw* **36** 76–82
[29] Li Z, Tian X, Qiu C W and Ho J S 2021 Metasurfaces for bioelectronics and healthcare *Nat. Electron.* **4** 382–91
[30] Wu Q and Zhang R 2020 Towards smart and reconfigurable environment: intelligent reflecting surface aided wireless network *IEEE Commun. Mag.* **58** 106–12
[31] Chandrasekhar V, Andrews J G and Gatherer A 2008 Femtocell networks: a survey *IEEE Commun. Mag.* **46** 59–67
[32] Olatinwo D D, Abu-Mahfouz A and Hancke G 2019 A survey on LPWAN technologies in WBAN for remote health-care monitoring *Sensors* **19** 5268
[33] Park J, Park S, Yang W and Kam D G 2019 Folded aperture coupled patch antenna fabricated on FPC with vertically polarised end-fire radiation for fifth-generation millimetre-wave massive MIMO systems *IET Microw., Antennas Propag* **13** 1660–3
[34] Shanmuga Sundar D, Sivanantharaja A, Sanjeeviraja C and Jeyakumar D 2016 Synthesis and characterization of transparent and flexible polymer clay substrate for OLEDs *Mater. Today Proc.* **3** 2409–12
[35] Thielens A, Deckman I, Aminzadeh R, Arias A C and Rabaey J M 2018 Fabrication and characterization of flexible spray-coated antennas *IEEE Access* **6** 62050–61
[36] Flexible antennas for wearable devices | Electronics360 https://electronics360.globalspec.com/article/17514/flexible-antennas-for-wearable-devices (accessed 24 April 2023)
[37] Sankaralingam S and Gupta B 2010 Determination of dielectric constant of fabric materials and their use as substrates for design and development of antennas for wearable applications *IEEE Trans. Instrum. Meas.* **12** 3122–30
[38] Dhanabalan S S *et al* 2022 Surface engineering of high-temperature PDMS substrate for flexible optoelectronic applications *Chem. Phys. Lett.* **800** 139692

[39] Sundar D S, Raja A S, Sanjeeviraja C and Jeyakumar D 2016 High temperature processable flexible polymer films *Int. J. Nanosci.* **16** 1650038

[40] Shanmuga sundar D, Sivanantha Raja A, Sanjeeviraja C and Jeyakumar D 2016 Highly transparent flexible polydimethylsiloxane films—a promising candidate for optoelectronic devices *Polym. Int.* **65** 535–43

[41] Trajkovikj J, Zurcher J F and Skrivervik A K 2013 PDMS, a robust casing for flexible W-BAN antennas [EurAAP corner] *IEEE Antennas Propag. Mag.* **55** 287–97

[42] Abbasi Q H, Rehman M U, Yang X, Alomainy A, Qaraqe K and Serpedin E 2013 Ultrawideband band-notched flexible antenna for wearable applications *IEEE Antennas Wirel. Propag. Lett.* **12** 1606–9

[43] Khaleel H R, Al-Rizzo H M, Rucker D G and Mohan S 2012 A compact polyimide-based UWB antenna for flexible electronics *IEEE Antennas Wirel. Propag. Lett.* **11** 564–7

[44] Raad H R, Abbosh A I, Al-Rizzo H M and Rucker D G 2013 Flexible and compact AMC based antenna for telemedicine applications *IEEE Trans. Antennas Propag.* **61** 524–31

[45] C. Plastics Supplier of PTFE Rod, PTFE (polytetrafluoroethylene) data sheet (at curbell plastics) (www.curbellplastics.com) (accessed 24 April 2023)

[46] An Introduction to PTFE | AFT Fluorotec (https://fluorotec.com/news/blog/an-introduction-to-ptfe/) (accessed 24 April 2023)

[47] Park M *et al* 2012 Highly stretchable electric circuits from a composite material of silver nanoparticles and elastomeric fibres *Nat. Nanotechnol.* **7** 803–9

[48] Liu H, Zhu S, Wen P, Xiao X, Che W and Guan X 2014 Flexible CPW-fed fishtail-shaped antenna for dual-band applications *IEEE Antennas Wirel. Propag. Lett.* **13** 770–3

[49] Masihi S *et al* 2020 Development of a flexible tunable and compact microstrip antenna via laser assisted patterning of copper film *IEEE Sens. J.* **20** 7579–87

[50] Rahman M A, Hossain M F, Riheen M A and Sekhar P K 2020 Early brain stroke detection using flexible monopole antenna *Prog. Electromagn. Res.* C **99** 99–110

[51] Kangeyan R and Karthikeyan M 2023 Implantable dual band semi-circular slotted patch with DGS antenna for biotelemetry applications *Microwave Opt. Technol. Lett.* **65** 225–30

[52] Ashyap A Y I, Zainal Abidin Z, Dahlan S H, Majid H A and Saleh G 2019 Metamaterial inspired fabric antenna for wearable applications *Int. J. RF Microw. Comput. Eng* **29** e21640

[53] Zaidi A, Awan W A, Hussain N and Baghdad A 2020 A wide and tri-band flexible antennas with independently controllable notch bands for sub-6-GHz communication system *Radioengineering* **29** 44–51

[54] Baran D, Corzo D and Blazquez G T 2020 Flexible electronics: status, challenges and opportunities *Front. Electron* **1** 2

[55] Li S, Da Xu L and Zhao S 2018 5G internet of things: a survey *J. Ind. Inf. Integr* **10** 1–9

[56] Marasco I *et al* 2022 A compact evolved antenna for 5G communications *Sci. Reports* **12** 1–11

[57] Yin A *et al* 2023 A highly sensitive and miniaturized wearable antenna based on MXene films for strain sensing *Mater. Adv.* **4** 917–22

[58] Li E, Li X J, Seet B C and Lin X 2020 Ink-printed flexible wideband dipole array antenna for 5G applications *Phys. Commun.* **43** 101193

[59] Jilani S F, Rahimian A, Alfadhl Y and Alomainy A 2018 Low-profile flexible frequency-reconfigurable millimetre-wave antenna for 5G applications *Flex. Print. Electron.* **3** 035003

[60] Dey S, Arefin M S and Karmakar N C 2021 Design and experimental analysis of a novel compact and flexible super wide band antenna for 5G *IEEE Access* **9** 46698–708

[61] Kirtania S G *et al* 2020 Flexible antennas: a review *Micromachines* **11** 847
[62] Kangeyan R and Karthikeyan M 2023 Miniaturized meander-line dual-band implantable antenna for biotelemetry applications *ETRI J.* **1** 1–8
[63] Kangeyan R and Karthikeyan M 2023 A novel wideband fractal-shaped MIMO antenna for brain and skin implantable biomedical applications *Int. J. Commun. Syst.* **36** e5509
[64] Nguyen D and Seo C 2023 An ultra-miniaturized circular polarized implantable antenna with gain enhancement by using DGS and Holey superstrate for biomedical applications *IEEE Access* **11** 16466–73
[65] Mosavinejad S S, Rezaei P, Khazaei A A and Shirazi J 2023 A triple-band spiral-shaped antenna for high data rate fully passive implantable devices *AEU—Int. J. Electron. Commun.* **159** 154474
[66] Salama S, Zyoud D and Abuelhaija A 2022 Modeling of a compact dual band and flexible elliptical-shape implantable antenna in multi-layer tissue model *Electron* **11** 3406
[67] Osman S A, El-Gendy M S, Elhennawy H M and Abdallah E A F 2022 Reconfigurable flexible inverted-F antenna for wireless capsule endoscopy *AEU—Int. J. Electron. Commun.* **155** 154377
[68] Ahmad S *et al* 2022 A wideband bear-shaped compact size implantable antenna for in-body communications *Appl. Sci.* **12** 2859
[69] Fernandez-Munoz M, Sanchez-Montero R, Lopez-Espi P L, Martinez-Rojas J A and Diez-Jimenez E 2022 Miniaturized high gain flexible spiral antenna tested in human-like tissues *IEEE Trans. Nanotechnol.* **21** 772–7
[70] De Cos Gómez M E, Álvarez H F, Andrés F L H, Valcarce B P, González C G and Olenick J 2019 Zirconia-based ultra-thin compact flexible CPW-fed slot antenna for IoT *Sensors* **19** 3134
[71] Kirtania S G, Younes B A, Hossain A R, Karacolak T and Sekhar P K 2021 CPW-fed flexible ultra-wideband antenna for IoT applications *Micromachines* **12** 453
[72] Ali E M *et al* 2023 A shorted stub loaded UWB flexible antenna for small IoT devices *Sensors* **23** 748
[73] Morales-Centla N *et al* 2022 Dual-band CPW graphene antenna for smart cities and IoT applications *Sensors* **22** 5634
[74] Saha T K, Knaus T N, Khosla A and Sekhar P K 2022 A CPW-fed flexible UWB antenna for IoT applications *Microsyst. Technol.* **28** 5–11
[75] Lavadiya S P, Patel S K, Ahmed K, Taya S A, Das S and Babu K V 2022 Design and fabrication of flexible and frequency reconfigurable antenna loaded with copper, distilled water and seawater metamaterial superstrate for IoT applications *Int. J. RF Microw. Comput. Eng.* **32** e23481

IOP Publishing

Advances in Flexible and Printed Electronics
Materials, fabrication, and applications
Shanmuga Sundar Dhanabalan and Arun Thirumurugan

Chapter 10

Advancement in flexible screen printing electrodes for medical and environmental applications

N Prabu and D Jeyakumar

Flexible printing technology plays a foremost role for the miniaturization of electronic devices and sensors systems with advancement of materials science. The study of flexible printing technologies and its contribution is more attractive and interesting for the young researchers in a productive manner. In the medical field, emergency care patients need live monitoring of the heart and lungs, and monitoring of blood components with the influence of gases is necessary for diagnosis. The assessment of corporeal surroundings are necessary for the benefit of living beings with the impact of various diseases. Environmental assessment also requires live monitoring in a high accuracy manner, so the analysers will be simple, effective and also stable with the atmosphere. In this chapter, the advancement of flexible screen printing techniques and their contribution to medicinal and environmental analysis may impact students and researchers to realize the current scenario of the flexible printing techniques in various applications.

10.1 Screen printing

Screen printing is a process used as a form of ancient art in many countries. Originally, designs were stencilled on a silk or nylon screen and ink was forced through the opening of the screen directly onto the substrate. Screen printing as a process of mass production was started during the nineteenth century. This process was developed and modified to reach all types of fields to print remarkable things. At present, screen printing is a well established and commercially exploited technique for the fabrication of printed electronics and electrodes on various applications in emerging fields.

The screen printing technique is a simple tool to make many useful things in a flexible and easily accessible manner. Interiors, floors and clothes were printed with

doi:10.1088/978-0-7503-5492-9ch10

definite structures and designs to modify the specific areas in a colourful manner. Advancements in screen printing equipment with precision control of specific areas with definite size and also computer controlled automatic/semi-automatic screen printers have a great impact in various advanced applications to fulfil all types of requirements. Especially via electronics and its miniaturization of the circuit boards and components, screen printing has tremendous impact for the creation of new products with precise accuracy. Most countries spend a maximum percentage of their funds on medicinal and defence research for the growth and survival of their nations. In this way screen printing shows extraordinary growth, and its applications influence huge changes in the world.

The influence of screen printing in various fields' development is as follows: from data storage and analysis on a room-sized computer to smart and super computers in pocket size and pen size, and from the medicinal field's internal medicine to targeted drug delivery and area-specific analysers like pacemakers etc. In the defence field miniaturization also happens thanks to screen printing techniques, turning huge tankers into smart devices the size of a spider or mosquito. In nature, many things are still unbelievable, but screen-printed circuit board miniaturization and low-weight devices and the advancement of sensors provide better understanding of and sort out issues in the environment to solve and minimize the natural disasters, but the disposal of these [1] devices and circuit boards is a big question mark.

10.2 Clinical diagnosis of blood electrolytes

Clinical diagnosis is among the most critical and important areas in biomedical research and results in the early diagnosis of ailments, thereby providing effective health care. High volumes of methodologies are available for various parameters in clinical diagnostics. Specifically, analysis of blood electrolytes and blood gases plays a vital role in health care for the emergency period. Sodium, potassium, chloride ions and pH organise the common blood electrolytes, and pCO_2 and pO_2 create the blood gases. In addition to this, the $[HCO_3]^-$ value is calculated from the pCO_2 and pH values of the blood samples [2]. Point-of-care monitoring of blood gases and blood electrolytes will be of interest to researchers. Nowadays, there are lot of pocket-sized products available to externally analyse the above mentioned blood components and blood gases, but to date the research is incomplete to accurately monitor the above components through live internal monitoring. This type of research was highly focused on screen printing techniques and its advancements with the help of nanomaterials.

The advanced level sensors can be utilized by two ways of sensing, which are potentiometric and amperometric concepts that sense components, and their respective changes were monitored by change in the potential or current. Preparing sensing electrodes or ion selective electrodes (ISEs) and their efficiency is highly dependent on the respective reference electrodes used for the analysis [3]. The development of screen-printed reference electrode (SPREs) plays a vital role for the ISE. The potential stability and the analysis condition will be highly motivated for the efficiency of the screen-printed potentiometric sensors. It is also a basic and vital criterion for the analysis of live monitoring of blood components and blood gas in emergency care units.

10.3 Blood electrolytes

Among the blood electrolytes sodium and potassium ions are continuously monitored for patients treated in the coronary care unit (CCU). Originally sodium and potassium ions were estimated using flame photometry from blood serum samples. Then electrochemical methods were found to be conducive and easy for the measurement of blood electrolytes and blood gases. Blood electrolytes and pCO_2 are analyzed using ISEs, and pO_2 by the amperometric method. ISE is an electrochemical analysing tool that converts the activity of an ion of interest in the presence of other ions into respective potential or current in low ranges up to nano (10^{-9}) or pico (10^{-12}) levels. The measured potential and concentration of the ion are related by the Nernst equation [4].

The Nernst equation is

$$E = Eo + \frac{RT}{nF}\ln[M^{n+}] \tag{10.1}$$

$$E = Eo + \frac{0.0592}{n}\log[M^{n+}] \tag{10.2}$$

where E is the potential developed in the presence of active ion,
Eo is the standard electrode potential,
R is the gas constant,
T is temperature,
F is the Faraday constant,
n is the number of electrons, and
$[M^{n+}]$ concentration of the active ion.

It can be seen from the equations (equations (10.1) and (10.2)) that the measured potential is related to the log concentration of the analyte. The ISE comprises an electrode coupled to an ion selective material directly related to the SPRE. ISEs can be classified into four different types depending on the nature of the ion selective material.

They are

(i) glass membrane electrode,
(ii) solid membrane electrode,
(iii) polymer membrane based electrode, and
(iv) carbon paste electrode.

In the case of the glass membrane electrode and polymer membrane based ISEs [5], an electrochemical sensing element is used for the reference electrode, which is usually a Ag/AgCl electrode with a chloride ion concentrator. The glass membrane and the polymer membrane comprised ISEs [6], and the ion selective element was based on the analyte. The pH electrode is a typical glass membrane electrode, which was working as a change in the potential of the electrode based on the concentration of H^+ ion in the solution. The stability and the effectiveness of the ISE is purely based on the reference electrode. In the glass membrane, pH electrode metal (Pt) wire may act as a reference (pseudo) electrode. Other than the reference electrode a lot of factors are affecting the stability and the reliability of this pH glass membrane electrode. The metallic reference (pseudo) electrode [7] required space and the environment for the maintenance of the pH electrode. These types of electrodes are risky to maintain and transport; all of the above factors were solved or minimized in the screen-printed ISEs (SP-ISEs) [8]. The advancement in the SP-ISEs may replace the traditional glass membrane electrode for pH sensing with remarkable deeds.

Main factors affecting the ISEs' stability and efficiency are

- temperature of the analysis,
- atmospheric pressure in the measuring chamber,
- concentration of the analyte or solution,
- nature of the solution or analyte,
- washing cycle or solution after/before analysis, and
- periodic usage of electrodes

10.4 Reference electrode

An ISE's potential is measured against a standard reference electrode as a function of concentration of the analyte using a potentiometer or ion meter. A reference electrode is an ideally non-polarizable electrode that has a stable and constant potential in a specific condition. A saturated calomel electrode is one of the examples for commercially and also universally accepted reference electrodes, which have the saturated chloride ion concentration in the electrode along with Hg_2Cl_2/Hg (calomel) and Ag/AgCl electrode with 0.1 M NaCl. Hence, in a practical system the reference electrode should be ideal i.e. free from potential drift as a function of time. A conventional potential electrode is quite large, and miniaturization is possible only to a limited extent. Similarly, ISE is also significantly large in size and miniaturization is rather difficult [9]. Hence, different approaches have been made that include field effect transistors (FETs) and screen-printed electrodes (SPEs). The research is focused on the above mentioned factors, and the basic need of ISEs for the specific analyte have to pave the way for the design and fabrication of SP-ISEs.

10.5 Screen-printed electrodes

The screen printing method, originally attached to artistic work, is a mature technology capable of manufacturing electrochemical transducers. Electrochemical transducers are electrodes that can be a single- or two-electrodes system or an array of electrodes. An interdigitated array sensor can also be prepared by screen printing [10]. It is basically a stencil cut technology. Screens are designed with the required pattern on screen mesh (may be polymer or silk or metal) with the help of the photo-polymerization process (figure 10.1).

Figure 10.1. Model of screen with embedding and masking of the inactive portion.

The customized pattern or design was embedded with the help computerised drawings, and the remaining exposed part of the screen mesh was polymerized: in this area the ink will not flow through. The screen printing was done in the specified substrates for the requirement of the applications and separated into individual parts for further modification. The effectiveness of the SPEs was based on the following factors [11].

Effectiveness of the SPEs includes

(i) design and fabrication of the screen,
(ii) selectivity of the substrate,
(iii) choice of the ink and the curing process,
(iv) thickness and the stability of the printing, and
(v) modification of the SPE.

(i) Design and fabrication of the screen: The major contribution of the efficiency of the SPE is the design of the screen, which confirms the stability and the reliability of the active participation of the sensing. The masking and printing were the secondary part of the design; the fabrication of the electrode and further modification based on the analyte (ionophore) could play the crucial role of the efficiency of the SPE. The fabrication of the SPE also includes the instrumentation tuning of the printing pressure, number of layers, and type of printing, which have improved the efficacy of the electrodes. The screen printing machine has different modes of operation and printing like ALT-Print, Flood-Print, Print-Print, Print-Flood. The above mentioned methods were given different thicknesses of screen printing, and the printings were pore free and easily durable.

(ii) Selectivity of the substrate: The substrate selectivity has a vital part in the SPE due to the impact and the usefulness of the electrode in various applications. The substrates are alumina, polyurethane sheets, polyvinyl chloride (PVC), and many other composite polymer materials. The substrate is not affected in the analysis environment, and also it can be made of cheap and easily available materials.

(iii) Choice of the ink and curing processes: The ink selection for the screen printing is also customised by the usage of the electrode in the applications. The basic selection of the ink was a strong adhesive; it will not be destroyed by the substrate, it has an easy curing process, and also the ink and the supporting materials will not be affected by the analysis environment.
(iv) Thickness and stability of the printing: The printing thickness or multilayer printing may affect the response time of the electrodes; the thickness can be optimised based on the substrate. The pinhole-free printing and the specific caution needs for multilayer printing helps the effectiveness of the thickness of the electrodes. The stability of the electrodes is highly focused on the working environment, so the choice of substrate, ink, and thickness is the strength of the electrodes.
(v) Modification of the SPE: The SPE was further modified by the ionophore or active materials based on the applications; the modification also improves and supports the stability of the ISEs. The effectiveness of the SPE in various applications is highly focused on the electrode modification with the help of active materials or sensing elements. Modification of the SPE shows the reliability and the actual value of the ISEs.

Layer by layer printing can be affected to have the desired pattern on a substrate and may be flexible or ceramic. Figure 10.2 shows a picture of the semiautomatic screen printer, AMI Presco 485 model. Electrically conducting inks comprising mainly silver, gold, and carbon are commonly used for electrode printing. Generally

Figure 10.2. Overview of the semiautomatic screen printing machine (AMI–PRESCO 485). Credit to AMI/Presco, Affiliated Manufacturers Incorporation, NJ, USA.

line printing of conducting materials is printed as the first layer. A dielectric or non-conducting/insulating layer of our choice is then printed to mask the conducting layer and expose the required area to constitute the SPE. Two electrode systems are mainly used for biosensors, with one working (active) electrode and another counter/reference electrode. The advancement in the printing techniques and the inks has improved dual activity in a single electrode and will be in the three-electrode system by the way of the working electrode; the reference electrode, the counter electrodes, and the development of the nanomaterial-based inks all have a solid-state SPE, which does not have any type of solution to maintain the effectiveness of the SPEs. This type of electrode has two active portions: one is measured by a change in the potential, and the other one shows a change in the current as potentiometric and amperometric ISEs.

SPE-based biosensors for monitoring glucose using screen printing technology in whole blood show tremendous commercial success. Three-electrode assemblies consist of a working electrode, counter electrode, and a reference electrode; these are all very popular for environmental monitoring of heavy metal ions (HMIs) in waste waters. Both of these transducers work on the principle of amperometry because of their high sensitivity and high selectivity for multiple sensing, and the transducers are modified with the mediators/enzymes or thin films of metals (gold, bismuth, mercury etc) for stripping voltammetry analysis depending on the nature of the analyte and its sensitive applications.

10.6 Screen-printed ISEs

During the fabrication of ISEs, the single electrode is used as a transducer that employs a Ag/AgCl ink for printing the contact portion or base of the active portion. It is mandatory to have the sensing element coupled to the electrochemical transducer for ion sensing. In the case of SPEs the ion sensing layer will be a polymer matrix comprising a polymer, a plasticizer, an ion (cation/anion) excluder to improve the sensitivity and a neutral carrier to impart ion selective faces [12] Synthetic as well as natural neutral carriers have been reported for different ions and are commercially available as ionophores.

In the case of alkali metal ions, more than one synthetic carrier is available with different selectivities and sensitivities for different ions. As an example, valinomycin is a natural carrier for the potassium ion, whereas bis benzo-15crown6 ethers are the synthetic analogue for the potassium ion. Similarly, the choice of polymers for giving support to the ISE also varies based on the natural and synthetic carriers. PVC, silicone polymer, polyvinyl acetate etc can be used as the polymers for creating the polymer matrix. In the case of plasticizers we also have different choices to use in different conditions and the temperature. It may be noted that the SPEs have a planar geometry, and hence the polymer has to be drop casted onto the electrode surface to mask the active material without masking its activity. There will be plenty of ongoing research on miniaturization and improving the efficiency of the screen-printed sensors and devices, but it has some faintness to reach the goal in that steadiness of the reference electrode pave the major contribution.

10.7 Screen-printed planar reference electrodes

As mentioned already, SPEs have planar configuration on the electrode surface. Hence, construction of the reference electrode will essentially produce a planar reference electrode. It is evident that a stable Ag/AgCl reference electrode requires constant activity of chloride ion in the system [13]. It will be rather difficult to achieve this, especially considering the dimensions of the SPE and the design of the printing electrodes [14]. Various attempts in this direction of research are ongoing for the effectiveness of the SPE-based sensors in various fields. Outline of the SPEs show that there are merits and demerits to reaching the expansions in flexible SPEs.

Merits and demerits of SPEs:

Merits of the SPE:

- Flexible printing is the main advantage of SPEs for customised design and modifications.
- The thickness of the printing may be tuneable to nanometer range.
- The stability and the efficiency of the SPE are incredible.
- It is easy to maintain the SPE's uniqueness for long time applications.

Demerits of the SPE:

- An advanced-level screen printer is required to prepare SPEs.
- High quality consumables like ink, substrate and other modification chemicals are high in cost.
- The reusability of the SPEs is miniscule; at maximum the SPEs are use and through type.

10.7.1 Applications/uses of SPEs

SPEs have applications in various aspects of electroanalytical methods including specifically

- biosensors
- ISEs
- heavy metal analysis
- antibody–antigen sensing devices.

10.7.2 Biosensors

Biosensors also sense in the same way as the usual chemical sensors like SPEs or FETs, only changing the active material or sensing element. In biosensors, the active part of the sensor is a biomaterial or a biologically active component, which will be active only in the specific environment or controlled atmosphere. Most biosensors are specialized sensors specifically for analysis in the laboratory.

10.7.3 Ion selective electrode

The name itself mentioning the ion-specific activity, it may be a single electrode or an array of electrodes; each portion is specific for particular ions [15]. The array ISEs can be used to evaluate the multivalent or multielement sensing activities. Mostly it is highly useful in the medical field [16] and also in the environmental era. ISEs are in two categories of sensor, potentiometric and amperometric; mostly it will be potentiometric sensors that change in the concentration [17]; the concentration of the ions changes in the potential of the ISE based on the change in the potential the printed circuit (PC)-controlled devices convert into direct values. The ISEs are a disposable type [18] of electrode, which may sense the ions present in biofluids, pollutants, and some grave places [19].

10.7.4 Heavy metal analysis

The analysis of HMIs in soil and water is evaluated by a metal–metal oxide probe electrode for periodic monitoring in most of the industries in the waste water disposal sector. Nowadays live monitoring is required for this type of analysis of heavy metals like lead, tin, aluminium, mercury, tungsten, sulphur and other free radicals, so every time calibration of the probe electrode is not possible on live monitoring, but in the case of the SP-ISEs for heavy metals/metal ion sensing, the calibration or activation of the electrode will be easy, and the surface-activated type of electrode favours the efficiency of the SPE [20, 21].

10.7.4.1 Biofluid analysis

Biofluids, especially blood components and various minerals in blood plasma, are critically monitored during the emergency period in intensive care units (ICUs) and CCUs. Impulsive changes in the blood components and other minerals present in the biofluids cause extraordinary changes in our body. Every collection of biofluids and analysis of components was difficult, so live monitoring of some of the components of biofluids, like electrolyte concentration, especially of sodium (Na), potassium (K), and chloride (Cl) ions in the blood plasma, may increase the pH of the blood and also increase the thickness of the blood, which may cause serious issues or, ultimately, death [15]. In emergency care patients, inhalation capability is very important. The heat functioning ability was evaluated by blood purification efficiency by means of oxygen and carbon dioxide percentage in the blood is also monitored by the analysis of blood gas periodically or continuously. The live monitoring of blood gas and electrolyte concentration is necessary for CCU patients; the above set of live monitoring is possible using multiarray SPEs in the name of point of care blood gas analysis PoCBGA [22]. Nowadays, a lot of health care industries are focused on this type of research.

The design and the selection of the active components and also the solid-state planar reference electrode (SSPRE) are the key criteria to achieve efficient live monitoring of the PoCBGA. Two types of analysis were required: potentiometric and amperometric analysis. In both of the analyses a constant stable potential SSPRE is required [9]; one it success all other portion is simple. But to date there is no SSPRE that is commercialized with constant potential in these specific conditions; the researchers are focused on the SSPRE to succeed in the live monitoring of the CBGA inclusive of the electrolyte test.

10.7.4.2 Care blood gas analysis

It is analysis of the blood components, especially the electrolyte portion and the partial pressure of the O_2 and CO_2 present in the blood volume, which was assessed for the functioning of the lungs, kidneys and heart. The basic components of the blood are red blood cells, white blood cells, platelets, and plasma. Other than these, electrolyte components also important for the proper functioning of various organs. Electrolytes are ions, minerals and salts such as ions of sodium (Na), potassium (K), calcium (Ca), magnesium (Mg), chloride and bicarbonate, which are found in the blood. The main role of the electrolytes present in the body is to conduct electrical impulses in the body. Body dehydration also causes electrolyte imbalance; dehydration is the basic symptom of electrolyte imbalance which causes alarm, in which our body has some imbalance which may be a malfunction of some of the primary internal organs like the liver, kidneys and heart. The concentration [23] of such electrolytes may vary in different timings due to functioning of the internal organs, of which the above are evaluated continuously in emergency care [24].

Generally the above components were analysed by collecting blood samples and assessed externally in the laboratory; the gas portion was monitored as separate equipment. The electrolyte-based blood gas pressure cannot be monitored, as it is only needed for the CCU patients. The advancement of this SPE-based ISE and ion selective field effect transistor (ISFET) electrodes ensured that the total analysis of all the CBGA was possible without collecting blood samples externally, and all the analyses and the results were received in one single column. A single chip-like array of SPE-modified ISE/ISFET contributes live monitoring of the above set parameters in different time intervals.

The total analysis of CBGA is on the basis of electrochemical analysis in which the measurement of change in the concentration of various ions and partial pressure of the gas portion reflects changes in the potential/current in the electrode, respectively [25]. The changes in the potential or current in the SPE for the specific changes in the concentration were on the order of 10^{-3}– 10^{-4} V A^{-1}; the tiny changes in the concentration of electrolyte components reflect massive changes in the body. The measured value can be multiplied properly using standard electrical components, and the real values were displayed with the help of computer-based display devices. The standard concentration values of various ions in the electrolytes for adults [21, 26–28] in the blood are as follows:

- **Sodium:** 136–144 mM
- **Potassium:** 3.7–5.1 mM
- **Chloride:** 97–105 mM
- **Bicarbonate:** 22–30 mM
- **Calcium:** 8.5–10.2 mM
- **Magnesium:** 1.7–2.2 mM
- **Phosphate:** 2.5–4.8 mM

The role of blood electrolyte components to our body health is as follows:

Sodium supports nerve and muscle functioning and also controls the level of body fluids [29].

Potassium supports functioning of the heart, nerves and muscle and also helps the metabolism [26].

Calcium maintains the bones and muscle strength, recovery of cells and nerves and controls the circulatory system [30].

Chloride supports and controls the blood pressure and also body fluids, especially electrolytes [23, 31].
Magnesium maintains the total skeletal system and the growth of bones and teeth [28].
Phosphate improves bone strength, healthy teeth and functioning of nerve and muscle.
Bicarbonate balances a fit steadiness of acids and basic compounds in blood. It helps move carbon dioxide through the bloodstream.

The major components we have to analyse from the blood electrolytes are sodium, potassium and chloride ions, which may state the overall functioning of the major internal organs like the kidneys, liver and heart. Apart from this the inhalations were also periodically monitored during the emergency time. The above five parameters were evaluated in single SPE with SSPRE supported by pc controlled display is the proper care blood gas analyser. The efficiency and the durability of the above mentioned array of electrodes for CBGA is based on the SSPRE and the design of the SPE. The detailed design of the SPE and the SSPRE stability of the array of ISEs for analysing care blood gas is explained as follows [22].

For the schematic design of the array of SP-ISEs with a constant potential reference electrode in the blood electrolyte condition, the detailed design and the fabrication of the SPE electrode is crucial because of its optimisation of trial and error basis. The substrate is a polyurethane sheet 1/0.5 mm thick; the first layer is a contact/conducting coating made by silver ink printing. The second layer coating is masked and creates the active sensing area of the electrode, as shown in figure 10.3; the first layer of dielectric coating or masking layer is slightly larger than the active area. The extra space in the first masking layer is a protective layer to avoid the seepage of the modified ion selective polymer coating. The dielectric layer coating was repeated thrice with increased pressure to minimize the coating thickness; after triple layer coating the active portion is filled by ion sensing polymer coating. The centre part of the electrode active area will be slightly larger than the SSPRE; the perfectly optimised composition of the polymer coating (depends on the condition and the analyte concentration, which may vary) is required for the constant potential during the analysis. The total setup acts as array of electrodes for Na and K ISEs. The five electrode array was used for the CBGA as mentioned in figure 10.4; the five electrode array is also available with carbon-based electrodes for analysing Na, K, PCO_2, and PO_2 with SSPRE. The array of five electrodes can be connected in a flow channel; the analysis was based on the socking time and response time of the electrode, which continuously monitors the electrolytes and the gases in the blood in the emergency care unit, especially in heart, liver and kidney functioning affected patients.

The steadiness of the pH of the blood in our body is very important for the healthy life of human beings. The change in the pH of the blood is not a usual change, but once it changes even slightly, it creates huge changes in our body's condition in a short time span. The pH analysis in the blood is very simple because its stability range is very small, in the range of 7.35–7.45 [32]. Commercially paper based strips are available to analyse this particular change, but all types of analysis

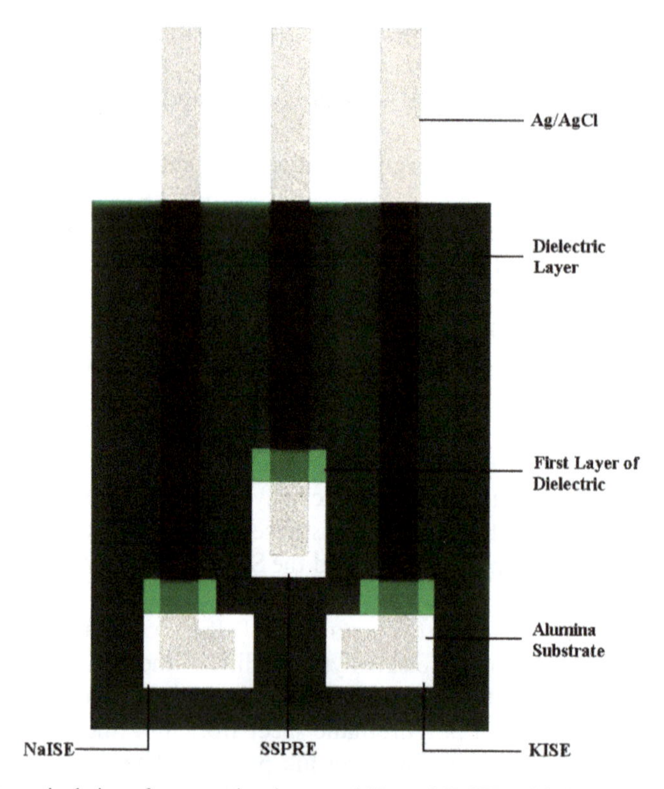

Figure 10.3. Schematic design of screen-printed array of Na and K ISEs with SSPRE (size: 20 × 40 mm). Reprinted from [22] with permission from IEEE, Copyright (2018).

Figure 10.4. Schematic design of screen-printed array of four analytes with SSPRE (size: 25 × 40 mm).

are external after collecting the blood samples. For patients in the ICU and CCU, monitoring the blood pH is always necessary, so this pH analysis [33] is also included in the CBGA. The multiarray of SPE with SSPRE is the only way to fulfil all the requirements of CBGA: in this screen-printed flexible printing plays a major role. The further advancement of the screen printing electrodes and the carbon-based ink coated paper based electrodes for various applications in the field of food industries also plays a major role. The monitoring of hazardous chemicals in food items and analysis of the percentage of food adulterants can be assessed using carbon-based SPE by electrochemical sensing. Figure 10.5 shows the simple way of carbon-based

Figure 10.5. Screen-printed carbon-based electrode with SSPRE (size: 15 × 40 mm).

SPEs to analyse the analyte for measuring changes in the potential or current as electrochemical sensors.

Health care issues commonly come from our environmental impact on pollution of the soil, water and air. It is necessary to evaluate the above set of parameters and pollutant levels in the atmosphere. There are a lot of sensing devices and probes available to monitor the pollutants in soil and water, but in the case of air pollutants monitoring in the atmosphere is lacking sensing devices because the active surface is severely affected by atmospheric changes. Self-cleaning surfaces based sensing devices are required to asses air pollutant monitoring, though this is not common to all air pollutants. Flexible screen printing based environment/atmosphere monitoring is one of the best ways to evaluate atmospheric pollutants by assessing through moisture [34, 35]. The basic assessment of the environment is moisture percentage or relative humidity monitoring [36], which gives basic information about the present environmental condition.

The purity of water and the soil is also essential for the proper functioning of living beings and their environment-benign life. In these the presence of HMIs like arsenic [37], antimony, lead, mercury, chromium, cadmium, tin etc [38] may cause severe effects in all living beings and can entirely change the ecosystem. A greater number of commercially available heavy metal analysis systems [39] are available on the market with portable sizes, and most of the available analysers are based on the SPE-based flexible sensing electrodes. The commercially available devices for analysing HMIs and other pollutants present in water and soil analysis were used in SPEs as active electrodes with the help of electroanalytical techniques. The researchers used noble metal based SPEs [39, 40] and also carbon-based electrodes [41] for evaluating the HMIs and other pollutants.

10.7.4.3 Enzyme inhibition-based biosensors SPEs for environmental monitoring
Enzyme based sensors are effective for particular conditions, especially temperature dependent environments; in this type of sensor, the active portions are enzymes or some other bioactive materials [42]. Biosensors that are based on the source of enzyme inhibition can be realistic for an extensive range of substantial analytes such as organochlorine organophosphorous, metal halides and derivatives of insecticides, pesticides, HMIs and alkaloids in a wide range of samples. Most of the biosensors are also used as flexible screen printing–based electrodes for the base electrode for monitoring environmental pollutants [13]. The advancement of flexible screen printing technology implies the futuristic view of chip-based sensors for evaluation of environmental impacts.

Acknowledgments

We thank the Department of Science and Technology (DST), India for the financial support in carrying out the screen-printed blood gas sensor–based work. One of the authors, N P, expresses their thanks to DST for the financial support. We thank the Director of the CSIR-Central Electro Chemical Research Institute for the support and encouragement and permission to publish the work. The authors thank Dr R

Kalidoss, Retired Senior Technical officer, CSIR-Central Electro Chemical Research Institute for sharing his ideas on designing part of the screen printing electrodes.

Acronyms

SPE	Screen Printed Electrodes
ISE	Ion Selective electrodes
mM	millimolar
SP-ISE	Screen printed Ion selective electrode
ISFET	Ion Selective Field Effect Transistor
SSPRE	Solid State Planar Reference Electrode
ICU	Intensive Care Unit
CCU	Coronary Care Unit
CBGA	Care Blood Gas Analysis
NaISE	Sodium Ion Selective electrode
KISE	Potassium Ion Selective electrode
PCO_2	Partial pressure of the Carbon dioxide
PO_2	Partial pressure of the oxygen
RH	Relative humidity
HMI	heavy metal ions
PC	Printed Circuit

References

[1] Xia H, Peng Z, Cuncai L, Zhao Y, Hao J and Huang Z 2016 Self-supported porous cobalt oxide nanowires with enhanced electrocatalytic performance toward oxygen evolution reaction *J. Chem. Sci.* **128** 1879–85

[2] Zhou Z-B, Liu W-J and Liu C-C 2000 Studies on the biomedical sensor techniques for real-time and dynamic monitoring of respiratory gases, CO_2 and O_2 *Sensors Actuators* B **65** 35–8

[3] Morf W E 1981 *The Principles of Ion-Selective Electrodes and of Membrane Transport* (Amsterdam: Elsevier)

[4] Bard A J, Inzelt G and Scholz F 2012 *Electrochemical Dictionary* ed A J Bard, G Inzelt and F Scholz (Heidelberg Dordrecht London New York: Springer Verlag) 2nd edition

[5] Sundar D S, Raja A S, Sanjeeviraja C and Jeyakumar D 2017 High temperature processable flexible polymer films *Int. J. Nanosci.* **16** 1650038

[6] Dhanabalan S S, R S, Madurakavi K, Thirumurugan A, M R, Avaninathan S R and Carrasco M F 2022 Flexible compact system for wearable health monitoring applications *Comput. Electr. Eng.* **102** 108130

[7] Moschou D, Trantidou T, Regoutz A, Carta D, Morgan H and Prodromakis T 2015 Surface and electrical characterization of Ag/AgCl pseudo-reference electrodes manufactured with commercially available PCB technologies *Sensors* **15** 18102–13

[8] Oesch U, Ammann D and Simon W 1986 Ion-selective membrane electrodes for clinical use *Clin. Chem.* **32** 1448–59

[9] Mamińska R, Dybko A and Wróblewski W 2006 All-solid-state miniaturised planar reference electrodes based on ionic liquids *Sensors Actuators* B **115** 552–7

[10] Li M, Li Y-T, Li D-W and Long Y-T 2012 Recent developments and applications of screen-printed electrodes in environmental assays—a review *Anal. Chim. Acta* **734** 31–44

[11] Economou A 2018 Screen-printed electrodes modified with 'green' metals for electrochemical stripping analysis of toxic elements *Sensors (Switzerland)* **18** 24

[12] Guenat O T, Generelli S, de Rooij N F, Koudelka-Hep M, Berthiaume F and Yarmush M L 2006 Development of an array of ion-selective microelectrodes aimed for the monitoring of extracellular ionic activities *Anal. Chem.* **78** 7453–60

[13] Turner A, Karube I and Wilson G S 1987 *Biosensors: Fundamentals and Applications* (New York: Oxford University Press)

[14] Narmada T and Lakshmaiah M 2017 Design and development of embedded system based heart beat sensor using raspberry Pi 2 *Int. J. Adv. Res. Electr. Electron. Instrum. Eng.* **6** 1931–7

[15] Diamond S W D 1997 Solid-state sodium-selective sensors based on screen-printed Ag/AgCl reference electrodes *Electroanalysis* **9** 1318–24

[16] Lee J S, Lee S D, Cui G, Lee H J, Shin J H, Cha G S and Nam H 1999 Hydrophilic polyurethane coated silver a silver chloride electrode for the determination of chloride in blood *Electroanalysis* **11** 260–7

[17] Simonis A, Lüth H, Wang J and Schöning M J 2004 New concepts of miniaturised reference electrodes in silicon technology for potentiometric sensor systems *Sensors Actuators* B **103** 429–35

[18] Mroz A 1998 Disposable reference electrode *Analyst* **123** 1373–6

[19] Idegami K, Chikae M, Nagatani N, Tamiya E and Takamura Y 2010 Fabrication and characterization of planar screen-printed Ag/AgCl reference electrode for disposable sensor strip *Jpn. J. Appl. Phys.* **49** 097003

[20] Oh B K, Kim C Y, Lee H J, Rho K L, Cha G S and Nam H 1996 One-component room temperature vulcanizing-type silicone rubber-based calcium-selective electrodes *Anal. Chem.* **68** 503–8

[21] Poplawski M E, B. Brown R, Lae Rho K, Yong Yun S, Jung Lee H, Sig Cha G and Paeng K 1997 One-component room temperature vulcanizing-type silicone rubber-based sodium-selective membrane electrodes *Anal. Chim. Acta* **355** 249–57

[22] Kalidoss R, Sivanantharaja A, Jeyakumar D and Prabu N 2018 Solid-state planar reference electrode with ion selective electrodes for clinical diagnoses *IEEE Sens. J.* **18** 8510–16

[23] Paciorek R, Bieganowski P and Maj-Żurawska M 2005 Miniature planar chloride electrodes *Sensors Actuators* B **108** 840–4

[24] Montiel F, Aimar P and Montoriol P 1995 Continuous monitoring of sodium concentration in blood during haemodialysis by a selective membrane and conductivity sensor *Sensors Actuators* B **27** 465–7

[25] Neelamegam P, Murugananthan K, Raghunathan R and Jamaludeen A 2009 ATmega8535 microcontroller based blood sodium analyzer using ise direct potentiometry *Instrum Sci. Technol.* **38** 63–71

[26] Pirovano P, Dorrian M, Shinde A, Donohoe A, Brady A J, Moyna N M, Wallace G, Diamond D and McCaul M 2020 A wearable sensor for the detection of sodium and potassium in human sweat during exercise *Talanta* **219** 121145

[27] Cranny A, Harris N and White N 2014 Screen printed potentiometric chloride sensors *Procedia Eng.* **87** 220–3

[28] Jarujamrus P, Meelapsom R, Naksen P, Ditcharoen N, Anutrasakda W, Siripinyanond A, Amatatongchai M and Supasorn S 2019 Screen-printed microfluidic paper-based analytical device (μPAD) as a barcode sensor for magnesium detection using rubber latex waste as a novel hydrophobic reagent *Anal. Chim. Acta* **1082** 66–77
[29] Liao W Y and Chou T C 2008 Development of a reference electrode chip for application in ion sensor arrays *Z. Naturforsch.* B **63** 1327–34
[30] Wang Y, Xu H, Yang X, Luo Z, Zhang J and Li G 2012 All-solid-state blood calcium sensors based on screen-printed poly(3,4-ethylenedioxythiophene) as the solid contact *Sensors Actuators* B **173** 630–5
[31] Tymecki Ł, Zwierkowska E and Koncki R 2004 Screen-printed reference electrodes for potentiometric measurements *Anal. Chim. Acta* **526** 3–11
[32] Kinlen P J, Heider J E and Hubbard D E 1994 A solid-state pH sensor based on a Nafion-coated iridium oxide indicator electrode and a polymer-based silver chloride reference electrode *Sensors Actuators* B **22** 13–25
[33] Mahinnezhad S, Emami H, Ketabi M, Shboul A A, Belkhamssa N, Shih A and Izquierdo R 2021 Fully printed pH sensor based in carbon black/polyaniline nanocomposite *2021 IEEE Sensors (Sydney, Australia, October 2021)* (IEEE) 1–4
[34] Barmpakos D and Kaltsas G 2021 A review on humidity, temperature and strain printed sensors—current trends and future perspectives *Sensors (Switzerland)* **21** 1–24
[35] Turkani V S, Maddipatla D, Narakathu B B, Saeed T S, Obare S O, Bazuin B J and Atashbar M Z 2019 A highly sensitive printed humidity sensor based on a functionalized MWCNT/HEC composite for flexible electronics application *Nanoscale Adv.* **1** 2311–22
[36] Seiyama T, Yamazoe N and Arai H 1983 Ceramic humidity sensors *Sensors Actuators* **4** 85–96
[37] Sanllorente-Méndez S, Domínguez-Renedo O and Arcos-Martínez M J 2010 Immobilization of acetylcholinesterase on screen-printed electrodes. Application to the determination of Arsenic(III) *Sensors* **10** 2119–28
[38] Barton J, García M B G, Santos D H, Fanjul-Bolado P, Ribotti A, McCaul M, Diamond D and Magni P 2016 Screen-printed electrodes for environmental monitoring of heavy metal ions: a review *Microchim. Acta* **183** 503–17
[39] Renedo O D and Julia Arcos Martínez M 2007 A novel method for the anodic stripping voltammetry determination of Sb(III) using silver nanoparticle-modified screen-printed electrodes *Electrochem. Commun.* **9** 820–6
[40] Song Y-S, Muthuraman G, Chen Y-Z, Lin C-C and Zen J-M 2006 Screen printed carbon electrode modified with poly(l-lactide) stabilized gold nanoparticles for sensitive As(III) detection *Electroanalysis* **18** 1763–70
[41] Honeychurch K C, Hawkins D M, Hart J P and Cowell D C 2002 Voltammetric behaviour and trace determination of copper at a mercury-free screen-printed carbon electrode *Talanta* **57** 565–74
[42] Amine A, Mohammadi H, Bourais I and Palleschi G 2006 Enzyme inhibition-based biosensors for food safety and environmental monitoring *Biosens. Bioelectron.* **21** 1405–23

IOP Publishing

Advances in Flexible and Printed Electronics
Materials, fabrication, and applications
Shanmuga Sundar Dhanabalan and Arun Thirumurugan

Chapter 11

The flexible and printed energy storage devices for foldable portable electronic devices applications

P Justin, H Seshagiri Rao, P Nagaraja, G Ramesh and G Ranga Rao

The progress in technology resulted in the development of flexible and foldable portable electronic devices (PEDs), and its performance is more sensitive to energy consumption. Rechargeable batteries and supercapacitors are the primary energy source of foldable PEDs. Flexibility in design, ensuring high safety, electrochemical properties, and efficiency of the energy storage device with high energy and power density are essential for flexible energy storage devices. However, developing fully flexible energy storage devices compared to conventional rigid energy devices is still a challenge. The chapter presents the development of flexible and printed energy storage devices for various smart foldable portable electronic gadgets. The mechanism of electrochemical energy storage, materials for energy storage devices, and current state of the art in making high-quality flexible and printed energy storage devices are discussed in detail. The challenges in the practical implementation of flexible and printed energy storage devices are also discussed, along with potent solutions.

11.1 Introduction

The portability and flexibility of electronic devices are highly desired by the current trend in humankind toward advancing communication, information technology, and consumer electronic applications as promising information-exchange platforms for real-time applications. Utilization of portable and flexible components in daily life became a common practice in managing and monitoring smart home devices, healthcare apparatus, safety systems, wearable displays, e-textiles, etc. Technological evolution attracts flexible and printable electronics due to the compactness and compatibility. Commercializing flexible and printable devices depends on the safety, flexibility, and efficiency of energy storage devices implanted. The increasing demand

doi:10.1088/978-0-7503-5492-9ch11

for advanced, flexible, and portable electronic devices drives the pursuit of more powerful electrochemical energy storage systems, meeting modern needs and promoting sustainable solutions. Researchers aim to enhance capabilities through innovation and eco-friendly practices. The electrochemical energy storage systems are the paramount energy source of flexible, portable electronic devices and hold the key to guaranteeing the desired performance stability. In the last three decades, advancements in electrochemical energy storage systems have driven the emergence of versatile flexible and portable electronic devices, meeting daily needs efficiently. Key to their success are Li-ion batteries and supercapacitors, offering design flexibility, safety features, and high energy density. These devices have become essential components in wearable technology, providing user comfort and longer battery life. Ongoing research promises further innovations, revolutionizing our technology interaction and enhancing daily experiences [1].

11.2 Types of electrochemical energy systems

Electrochemical energy storage devices have become vital technologies for powering modern portable electronic devices. The rising demands for higher capacity, smaller size, and longer operating times pose challenges to current systems. Researchers are actively seeking innovative solutions to address these needs and shape a sustainable future. The quest for superior performance and efficiency in energy storage reflects the importance of this field in technological advancements. By investing in research and embracing creativity, we aim to develop integrated and high-performing electrochemical energy storage systems to meet the dynamic needs of society, revolutionizing our technological landscape. Hence, the continuous pursuit of creating innovative electrochemical energy storage devices is crucial for advancing future flexible, portable electronic devices. For years, electrochemical energy storage systems, notably Li-ion batteries and supercapacitors, have served as the primary energy sources for these portable gadgets.

The two energy-storage systems are efficient and reliable in their fabrication, operation, and storage. Li-ion batteries store energy through a diffusion-controlled redox process throughout the bulk, providing a high energy density. In contrast, supercapacitors store energy through surface-controlled adsorption/desorption of ions or faradaic reactions of electrolyte ions on or near the surface, providing a high-power density. The electrochemical performance of Li-ion batteries is characterized by high energy density, low power density, and longer charge/discharge times (minutes to hours). At the same time, supercapacitors are characterized by high power density, low energy density, and rapid charge/discharge times (seconds to minutes) [2]. Structurally, these energy storage devices constitute electrodes, electrolytes, separators, and current collectors. The chemistry, flexibility, and stability of each component is crucial to the device's overall performance.

11.2.1 History of supercapacitors

The history of energy storage started in a 'Leyden jar' invented by the German von Kleist (1745) and Dutch scientist van Musschenbroek (1746) independently in which the static electricity is stored in a glass jar on the surface of two metal plates dipped in an electrolyte. This groundbreaking experiment paved the way for the revelation of storing static electricity at the interface of a solid electrode and a liquid electrolyte. However, the nature of static electricity was unclear during the initial stages of discovery. In 1853, another German, van Helmholtz, provided the first insights into capacitors' energy storage mechanism through his colloidal solutions studies. The Helmholtz theory is the simplest model explaining the charge distribution at the electrode and electrolyte interface. According to this model, the charge is stored by the accumulation of electrolyte ions near the surface of the electrode to counterbalance the charge on the metal plate. Later on, the work of pioneer electrochemists, namely Gouy, Chapman, and Stern, developed the modern views of charge storage in the capacitor. Gouy-Chapman model assumes the diffusion of electrolyte ions rather than rigid layer through a distance, δ, from the surface. The concentration of ions in the diffuse layer follows the Boltzmann distribution. The thickness of the diffusion layer partially depends on the kinetic energy of the ions in the solution. This model provides a better insight than the Helmholtz model but still suffers from a few shortcomings. The stern model is the hybrid form of the Helmholtz and Gouy–Chapman model that assumes a fixed, compact layer on the surface of the metal called the Stern layer or inner Helmholtz layer that contains specifically adsorbed ions. In contrast, the counter layer is non-specific, diffused, and often called the outer Helmholtz layer. These models collectively give a satisfactory explanation of the electric double-layer formation on the surface of the electrode. In 1954, a noncommercial patent was filed by H I Becker at General Electric. In 1978, the Japanese company Nippon Electric Corp. commercialized the first electrochemical capacitor developed by Robert Rightmare at Standard Oil Co. of Ohio (SOHIO) in the name of 'Super-Capacitor.' These supercapacitors are successfully being used in backup memory devices and many electronic devices [3].

11.2.2 Historical milestones of Li-ion batteries

The story of Li-ion batteries is much younger than supercapacitors. The first Li-based battery was demonstrated in 1970 as non-rechargeable and used in watches, calculators, and implanted medical devices. Later it was found that inorganic compounds react with Li metal reversibility through the intercalation process. Based on this observation, in 1976, Whittingham at Exxon proposed an electrochemical cell composed of Li metal as an anode due to its high energy density and layered TiS_2 as a cathode. Dendrite formation by Li metal during the cycling process and leading to the short circuit was the primary limitation in commercializing the concept. Goodenough group (1980) overcame this limitation by using layered chalcogenides (sulfides and selenides) as cathode materials and Al alloy with Li as an anode. However, they failed to stabilize the oxidation states of transition metal ions, limiting them to achieving higher voltages (>2.5 V) versus Li/Li^+. Later, several transition metal-oxide–based materials with the formula Li_xMO_2 (M = Co, Ni, and Mn) were identified as the best candidates for cathode materials and are currently used as cathode materials in commercial portable electronic devices [4].

Besenhard, Yazami, and Basu explored that layered graphite can reversibly intercalate the Li^+ ions and be used as the stable anode. The invention of intercalation materials opened a breakthrough that circumvented the safety issues associated with anode materials in the early 1990s. In 1987, the Yohsino group made a prototype Li-ion battery using graphite as anode and $LiCoO_2$ (LCO) as cathode and filed a first patent with assured safety. Finally, in 1991, Sony commercialized the first Li-ion battery composed of graphite as an anode that acts as a host for Li^+ ions through the intercalation process and LCO as a cathode having stable and high voltage oxide material that provides a voltage of 3.8 V. This most successful battery empowers the entire consumer and mobile electronic technology. The Li-ion battery technology was further progressed by the invention of a low-cost cathode $LiFePO_4$ (LFP) in 1996 by the Goodenough group and a high-capacity anode C–Sn–Co in 2005 by Sony [5, 6].

11.3 Thermodynamics and kinetics

The thermodynamics of a battery involve both cathodic half-cell and anodic two-half-cell reactions. In general, the cathode is usually made of metallic oxides or sulfides, and the anode is a metal that would be electrochemically oxidized to form electrolyte-soluble metal ions. The electrolyte is one in which the ions must move from the anode to the cathode and vice versa. There should not be any electronic conductance through the electrolyte to avoid internal short-circuiting.

As given below, let us consider a general expression for anodic and cathodic reactions:

at cathode,

$$aA + ne^- \rightarrow cC,\ E_C^o, \tag{11.1}$$

at anode,

$$bB \rightarrow dD + ne^-,\ E_B^o, \tag{11.2}$$

overall reaction,

$$aA + bB \rightarrow cC + dD,\ E_{\text{cell}}^o. \tag{11.3}$$

The change in standard free energy is given by

$$\Delta G^o = -nFE^o = -nF(E_{\text{Cathodic}}^o - E_{\text{Anodic}}^o), \tag{11.4}$$

where E° refers to the standard cell potential. At conditions other than standard conditions, E can be given using Nernst Equation:

$$E = E^o - \frac{RT}{nF}\ln\frac{[a_C]^c[a_D]^d}{[a_A]^a[a_B]^b}, \tag{11.5}$$

$$\text{Thus } \Delta G = -nFE = -nF\left(E^o - \frac{RT}{nF}\ln\frac{[a_C]^c[a_D]^d}{[a_A]^a[a_B]^b}\right). \tag{11.6}$$

Thus, the change in Gibbs's free energy drives a battery to deliver electrical power. Thermodynamic data can be used to check the feasibility of a cell reaction and the theoretical cell potential. Notably, charge transfer is always a limiting factor in electrochemical cells. Hence, to better assess actual cell potential, it is necessary to consider the kinetics of cell reactions.

Charge deposited on the electrode placed in an electrolyte attracts the oppositely charged ions, leading to electrical double-layer formation. Electrochemical reactions occur on this layer and thus demand movement of ions from bulk to the electrode surface through the formed double layer. Various kinetic factors influence the ability to pass through this layer, indicating that the electrochemical reaction is also kinetically controlled.

According to Arrhenius's theory of reaction rates, the rate of a reaction is a given as

$$k = A.\, e^{\frac{-\Delta G}{RT}}. \tag{11.7}$$

The rate of an electrochemical reaction can be measured by the current produced since current can be said as charge per unit of time; in other words, several electrons are transferred per unit of time.

Thus, a dynamic equilibrium is to be established, with a change in concentrations of charge carriers failing below or above the dynamic equilibrium leading to the deviation from the equilibrium potential will result. This deviation from equilibrium potential is known as overpotential:

$$\text{Overpotential}\ (\eta) = E - E^{o}. \tag{11.8}$$

Let us consider the general equation for metal oxidation reaction at the anode:

$$M \rightarrow M^{z+} + ze^{-}. \tag{11.9}$$

The rate of this reaction

$$k_a = K.\, e^{\frac{-\Delta G}{RT}}, \tag{11.10}$$

where K is a constant.

From Faraday's Law,

$$i_o = zFk_a = zFK.\, e^{\frac{-\Delta G}{RT}}. \tag{11.11}$$

If an overpotential is applied in the anodic direction, the activation energy of the reaction is reduced to($\Delta G - azF\eta$), where a is the symmetry factor of the double layer.

Thus, the anodic current is given by the Tafel equation:

$$i_a = zFK.\, e^{\frac{-(\Delta G - azF\eta)}{RT}}, \tag{11.12}$$

$$i_a = zFK.\, e^{\frac{-\Delta G}{RT}}.\, e^{\frac{azF\eta}{RT}}, \tag{11.13}$$

$$i_a = i_o.e^{\frac{azF\eta}{RT}}. \tag{11.14}$$

This equation can be written as

$$\eta = \left(\frac{RT}{azF}\right)\ln\ln\left(\frac{i_a}{i_o}\right). \tag{11.15}$$

In terms of electrode potentials,

$$E = b_a \log\log\left(\frac{i_a}{i_o}\right) + a_a, \tag{11.16}$$

where b_a is the anodic slope.

Similarly, the cathodic equation is given as

$$E = b_c \log\log\left(\frac{i_c}{i_o}\right) + a_c. \tag{11.17}$$

Overall polarization is governed by the electrochemical reaction kinetics. Charge transfer reactions kinetics occurring on the electrode/electrolyte interface is attributed to activation polarization, whereas resistance offered by all the cell components and contact junctions is mentioned as ohmic polarization. Mass transport limitation during charge-discharge is ascribed as concentration polarization. Cumulative increase in overpotential increases internal resistance and thus reduces the overall performance of a battery.

Based on the thermodynamic calculation, a Li-ion battery can have a nominal cell potential between 3.2 and 3.85 V, depending on the cathodic material. Since the potential window is beyond 1.23 V, we cannot use aqueous electrolytes, i.e., nonaqueous solvents are mandatory for Li-ion batteries. Nonaqueous electrolytes are electrochemically stable even beyond 4 V. With this, Li-ion batteries can store large amounts of energy.

11.4 Mechanism of energy storage

11.4.1 Mechanism of energy storage in Li-ion battery system

A conventional battery comprises a cathode, anode and an electrolyte and plays a crucial role in storing energy and functioning of batteries. Batteries generally store energy through oxidation, and reduction reaction (redox reaction) occurs at the anode and cathode, respectively. A primary battery involves spontaneous redox chemistry in only one direction; hence, it can be exhausted after a specific lifetime. On the contrary, a secondary battery involves redox reactions in both directions, which are spontaneous in one direction and non-spontaneous in another. For instance, the Li-ion battery, the most successful rechargeable battery that empowers almost all electronic devices, comprises a graphite anode and LCO as a cathode material with a non-aqueous electrolyte. During charging (storing), Li^+ ions migrate from cathode lattice sites to the anode layers; while discharging (powering), Li^+ ions are diffused back to cathode lattice sites through an electrolyte. The cell capacity (mAh g^{-1}) usually depends on the number of Li^+ ions that can be deintercalated from a cathode and the number of ions that can be hosted/stored by the anode. A separator separates the anode and cathode electrically, preventing the shorting of electrodes, but it allows the movement of ions across it [8] (figure 11.1).

$$LiCoO_2 + C_6(\text{graphite}) \leftrightarrow Li_l - xCoO_2 + xLiC_6(\text{graphite}). \quad (11.18)$$

11.4.2 Mechanism of energy storage in supercapacitor system

Mechanistically, there are three kinds of capacitors: electric double-layer capacitors (EDLC), pseudocapacitors, and hybrid capacitors. In the EDLC type, energy is mainly stored through the adsorption of ions on the electrode's surface. The storage solely depends on the extent of surface area available on the electrode material. Due to the confinement of charge on the surface, the charge/discharge process is almost instant with infinite cycles. For example, porous or activated carbon materials are the best-known EDLC materials. In contrast, pseudocapacitors store the charge through a redox reaction (Faradaic process) at or near the surface of the electrode by accessing multiple oxidation states of transition metal ions. The amount of charge that can be stored depends on the active surface area, porosity, redox capability, and morphology of metal electrodes. Due to rich redox chemistry, pseudocapacitors are capable of storing a larger amount of energy than EDLC. Since the process is surface-controlled and Faradaic, the charge/discharge times are slightly longer, and cyclic stability is shorter than EDLC. Metal oxides and conducting polymers are the best choice of materials for pseudocapacitance. For example, RuO_2, MnO_2, Co_3O_4, NiO, etc are well-known metal oxide pseudocapacitive materials; polyaniline, polypyrrole, etc are well-studied conducting polymers. Further, in pseudocapacitance, two modes of energy storage can be observed, i.e., redox pseudocapacitance and intercalation pseudocapacitance. In redox pseudocapacitance, energy is stored via Faradaic reactions on or near the surface of the electrode material. In the intercalation type, the charge is stored via the rapid diffusion of ions into and out of the ion-conduction channels or layers of the electrode material without any phase transformation. Hybrid capacitors result by amalgamating an EDLC and battery-type electrodes in a single device, usually called supercapatteries. They store energy in two ways; one is electrostatically on an EDLC-type electrode that provides a rapid charge/discharge process and a longer life cycle. The other electrode stores charge Faradaically through a redox reaction that offers high energy density. Thus, hybrid capacitors can provide power and energy density in one component [2, 9] (figure 11.2).

11.5 Materials for Li-ion battery system

A Li-ion battery is mainly composed of three components, namely a cathode, an anode, and an electrolyte. The intrinsic properties of a battery, such as cell potential, capacity or energy density, etc, are dependent on the chemistry of battery materials. The lifetime and cyclability are related to the nature and phenomena occurring at

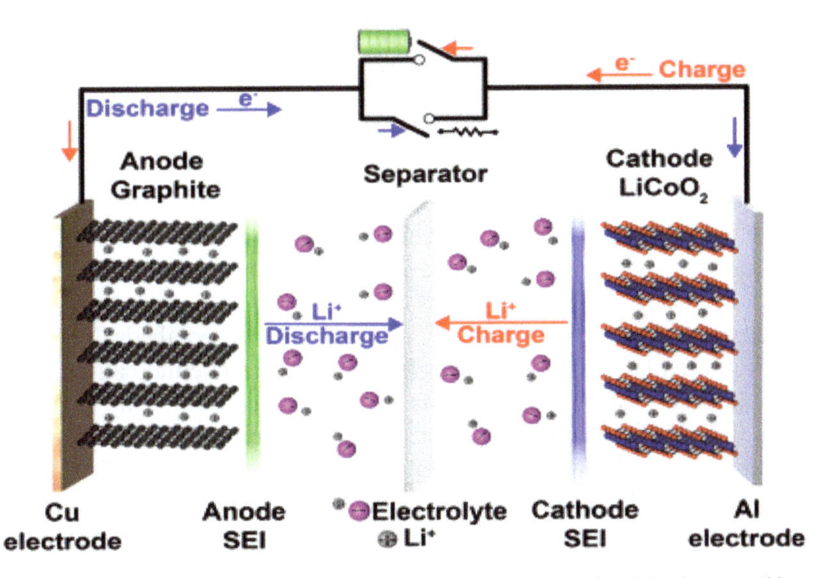

Figure 11.1. Schematic illustration of charging and discharging mechanism in Li-ion battery taking graphite anode and $LiCoO_2$ as a cathode [8]. Reprinted from [8] with permission from the American Chemical Society, Copyright (2017).

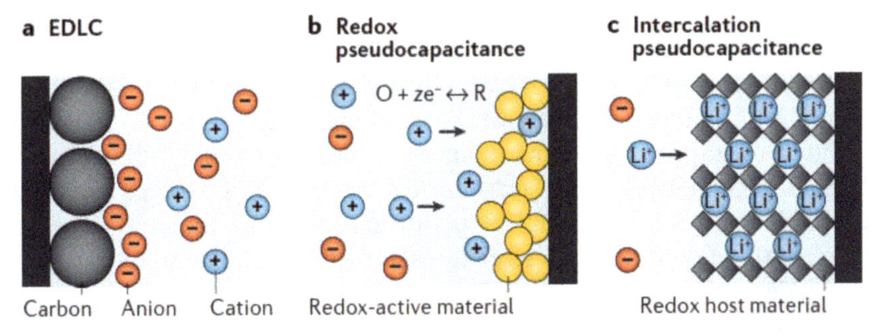

Figure 11.2. The schematic illustration of energy storage mechanism in (a) EDLC, (b) redox pseudocapacitor, and (c) intercalation pseudocapacitance [9]. Reprinted from [9] with permission from Springer Nature, Copyright (2020).

the electrode-electrolyte interface, and the safety is a function of the stability of electrode materials. Thus, the development of battery technology solely depends on the active chemistry of positive and negative materials and their interfacial phenomena. Various kinds of materials and their merits and demerits are discussed below.

11.5.1 Positive (cathode) electrode materials

In cathode chemistry, three types of materials have been explored as best suitable in Li-ion batteries. The cell capacity, rate capability, cell voltage, and cost of Li-ion battery mainly depend on the chemistry of the cathode material. The current Li-ion battery technology mainly focused on two types of cathode materials, namely intercalation-type and conversion-type cathodes.

Intercalation cathode materials consist of a solid network that can host guest (Li^+) ions. The intercalation compound intercalate/deintercalate Li^+ ions reversibly through the host network of compounds. LCO is a widely used cathode material in commercial Li-ion batteries due to its layered structure, high theoretical capacity (274 mAh g^{-1}), good rate capability for efficient charging and discharging, low self-discharge for prolonged charge retention, and excellent cycling stability for long-lasting performance.

The layered structure contains Co^{3+} and Li^+ ions in octahedral sites of oxide lattice in alternate planes that facilitates fast intercalation/deintercalation of Li ions at a cell voltage of 3.8 V. However, its wide utility is limited by high cost, low thermal stability, and fast capacity fading at high currents. The substitution of Co metal with some metals such as Mn, Cr, Al, and Fe demonstrated promising but limited performance. Further coating of LCO with some metal oxides (Al_2O_3, B_2O_3, TiO_2, ZrO_2) could enhance the stability and performance of electrode material during deep cycling due to high mechanical and chemical stability.

$LiNiO_2$ (LNO), is another positive material with the same characteristics as LCO in terms of the layered structure and theoretical capacity (274 mAh g^{-1}). The relatively cheaper cost of Ni compared to Co attracts LNO cathode material by the battery community. However, pure LNO suffers from the limitations of poor structural stability due to the insertion of Ni^{2+} ions in Li^+ ion sites and also blocks the Li^+ ion diffusion path. LNO is thermally less stable than LCO due to the ease of reduction of Ni^{3+} than Co^{3+}. Therefore, several metal substitutes (Ti, Al, Mg) were demonstrated to improve the performance characteristics of LNO. For example, partial Co-doping decreases the cationic disorder, and Al-doping improves the thermal stability and electrochemical performance of the material. The resulted $LiNi_{0.8}Co_{0.15}Al_{0.05}O_2$ is a commercially successful cathode used in Panasonic batteries made for Tesla electric vehicles.

$LiMnO_2$ (LMO) is another alternative positive electrode material to LCO and LNO due to its high theoretical capacity (285 mAh g^{-1}) and good rate capability. Moreover, LMO is relatively cheaper than LCO and environmentally benign. Although the chemical formula of LMO is identical to LCO, it is thermodynamically unstable and results in a poor cycle life due to phase transition from layered to spinel ($LiMn_2O_4$) during Li^+ ion insertion/extraction. A mixed transition metal oxide comprises of Co, Ni, and Mn is prepared to benefit the assets of low cost, high capacity, and good structural stability. The mixed metal oxide $LiNi_{0.33}Mn_{0.33}Co_{0.33}O_2$ has a layered structure and is popularly known as NMC cathode. The combination of Ni, Mn, and Co metals in the layered structure benefits from the high charge capacity of LNO, rate capability of LCO, and inexpensiveness of LMO. NMC provides a high discharge capacity of 160 mAh g^{-1} in the voltage range of 2.5–4.4 V. The low content of Co reduces the cost of the material and improves the structural stability by suppressing the Jahn-Teller distortion stemming from the Ni^{3+} cation in the layered structure. As a result, MNC is utilized as a successful cathode material in the battery market. However, the formation of an unstable solid electrolyte interphase (SEI) layer on the electrode surface with a decomposition of transition metals incurs sluggish electrode kinetics and affects the cycling and rate capability of the electrode. Recent research works showed that macroporous NMC can achieve a charge capacity of 234 mAh g^{-1} with good cycling stability even at high temperatures of 50 °C. Several non-stoichiometric combinations of NMC were demonstrated to show higher energy density, power density, better cycle life, and safety [10].

11.5.2 Polyanion compounds

Olivine-based polyanions are a new class of cathode materials having XO_4^{3-} (X = P, Si, As, Mo, W) anion unit in which X occupies tetrahedral sites in the slightly distorted hexagonal close packed array of oxide ions. LFP is a typical polyanion material that contains Li^+ and Fe^{2+} ions in octahedral sites, while P occupies a tetrahedral site. The presence of a strong P-O bond in the PO_4^{3-} structure lowers the cathode redox potential and improves the structural stability of the material. Thus, LFP is known for its strong thermal and electrochemical stability, high power capability, eco-friendliness, and inexpensive. However, it suffers from the weaknesses of poor Li^+ ion diffusion, low average potential, and low electrical and ionic conductivity. Research efforts were focused on the increase in the performance of LFP through the reduction in particle size, carbon coating, and cationic doping, which are a few strategies that improved the electrochemical performance of LFP. The polyanion class of compounds performs better thermal stability and safety than the layered and oxide-based cathode materials due to strong X–O (X = P, S, Si) covalent bonds [6, 11].

11.5.3 Negative (anode) electrode materials

A stable and Li-free anode is essential for Li-ion batteries for the successful storage of energy. In current battery technology, anode material act as a host for Li ions. Initially, Li metal anode is used as anode material with a layered TiS_2 as cathode material that can reversibly react with Li metal. However, the utilization of Li metal as an anode causes severe safety issues due to dendrite formation on cycling and its thermal run-away reaction at the cathode. Several types of materials are explored as anodes, and they are discussed below.

11.5.3.1 Carbon-based materials (intercalation-type anodes)

Carbon materials are more popular in battery chemistry due to low cost, low redox potential, and high thermal, chemical, and electrochemical stability. The usage of carbon-based materials or coatings as anode retards the side reactions of electrodes with electrolytes, thus reducing the thickness of the SEI layer. The reduction of SEI layer thickness greatly enhances the Li^+ ion diffusion and eliminates safety issues.

Carbon materials are of two types, soft and hard carbons, based on the degree of crystallinity and carbon atoms stacking. Soft carbon materials are most commonly found in battery technology as they show high reversible capacity (350–370 mAh g^{-1}), long cycling life, and good coulombic efficiency (>90%). The current model of

Li-ion battery uses graphite as successful anode material with a specific capacity of 372 mAh g^{-1} that powers most of the portable electronic gadgets today. However, the capacity is still limited to realize the complete electrification of the transportation sector. Therefore, several carbon-based materials such as porous carbons, carbon nanotubes (CNTs), nanofibers, and graphenes are extensively studied with novel properties to achieve larger specific capacity and high rate capability that resulted in excellent energy performance.

Hard carbons are non-graphitizable and possess disordered carbon layers. They were also explored as anode materials for Li-ion batteries due to their high specific capacity (>500 mAh g^{-1}) in the potential range of 0–1.5 V (versus Li/Li+) and hence tested as an alternative to soft carbons. Yet, the accommodation of Li ions in a random and non-uniform manner results in a very slow Li^+ ion diffusion and hence poor rate capacity. Many research efforts are still focused by battery communities to overcome the limitations of hard carbon toward the utilization of the high reversible capacity of hard carbons as anodes for use in electric vehicles.

Ti oxide-based materials significantly attract the battery community due to low cost, low toxicity, low volume variation, and reasonably good reversible capacity (175–330 mAh g^{-1}). However, the electrochemical performance of Ti-based materials depends on the structure, morphology, and size [12].

$Li_4Ti_5O_{12}$ (LTO) spinel is the most popular intercalation anode due to its excellent Li-ion reversibility at a high operating potential of 1.55 V versus Li/Li^+. Thus, LTO offers better safety properties than graphite-based materials. The insertion of Li into LTO spinel results in a stable rock salt type $Li_7Ti_5O_{12}$ structure that remains unaltered during Li insertion/de-insertion and hence provides a stable performance with the outstanding power density and excellent cycle life. Nevertheless, its commercialization is hindered by its low reversible capacity (175 mAh g^{-1}) and low electronic conductivity (10^{-13} S cm^{-1}). Several research strategies were employed to improve the electronic conductivity such as growing nanowires directly on the current collector (Ti plate), increasing the Li-ion diffusion, and showing remarkable enhancement in electrochemical performance.

TiO_2 is another promising anode material in the Li-ion batteries field. TiO_2 has the advantages of high electro-activity, good chemical stability, high natural abundance, and structural diversity. Titania can take 1 mole of Li per 1 mole of TiO_2 and can achieve a maximum theoretical capacity of 330 mAh g^{-1}. The quantity of Li insertion/extraction per titania depends on its crystallinity, particle size, phase structure, and surface area. Among various allotropic forms of titania, rutile, and anatase forms are most commonly explored due to their high electro-activity. The Li storage property of these titania forms depends on size and morphology; the decrease in size of the particles enhances the capacity and rate capability of anode material due to reduced path length for Li^+ ion diffusion and increased electrode kinetics [12, 13].

11.5.3.2 Alloy/de-alloy-type anode materials

The high energy demand of the current smart cities and electrification of automobile vehicles can fulfill by the materials with high specific capacity. Anode materials that can store Li through alloy/de-alloy mechanism are believed to satisfy the high energy demand due to high theoretical specific capacity ranges from 783 to 4211 mAh g^{-1}. The alloying mechanism between Li ions and metals can be presented by the general equation. $xLi^+ + xe^- + M \rightarrow Li_xM$. Even though their practical utilization is mainly restricted by large volume variations during Li insertion/extraction and a large irreversible capacity loss at the initial cycles. These issues can be addressed with nanostructured alloy materials with various morphologies such as nanowires, and nanotubes that facilitate faster Li diffusion with shorter path lengths and reduced volume changes.

Si is the second-most abundant element in earth's crust, having high gravimetric (4200 mAh g^{-1}) and volumetric specific capacity (9786 mAh cm^{-3}) that greatly attracts the battery community. However, the large volume variations (400%) cause poor cycling life and larger irreversible capacity loss during the charge/discharge process. The researchers employed x-ray diffraction (XRD), nuclear magnetic resonance (NMR), and transmission electron microscopy (TEM) techniques to study capacity fading in Si-based anode materials during charge and discharge cycles. XRD analysis revealed changes in the crystal structure due to Li-alloying reactions, leading to volume changes and capacity fading. On the other hand, NMR provided valuable insights into the interactions between Li ions and the anode material, influencing the battery's reversible capacity.

TEM visualization showcased microstructural defects, like cracks and voids, resulting from volume changes and also shed light on the electrical contact between the anode material and current collector, which affected battery performance. By integrating data from XRD, NMR, and TEM, the researchers gained a comprehensive understanding of the capacity fading processes. This knowledge holds significant value in the design of more efficient and durable Li-ion batteries, because it identifies critical factors that must be addressed to mitigate capacity fading and enhance battery performance [14–16]. Researchers are addressing challenges in Si-based anodes for Li-ion batteries by focusing on nanostructured Si materials like nanowires, nanotubes, and nanospheres. These structures provide space for Si expansion during charge and discharge cycles, reducing mechanical stress and capacity fading. Si nanowires and nanotubes grown on a current collector exhibit a reversible capacity of 2000 mAh g^{-1} and good cycling stability, making them potential high-performance anode materials. Two effective strategies to enhance electrochemical performance and extend cycling life are carbon coating and metal doping. Carbon coating acts as a protective barrier, reducing capacity fading, while metal doping stabilizes the Si structure, improving cycling performance. These advancements contribute to the development of long-lasting Li-ion batteries for various applications.

For example, B-doped Si nanowires and carbon-coated nanotubes showed an impressive larger reversible capacity of 3247 mAh g^{-1} and a high coulombic efficiency of ~89%. Si nanotubes are successfully tested as anode material for commercial Li-ion battery technology and exhibited a higher capacity than graphite anode (~10 times higher) even after 200 cycles. Similarly, Ge, Sn, and P also reversibly react with Li ions through the alloying mechanism and can have high specific capacitance, still suffering from large volume variation issues. The volume changes during Li insertion and extraction result in mechanical fractures of electrode particles present on the interface leading to an unstable SEI layer. The swelling of individual particles of the electrodes challenges the cycling life and capacity of the anode material. The fabrication of nanosized metal particles or coating with carbon is a viable solution to improve the mechanical stability and the subsequent electrochemical performance of the materials [12, 17].

11.5.3.3 Conversion-based anode materials

A variety of metal oxides (MO; M = Co, Ni, Fe, Cu, Mn, etc) reacts with Li reversibly to form metal nanoparticles (2–8 nm) that are embedded in lithia (Li_2O) matrix. These reactions are generally called conversion reactions. A typical form of conversion reaction may be represented as $MO + 2Li^+ + 2e^- \leftrightarrow M + Li_2O$. These reactions involve the complete transformation of electrode material into a different structure and composition, and hence, the stable materials that can regenerate their initial form in a reverse reaction and can act as good conversion materials providing a high Li-storage capability and stable performance. These materials provide high specific capacity (500–1000 mAh g^{-1}) due to the participation of more electrons. Although, conversion materials are mainly suffering from the limitation of poor electrode kinetics associated with the diffusion of Li ions into the interfacial of nanodomains that cause significant voltage hysteresis during charging and discharging and hence effects the energy performance of the device. This limitation can be manifested by hybridizing metal oxides with CNTs/graphene sheets that could effectively enhance the diffusion rate through short path lengths, control the agglomeration of nanoparticles during first charging, and impart better rate performance [10].

In general, the choice of cathodic material, anodic material, electrolyte, design, and fabrication of a battery should meet the requirements—the ability to recharge, realistic voltage drawn, constant voltage discharge, capacity, energy density, power density, temperature dependency, physical requirement, cycle life, cost, and the ability to deep discharge, etc.

11.6 Development of flexible materials for energy storage applications

Unprecedented progress in development and fabrication of flexible energy storage devices over conventional rigid batteries and supercapacitors is due to their higher degree of foldability, stress-strain tolerance, and flexibility in design. The crucial steps in achieving a functional flexible energy storage device are (a) the development of new electrode material and packing material, (b) understanding of the interactions between isolated components, and (c) checking of the key parameters to meet the requirement. Thus, with a series of iterations in the above-mentioned steps, along with a constant introspection of the modifications made in each step and consequent changes in the desired property, a fully functional flexible energy storage device can be achieved.

Numerous portable energy storage materials have been documented, yet Li-ion batteries remain at the forefront as pioneers, boasting high energy density and extended cycle life. Hence, more focus is being paid toward establishment of flexible Li-ion batteries. Basically, flexibility in a device can arise if all the material counterparts are flexible and these materials can be classified as intrinsically flexible devices, whereas if a specific structure of a device offers flexibility, it can be called a structure-dependent flexible device. In comparison, fabricating an intrinsic flexible device is much more laborious and demands the skill of the art, starting from the choice of the material to the expected outcomes of the device fabrication.

Flexible devices can be fabricated mainly in three ways:

(a) Making an inherently rigid conducting substance elastic via patterning: Pre-straining is practiced in designing electrodes to minimize actual deformations of the electrode microstructure during large tensile strain on the battery. Spring-like coiled electrodes and wave-structured electrodes involve pre-strain application and can withstand strain up to 400% without failure. Amorphous materials can withstand strain, but binders are needed to rebound on relieving the stress. Kirigami-based stretchable Li-ion batteries show a tensile strain limit of 150%, whereas thin-film Li-ion batteries can withstand bending radii of 14 mm, and elastic foam-based Li-ion batteries can be compressed up to 90%.

(b) Introducing some rigid conductive fillers into an elastomeric composite: Crystalline, stiff, and rigid materials are being used currently in commercial-scale batteries. Bending and stretching of the batteries with these materials may lead to short circuits or broken circuits and cell damage. Structural integrity between current collectors and rigid active material can improve flexibility and battery durability under repeated cyclic stress. Most of the rigid components like electrode material, metallic components, packing material, and electrolytes are being replaced with advanced flexible materials. Nanomaterials are chosen for binding current collectors and active material due to their high surface area. However, it is not an exception to improving mechanical interaction between the current collector and active material to enhance flexibility under large deformations.

(c) Using conducting substances that are inherently stretchable: Bendability in the electrode material is achieved using thin-film–based materials. Deformations due to strain on the electrode and active material can be minimized by fabricating them to thin and ultrathin material. Thus, even small bending radii can bring about insignificant strain in ultrathin batteries. Composite thin-film electrodes are generally composed of 1D material like CNTs, carbon nanofibers, and 2D material like graphene, which offer high conductivity, flexibility, and stability to the electrochemical reactions and temperature variations occurring during the charge-discharge processes. Conductive paper, carbon cloth, etc, are also being used in recent times as

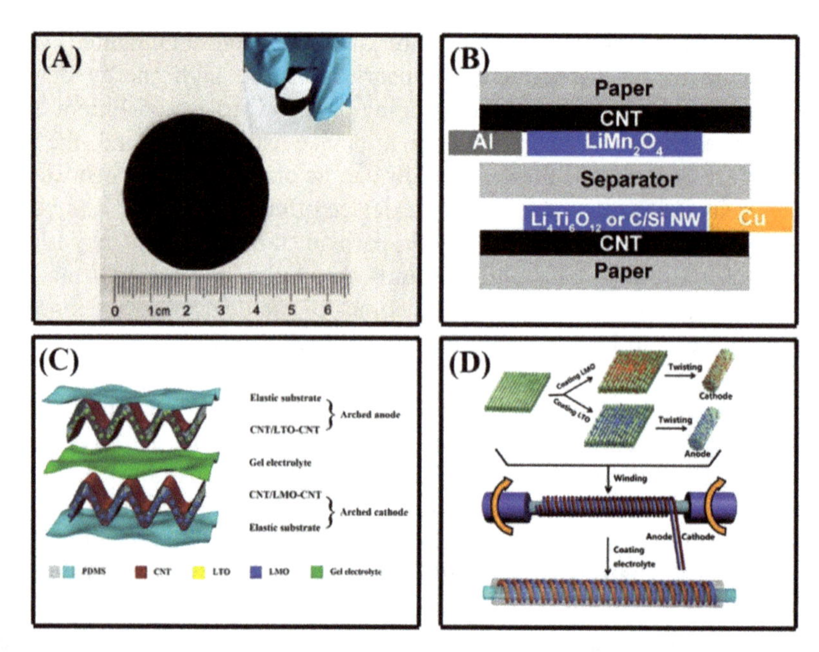

Figure 11.3. (a) Free-standing network film, (b) cross-section of the electrode material, (c) structure of the super-stretchy Li-ion battery, and (d) multiwalled CNTs used in making super-stretchy material [19]. Reprinted from [19] with permission from Elsevier, Copyright (2019).

they offer a high surface area to accommodate more active material owing to its high porosity and flexibility.

Although the flexibility of Li-ion batteries has been improving other parameters like energy density, cell cycle, cost, and safety of the devices are not fully replenished. Computational modeling is being developed to check the deformation mechanism, and this enables to development a better design [18]. Figure 11.3 represents the internal layering in a network like super-stretchable Li-ion battery, and figure 11.4 represents the theoretical model with strain behavior of a layered battery material.

In recent times development of composites containing cellulose, conducting polymers, and CNTs are also being tried to develop paper-based flexible batteries and supercapacitors. The electronically conducting polymers, including polypyrrole, polyaniline, polythiophene, etc, are being investigated to induce flexibility and high conductivity in the device.

11.7 Electrolyte material for flexible energy storage applications

Ionic conductivity occurs through the electrolyte in an electrochemical device. The mobility of ions toward their respective electrodes is to be facilitated, and moreover, the electrode separation is also of extreme importance. Primary research in the development of flexible energy storage devices was regarding the development of flexible electrodes, and hence, very few research reports are being noted on electrolyte material for flexible energy storage devices. Gel-based electrolytes are preferred

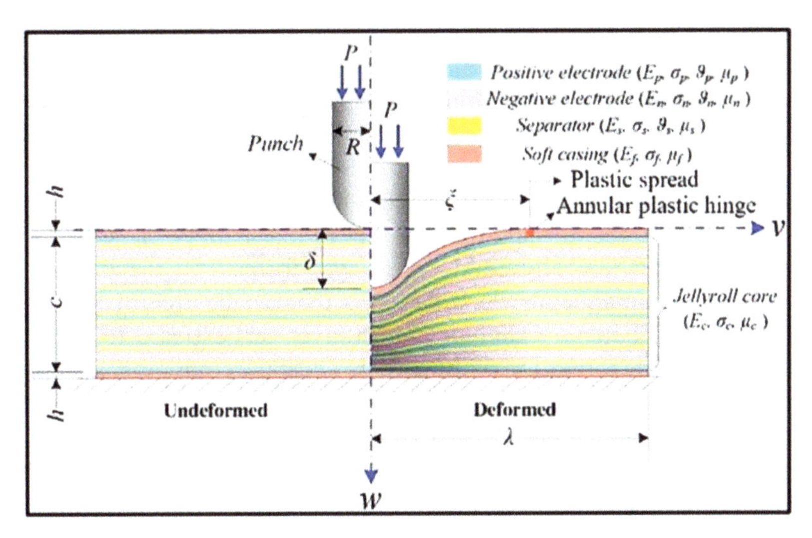

Figure 11.4. Theoretical model representing a rigid wall multilayered material (both deformed and undeformed structures in presence and absence of force P applied in normal to the surface) [19]. Reprinted from [19] with permission from Elsevier, Copyright (2019).

over direct liquid electrolytes in flexible devices because there is a chance of short circuits or leakage during the folding process. A flexible electrolyte needs to have high ionic conductivity, self-healing properties, and mechanical support to achieve overall cell safety. In general, polymer-based substrates are chosen as a flexible electrolyte due to their stretchability and porosity to facilitate ion movement. Mixing plasticizers, nanofillers, ionic liquids, crosslinkers, etc as additives can improve the electrochemical ion conductivity and mechanical performance of the device. Nanocomposite polymer electrolyte and gel polymer electrolyte are the two major classes of flexible electrolytes in which nanofillers are used in the former class to enhance ion mobility, whereas plasticizers are added in the latter case to improve ion mobility. Interpenetrating polymeric networks such as hydrogels are being used in stretchable aqueous cells, whereas ionogels composed of ionic liquids and polymers are being used for broad potential window applications.

11.8 Challenges in practice and possible solutions

Li ion battery has made a remarkable milestone in the path of energy storage, but a few challenges still exist on the way to develop next-generation flexible energy storage devices. Vast improvements are being made in developing potential anode and cathode material. Fabricated small and thin cathodes can minimize the distance between the electrodes and improve Li ion mobility. Expensive and toxic cathodic material like LCO and LFP are to be replaced with affordable, nontoxic, eco-friendly material with better specific capacity and durability. To minimize dendritic growth modified graphite-based anodes, Si- and Sn-based anodes are being developed. But large-scale production of Si nanomaterials is still a challenge. Thus, suppressing dendritic growth had been a challenge, and various techniques like electrode modification; implementation of solid, diverse electrolytes including solid gel electrolytes; and finally changes in cell design are being reported effective in suppressing the dendritic growth. Thus, the strategies applied to suppress dendritic growth can be captioned as mentioned below.

(a) *In situ* development of stable solid electrolyte interface formation: As mentioned in the mechanism of the Li-ion battery, a solid electrolyte interface is an ionically conducting layer formed by the interaction of a metal with non-aqueous electrolyte. As the surface area of the Li dendrite grows during battery recharge, a fresh layer gets exposed to the electrolyte and leads to the formation of more SEI. Thus, the coulombic efficiency of the cell drastically decreases and cell resistance increases. A highly conductive, more intense, thin, and elastic SEI formation could reduce the loss. This can be achieved by checking the combinations of organic solvents, Li salt, and added functional additives [20, 21].

(b) *Ex situ* formed surface coatings: Stable protective coatings on Li metal using amorphous carbon spheres had shown a significant stabilization of SEI formed, and dendritic growth was hindered. These coatings get destroyed during reversible reaction cycles, and later dendritic growth cannot be arrested.

(c) Solid-state polymer electrolyte: Solid-state electrolytes with shear modulus twice higher than Li metal anode can suppress dendritic growth because they have the good mechanical strength to arrest dendritic protrusion. Polymer electrolytes like polyethylene oxide are thermally viable and chemically inert and are gelled with polypropylene oxide, polyethylene glycol, polyacrylonitrile-like polymers, and ceramic particles like nanostructured silica as inorganic fillers. These combinations offer high ionic conductivity and good mechanical strength and thus can hinder dendritic growth. Figure 11.5 represents the schematic representation of a solid electrolyte composite material.

Li ion batteries also suffer from parasitic reactions occurring at the anode, which lead to the formation of a passivation layer on the electrode surface in a long run and induce capacity fade. Limiting electrode area and choosing electrolytes with high ionic conductivity and low viscosity can minimize side reactions occurring in the cell. It is mentioned that low viscosity, high ionic conductivity, wide potential window, good chemical stability, and reliable safety are the ideal characteristics of an electrolyte in a battery, but practically there is no electrolyte with all these ideal characteristics in it. Hence, a mixture of solvents and addition of additives are being tried, and unpretentious performance was noticed [22].

Problems with the dendritic growth can be mitigated by any of the above-mentioned methods, but on the other hand, limited Li availability and lack of

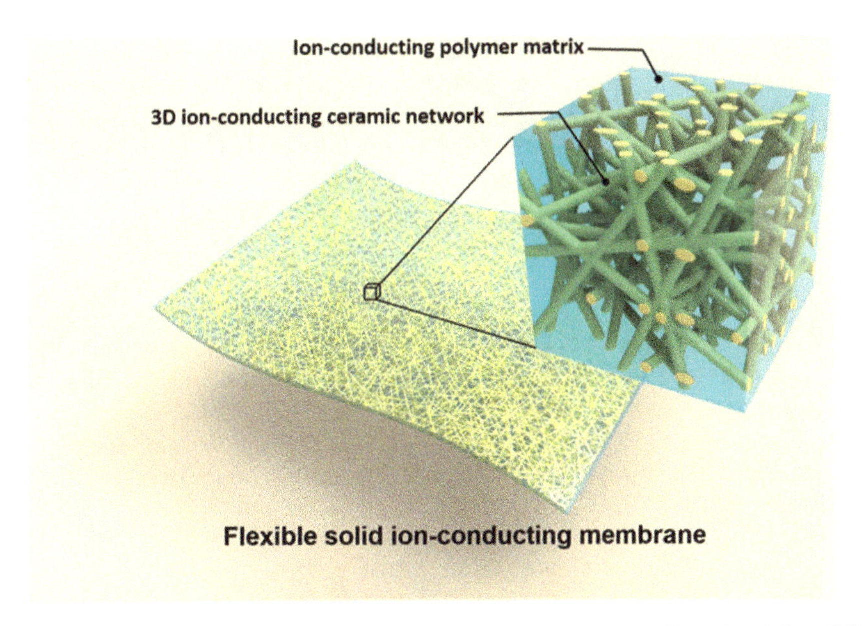

Figure 11.5. Schematic representation of solid-state composite electrolyte [22]. Reproduced from [22] with permission from PNAS, Copyright (2016).

development in the recycling process lead to the development of Na-ion, Mg-ion, Ca-ion, and F-ion–based batteries [23].

All the mentioned above are the current challenges in the development of Li-ion batteries, and thus, flexible Li-ion batteries face new challenges during onsite operation. Size elongation of the device on applying strain is always accompanied by elongation of the pore size, through which the moisture resistance of the device will be altered. To overcome this, water repellent material is being opted in device fabrication. Flexible packing materials like graphene composites that can effectively prevent moisture penetration are being developed but these materials create discomfort in the users on continuous operation due to sweating. On the other hand, rigid electrode materials minimize the flexibility in design to restore structural integrity. Hence, alternative electrode materials like high-capacity Si-based anode are being used. The issue of volumetric expansion of the battery during usage is addressed by, limiting the passive materials like polyvinylidene fluoride (PVDF) in battery construction and reducing the thickness of the electrode, thus improving the energy density [24]. Alternatively, Zn-ion batteries and Zn-air batteries are also being encouraged for better safety.

11.9 Summary

Flexible energy storage devices are a promising tool for fourth-generation industries with demanding lightweight, higher flexibility, safety, and long durability. Li-ion batteries are the better choice of material due to their high energy density and cyclic stability. Dendrite formation is the major hurdle in Li-ion batteries and novel methods are being investigated to arrest the issue. But the fabrication of flexible Li-ion batteries faces a few new challenges with teething problems. Alternatives in electrode composition, electrolyte, and packing material are being rapidly investigated to suit the need; however, a few of the outstanding qualities of Li-ion batteries like high energy density and cyclic stability are being compromised at the early stages of development. Nevertheless, flexible Li-ion batteries open a wide arena in portable electronics with esteemed applications.

References

[1] Dhanabalan S S, R S, Madurakavi K, Thirumurugan A, M R, Avaninathan S R and Carrasco M F 2022 Flexible compact system for wearable health monitoring applications *Comput. Electr. Eng.* **102** 108130

[2] Fleischmann S, Mitchell J B, Wang R, Zhan C, Jiang D E, Presser V and Augustyn V 2020 Pseudocapacitance: from fundamental understanding to high power energy storage materials *Chem. Rev.* **120** 6738–82

[3] Shao Y, El-Kady M F, Sun J, Li Y, Zhang Q, Zhu M, Wang H, Dunn B and Kaner R B 2018 Design and mechanisms of asymmetric supercapacitors *Chem. Rev.* **118** 9233–80

[4] Xie J and Lu Y 2020 A retrospective on lithium-ion batteries *Nat. Commun.* **11** 9–12

[5] Kim T, Song W, Son D Y, Ono L K and Qi Y 2019 Lithium-ion batteries: outlook on present, future, and hybridized technologies *J. Mater. Chem.* A **7** 2942–64

[6] Manthiram A 2020 A reflection on lithium-ion battery cathode chemistry *Nat. Commun.* **11** 1550

[7] Lu J, Chen Z, Pan F, Cui Y and Amine K 2018 High-performance anode materials for rechargeable lithium-ion batteries electrochem *Energy Rev.* **1** 35–53

[8] Chen K-S, Balla I, Luu N S and Hersam M C 2017 Emerging opportunities for two-dimensional materials in lithium-ion batteries *ACS Energy Lett.* **2** 2026–34

[9] Choi C, Ashby D S, Butts D M, DeBlock R H, Wei Q, Lau J and Dunn B 2020 Achieving high energy density and high power density with pseudocapacitive materials *Nat. Rev. Mater.* **5** 5–19

[10] Nitta N, Wu F, Lee J T and Yushin G 2015 Li-ion battery materials: present and future *Mater. Today* **18** 252–64

[11] Armand M and Tarascon J-M 2008 Building better batteries *Nature* **451** 652–7

[12] Goriparti S, Miele E, De Angelis F, Di Fabrizio E, Proietti Zaccaria R and Capiglia C 2014 Review on recent progress of nanostructured anode materials for Li-ion batteries *J. Power Sources* **257** 421–43

[13] Srivastava P R and K S 2015 Nanostructured anode materials for lithium ion batteries *J. Mater. Chem. A Mater. Energy Sustain.* **3** 2454–84

[14] Misra S, Liu N, Nelson J, Hong S S, Cui Y and Toney M F 2012 *In situ* x-ray diffraction studies of (de)lithiation mechanism in silicon nanowire anodes *ACS Nano* **6** 5465–73

[15] Key B, Bhattacharyya R, Morcrette M, Seznéc V, Tarascon J-M and Grey C P 2009 Real-time nmr investigations of structural changes in silicon electrodes for lithium-ion batteries *J. Am. Chem. Soc.* **131** 9239–49

[16] McDowell M T, Lee S W, Harris J T, Korgel B A, Wang C, Nix W D and Cui Y 2013 *In situ* TEM of two-phase lithiation of amorphous silicon nanospheres *Nano Lett.* **13** 758–64

[17] Zhang X, Cheng X and Zhang Q 2016 Nanostructured energy materials for electrochemical energy conversion and storage: a review *J. Energy Chem.* **25** 967–84

[18] Mohanta J, Kang D W, Cho J S, Jeong S M and Kim J K 2020 Stretchable electrolytes for stretchable/flexible energy storage systems—recent developments *Energy Storage Mater.* **28** 315–24

[19] Zeng L, Qiu L and Cheng H-M 2019 Towards the practical use of flexible lithium ion batteries *Energy Storage Mater.* **23** 434–8

[20] Song W J, Lee S, Song G, Son H B, Han D Y, Jeong I, Bang Y and Park S 2020 Recent progress in aqueous based flexible energy storage devices *Energy Storage Mater.* **30** 260–86

[21] Dhanabalan S S, Arun T, Periyasamy G, D N, C N, Avaninathan S R and Carrasco M F 2022 Surface engineering of high-temperature PDMS substrate for flexible optoelectronic applications *Chem. Phys. Lett.* **800** 139692

[22] Fu K *et al* 2016 Flexible, solid-state, ion-conducting membrane with 3D garnet nanofiber networks for lithium batteries *Proc. Natl Acad. Sci. USA* **113** 7094–9

[23] Deng K, Qin J, Wang S, Ren S, Han D, Xiao M and Meng Y 2018 Effective suppression of lithium dendrite growth using a flexible single-ion conducting polymer electrolyte *Small* **14** 1–10

[24] Shanmuga sundar D, Sivanantha Raja A, Sanjeeviraja C and Jeyakumar D 2016 Highly transparent flexible polydimethylsiloxane films—a promising candidate for optoelectronic devices *Polym. Int.* **65** 535–43

Chapter 12

Flexible piezoelectric, triboelectric and hybrid nanogenerators

Durga Prasad Pabba, Mani Satthiyaraju, Radhamanohar Aepuru, Ali Akbari-Fakhrabadi, Viviana Meruane, Hari Prasad Sampatirao and Arun Thirumurugan

Environmental degradation and the depletion of fossil fuels have emerged as pressing issues that must be addressed for progressing sustainable life. Reducing the usage of the fossil fuels and exploring different eco-friendly resources and energy are in highly valued. The abundant mechanical energy is the most promising source of accessible alternative energy in our daily lives. Harvesting energy from bodily movements such as running, jogging, shaking, and so on enabled to power low-power electronics using nanogenerators (NGs) and has recently emerged as an important area of study due to their diverse applicability in practical scenarios. To obtain highest mechanical energy optimization and greater voltage outputs from NGs, scientists all over the globe have put forward potential options of hybrid NGs with numerous and unique designs utilizing a combination of more than one of the following pyroelectric, triboelectric, piezoelectric, thermoelectric, and other NGs. In large-scale operations, hybridization of NGs has proven to be an effective approach for mechanical energy harvesting and sensing applications. This book chapter focuses on major types of NGs for energy harvesting, such as piezoelectric NG (PENG), triboelectric NG (TENG), and hybrid NGs. The emphasis is primarily on the fundamental working mechanisms of three types of NGs, newly developed materials, and typical applications for energy harvesting in fields such as wearable and human monitoring, implementable devices, and artificial intelligence. Finally, we summarize recent advancements in this field and describe future opportunities for improving energy harvester performance.

12.1 Introduction

Humanity needs a novel technology for a higher standard of living in light of the social globalization's rapid rise [1–5]. Numerous studies have been conducted in

doi:10.1088/978-0-7503-5492-9ch12

search of alternative green and renewable energy sources because of the rapidly depleting supply of fossil fuels and the high pollution issues they cause, both of which pose significant challenges to modern society [6–11]. In general, a conventional battery is a crucial component of a smart system's power supply, but it is relatively big, heavy, and rigid. With regards to battery deterioration, the hazardous risk from lead, cadmium, and lithium during replacement raises issues for the environment and public health. Therefore, there is an urgent need for the creation of self-powered harvesting and sensing tools that offer high performance, adaptability, portability, and eco-friendliness, as well as the capacity to interface with sophisticated electronics [12]. Among the various energy transformation apparatuses, NGs have the potential to transfer a low-frequency mechanical energy to electrical energy by using different phenomena such as triboelectric and piezoelectric [13, 14]. A TENG has gradually evolved to seek widely dispersed mechanical energy for independently powered sensing systems used for a variety of physical exertions. Furthermore, modern TENGs research is developing techniques to connect humans with robots, offering up exciting opportunities in the fields of artificial intelligence [15–17].

The most appealing aspects of PENG technology is its reliability, ease of processing, greater power density, and cleanliness. Piezoelectricity is a property of certain materials that allows them to generate an electric charge when subjected to mechanical strain and vice versa. Natural and synthetic piezoelectric materials are the two primary categories of piezoelectric materials. Furthermore, ceramics, polymers, and composites are three subcategories of synthetic piezoelectric materials. The earliest naturally occurring substance with piezoelectric characteristics is Rochelle salt [18, 19]. Other natural piezoelectric material includes sugar, quartz, bone, etc. Similarly, various ceramics materials such as $BaTiO_3$ [20], lead zirconium titanate (PZT) [21], $Bi_{0.5}(Na_{1-x}K_x)_{0.5}TiO_3$ [22], $(1-x)Ba_{0.95}Ca_{0.05}Ti_{0.95}Zr_{0.05}O_3$-$(x)$ $Ba_{0.95}Ca_{0.05}Ti_{0.95}Sn_{0.05}O_3$ [23], and polymers such as polyvinylidene difluoride (PVDF), poly(vinylidene fluoride-trifluoroethylene), nylon-11, polypropylene, polylactic acid, and other materials that are exhibiting piezoelectricity [24]. PENG may turn tiny vibrations from our environment, human body motions, and other sources of mechanical energy into usable electrical energy. PENGs are highly flexible and resistant to moisture and dust; however, they have several limitations such as low output power and narrow applications. Wang's group developed the TENG in 2012 based on the combined effect of electrostatic induction and triboelectrification, which was proved to harvest biomechanical energy with a wide frequency distribution [25, 26]. The TENG has demonstrated several outstanding benefits, including low cost, light weight, simple structure, biocompatibility, high electrical outputs, a diverse range of material options, and high adaptability for sound wave energy harvesting, wind energy harvesting, blue energy harvesting, and vibration energy harvesting. According to most of the new and revolutionary research, it can also be flexible, elastic, humidity-proof, shape-adaptable, and washable. However, TENG suffers from significant humidity and dust dependencies due to the transient triboelectric charges [27, 28].

Achieving high output performance in a single energy-harvesting technology has remained a significant issue. As a result, the hybrid technique of PENG and TENG can overcome these constraints while enhancing the unique characteristics of each NG, such as high flexibility, high electrical outputs, high resistance to moisture and dust, and broad applicability. PENG and TENG hybrid concepts are realized by using a single piezoelectric material and polymer-composite piezoelectric nanostructures operated via the TENG technique. Outstanding characteristics of hybrid NGs improved the performance and have significant potential for human monitoring, self-powered sensors, etc. PVDF and its co-polymers are currently being used to manufacture flexible NGs due to its inherent piezoelectric properties, easy processing, low cost, non-toxic nature, and very flexible nature. PVDF's piezoelectric behavior is essentially determined by its β-phase content, which can be increased by a variety of techniques including stretching, electric poling, electrospinning, etc [29, 30]. Despite its piezoelectricity and flexibility, piezoelectric output is still limited for a wide range of practical applications. Incorporating piezoelectric nanofillers such as $BaTiO_3$, KNN, ZnO, PZT, and BCZT into the PVDF matrix is an efficient way to get a combined effect that improves β-phase and intrinsic piezoelectric properties [31–34]. As a result, PVDF could be a promising polymer for mechanical energy-

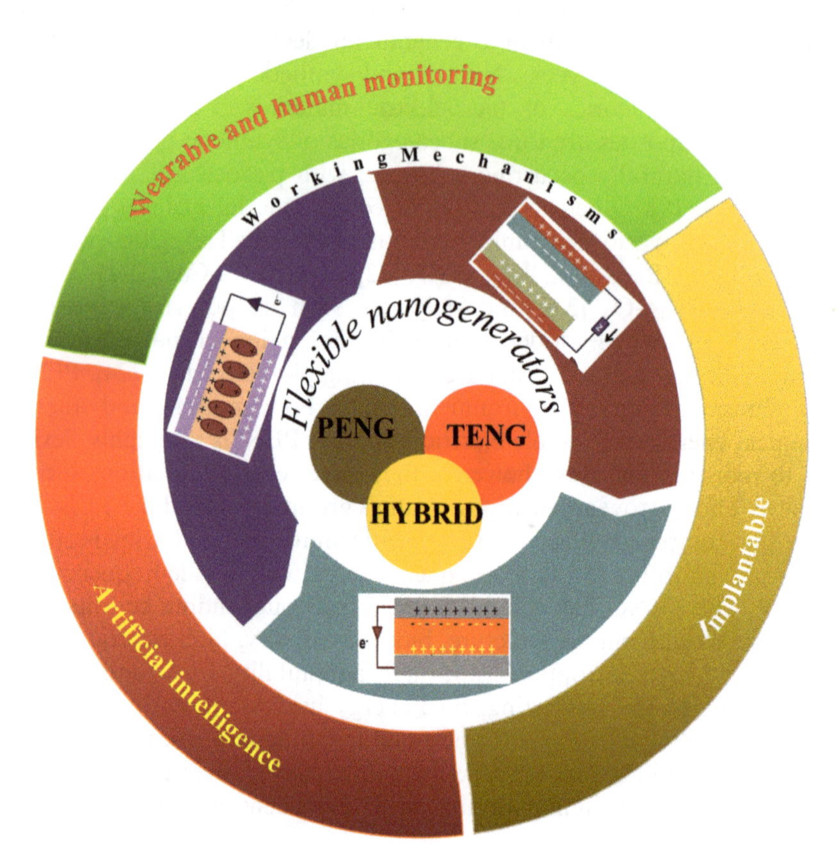

Figure 12.1. Schematic overview of NGs, working mechanisms and applications.

harvesting devices. Many of the transition metal dichalcogenides (MoS_2, $MoSe_2$, etc) and other semiconducting materials (ZnO, $CsPbBr_3$, PZT, $ZnSO_3$) have also been used as nanofiller in the PVDF matrix to improve β-phase, piezoelectric properties, dielectric properties, and surface roughness, resulting in enhanced TENGs output. As a result of the combined impact of adding the nanofiller to the PVDF, the possibility of producing hybrid NGs has emerged, in which both the piezoelectric and triboelectric effects can be utilized to magnify the NG performance for powering a wide range of electronic devices.

The current chapter discusses in detail regarding the working mechanisms of flexible NGs such as PENG, TENG, and hybrid NGs, as well as their applications in fields of human monitoring, implanted devices, and artificial intelligence. Figure 12.1 shows the scheme to reflect the overall content of the present book chapter.

12.2 Working mechanisms

12.2.1 PENG

Figure 12.2 depicts the general device structure of the flexible PENG harvester, which comprises of a piezoelectric material sandwiched between two metal electrodes. The PENG mechanism is stated as follows: in the initial condition, no electrical signal is formed due to the absence of any mechanical stimuli. When an external force is applied to the PENG, the piezoelectric material becomes distorted because it possesses a non-centrosymmetric crystal structure. As a result, the material's positive and negative charge centers move, resulting in a flow of electrons through the external circuit driven by stress-induced polarization [35]. Once the external force is removed, the compressed piezoelectric material is released, resulting in a current

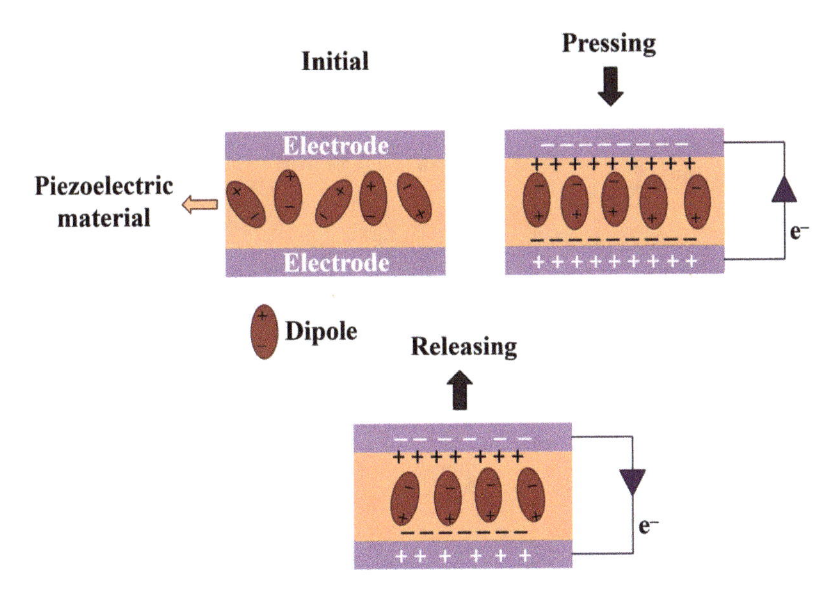

Figure 12.2. PENG working mechanism.

signal flow in the opposite direction. Despite the fact that PENG manufacture is simple and has high flexibility, the output signals produced is typically minimal due to a low charge density. This restricts practical applications for driving numerous electronic gadgets. As a result, the fundamental applications of piezoelectric materials are mostly centered on active sensors utilized in, for example, healthcare, human monitoring, environment monitoring, speed tracking, etc.

12.2.2 TENG

From diverse TENG constructions and applied force directions, TENGs normally operate in four main working modes, including vertical contact-separation, lateral sliding, single-electrode, and free-standing. The vertical contact-separation mode is highly common in TENG structures and was first observed by Fan *et al* in 2012 [36]. Because of the simple device configuration and low surface abrasion, this mode has a long durability of up to 10^5 cycles and has been extensively used in biomechanical energy harvesting, self-sustaining sensors, implantable devices, wearable technology, and large-scale blue energy harvesting [37]. In this mode of operation, a pair of two triboelectric materials, one tribopositive and the other triboneagative, are brought into contact and separated in vertical configuration. When a vertical compression force is applied to the device, the top and bottom layers come into contact, resulting in the formation of triboelectric charges on each layer surface. Upon release, electrostatic induction occurs, resulting in the formation of an electric field across two electrodes connected to the two triboelectric layers as shown in figure 12.3. Output current is formed when the generated triboelectric charges on each material surface encourage the transmission of its opposing charges to the opposite end of the material. At the electrical equilibrium, however, no current flow is visible. The induced charges are driven back when the TENG device is pressed again.

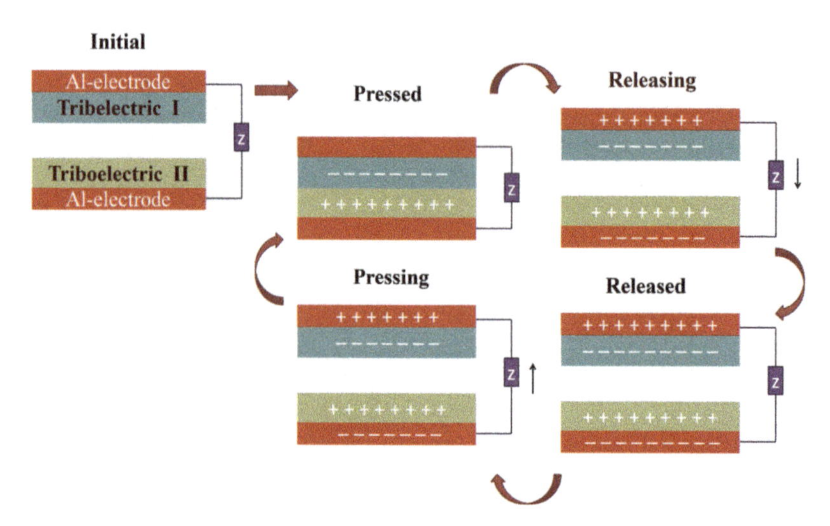

Figure 12.3. TENG working mechanism.

The alternating electrical outputs will be recorded while repeatedly operating this mechanism.

12.2.3 Hybrid NG

The triboelectric surface charge density is the most important component in influencing the output performance of a TENG. A range of characteristics, including triboelectric material selection, surface roughness, and dielectric characteristics, can influence surface charge density. TENG performance can be further improved by integrating piezoelectric and triboelectric effects. Yang *et al* present the multi-effects linked NG for the first time in order to extract more power from single structure-based diverse energy scavenging devices, which is critical for creating future coupled energy-harvesting devices [38]. The most typical approach for creating a piezo-triboelectric hybrid NG is to encapsulate a piezoelectric material, such as ZnO, BZT-BCT, $ZnSO_3$, PZT, and $BaTiO_3$, in a non-piezoelectric polymer, such as polydimethylsiloxane (PDMS), PTFE, nylon, etc. The piezoelectric nanomaterials incorporated inside the polymer produce piezoelectric output, while the contact-separation between the two tribo series materials produces triboelectric output. However, if we use a piezoelectric polymer, such as PVDF or its copolymers, and can improve both the piezoelectric and triboelectric properties of PVDF, we can benefit from the dual enhancement effect, and the composite material could be used as an innovative material for piezo-tribo hybrid generators [39].

In general, the bi-functional performance of hybrid NGs has a strong connection to the design of their structure. Hybrid NGs are classified into two major electrode configurations in the structure: three-electrode modules and two-electrode modules as shown in figure 12.4(A–B). The piezoelectric and triboelectric functional layers are introduced individually in the three-electrode module architecture. These layers are typically separated by inserting an electrode between them, with two triboelectric layers operating in the contact separation mode. A structure with two-electrode modules has only one functional layer placed between the electrodes that combine both triboelectric and piezoelectric capabilities. Moreover, a spacer requirement is essential that often separates one of the electrodes and the functional layer [40].

The two-electrode hybrid NG's operation is classified into four stages, as shown in figures 12.4(C)(a)–(d). Initially, the top electrode of the pre-rubbed NG approaches the piezoelectric layer until they touch. Positive charges flow from the bottom to top electrode through the external circuit as the electrostatic induction between the piezoelectric layer and the top electrode gradually increases, generating a triboelectric current from top to bottom. Henceforth, piezoelectric-bound charges are created inside the piezoelectric layer as the applied pressure increases. The negative charges in the top electrode flow to the bottom electrode due to electrostatic induction of the piezoelectric charges, causing a piezoelectric current to flow from top to bottom in the external circuit. When the pressure is released, the piezoelectric charge vanishes, and the charges created by the piezoelectric effect in the electrode are reversed to their original positions, restoring electrostatic balance. This procedure generates a piezoelectric current in the opposite direction of the first

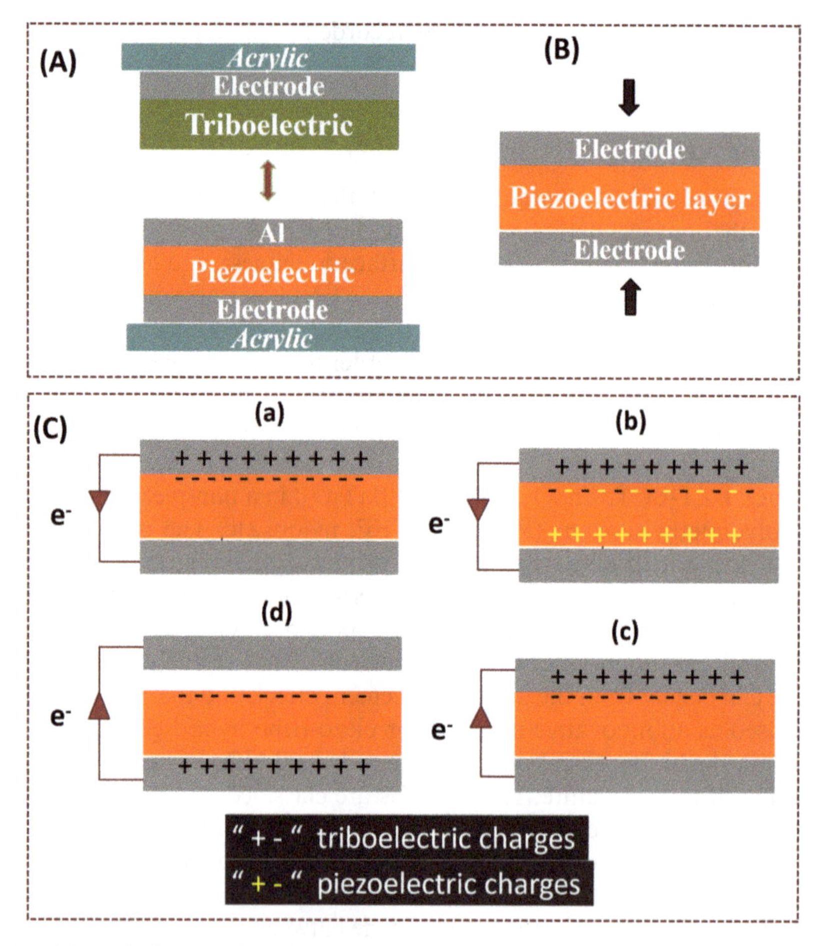

Figure 12.4. Schematic diagram of (A) three electrode modules, (B) two electrode module Hybrid NGs, and (C) working mechanism of two electrode module hybrid NG.

piezoelectric current. At last, when the separation between the top electrode and the piezoelectric layer grows, the electrostatic induction between them weakens. The extra positive charge flows from the higher electrode to the lower electrode, creating a triboelectric current in the external circuit opposite to the initial triboelectric current. Qian *et al* recently developed a hybrid NG with a two-electrode module made of PDMS/ZnO nano flakes/3D-graphene heterostructures. The hybrid energy harvester demonstrated open-circuit voltage (V_{OC}), current densities and power density of up to 122 V, 51 mA cm^{-2}, and 6.22 mW cm^{-2}, repetitively [41].

12.3 Applications

With the increase of microelectronic devices, flexible polymer-based devices are being regarded as viable micro battery power alternative solutions. Besides the continuous growth of micro technology in recent times, new challenges for low-power electronic devices have been reported in terms of energy supply components especially batteries. In general, batteries are primarily used as energy supply units to power the conventional miniature medical devices because of their high energy supply density and stability. However, because of the frequent maintenance of the batteries, they are unable to meet the demands of complex application environments. Methods overcome this include harvesting the energy in the surroundings by converting mechanical energy in physical movement into electrical energy for uninterrupted power supply and forming a self-powered system with wearable/implantable medical devices, and a number of problems, like environmental damage and chemical battery replacement problems, could be fixed. As a result, it is extremely desirable for the development of a long-term green power harvester, such as a self-powered PENG. In the subsequent sections, we discuss the use of a flexible polymer-based energy harvester in three different contexts: wearable NG, IPNG, and artificial intelligence [42, 43].

12.3.1 Wearable and human monitoring

The best approach to the structural design of a wearable energy harvester would be to obtain the energy caused by vibration while not interfering with the physical movement of the body or causing a hardship. Numerous theoretical and experimental approaches have been reported on the flexible piezoelectric polymer–based wearable energy harvesters due to the significant possibilities of generating power from human movements. Wang's group was the earliest to report the concept of human-motion–based energy harvesting in 2012 by demonstrating a flexible ZnO NWs/PVDF PENG for transforming low-frequency physical action into electricity at different degrees of deformation [44]. Recently, Pe *et al* [45] developed a strategy by combining solid-state shear milling (S^3M) and fused filament fabrication (FFF) 3D-printing technology in order to fabricate an outstanding performance biomimetic wearable piezoelectric energy harvester using fish-scale–like metamaterial PVDF/tetraphenylphosphonium chloride/$BaTiO_3$ nanocomposite. The S^3M technology has the potential to significantly improve $BaTiO_3$ sub-micrometer particle dispersion and compatibility in the polymeric matrix, resulting in better processability and piezoelectric performance of nanocomposites. The highest piezoelectric outputs, with a V_{OC} of 11.5 V and a short-circuit current (I_{SC}) of 220 nA; at 30%, loading of $BaTiO_3$ was recorded for the FFF 3D-printed energy harvester.

The flexible biomimetic energy harvester device can be easily worn on the wrist as shown in figures 12.5(A)(a). The device exhibited a maximum I_{SC} 40 nA while tapping a table, demonstrating the tremendous potential in wearable technology. Further, the FFF 3D-printed flexible device was mounted on various positions of a bicycle to harvest mechanical energy, as depicted in figures 12.5(A)(b). The devices undergo various mechanical stimulus such as tire rolling and human motion because of the applied forces and mechanical deflections producing corresponding piezoelectric outputs (figures 12.5(A)(c–e)). The inset diagrams depict the variations in the current produced by the energy harvester device under various motion circumstances. The I_{SC} generated by the energy harvester assembled on the handgrip and seat attained~ 45 and 85 nA, respectively, by hand pressing. As shown in the

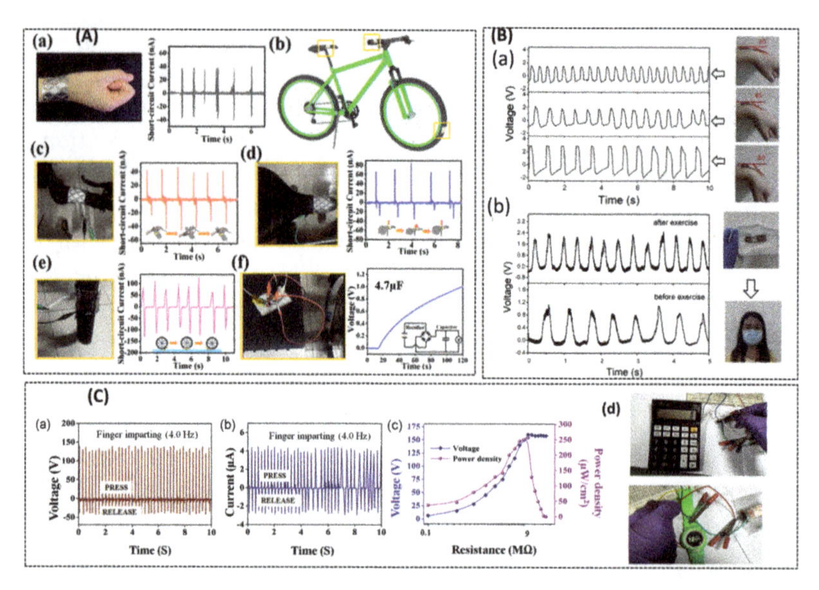

Figure 12.5. (A) (a) Digital photograph of the fish-scale–like energy harvester placed on a wrist and the current signals. (b) Schematic representation of the fish-scale–like energy harvesters mounted on a bicycle, as seen in the photograph, and output signals generated by (c) pushing the handgrip, (d) pressing the seat, (e) rolling the tire, and (f) charging a capacitor while rolling the tire. Reprinted from [45] with permission from the American Chemical Society, Copyright (2022). (B) Output voltage signals generated while (a) finger pressing, (b) bending movements at various bending angles, and (c) breathing with a mask. Reprinted from [46] with permission from Elsevier, Copyright (2022). (C) The generated (a) V_{OC} and (b) I_{SC} for (c) hybrid NG output voltage and power density under various load resistances; (d) photograph of powering of digital calculator and wristwatch. Reprinted from [47] with permission from John Wiley & Sons, Copyright (2023).

figures 12.5(e) and (f), tire rolling generated an I_{SC} with an average value of 130 nA. The harvested electric energy could successfully charge a 4.7 μF capacitor. Li *et al* [46] designed a highly sensitive piezoelectric sensor for human health monitoring using polythiophene (PT) to enhance the β-phase of PVDF. The feasibility of a PVDF/PT sensor in human health monitoring was investigated by measuring voltage output under various physiological input modes. The applied external force due to the manual tapping of the sensor generated a positive output voltage as shown in figure 12.5(B)(a). At wrist bending angles of 30°, 45°, and 60° (figure 12.5(B)(a)), the voltage output values of 1.7, 2.1, and 2.9 V were obtained, respectively. The force exerted on the sensor increased as the wrist bending angle increased, resulting in an increase in sensor output voltage. This result indicates that the sensor was capable of monitoring the human joint bending. Further, they attached the sensor to the mask to monitor the intensity and frequency of breathing to further validate the sensor's practical capability for human health monitoring. Before exercise, the sensor responded to respiration at a heart rate of 88 bpm (figure 12.5(B)(b)), and after exercise, the participants' heart rates increased to 134 bpm, and the sensor response frequency increased. The sensor demonstrated an incredibly stable and sustained response to respiration. Ojha *et al* [47] recently produced a MoS_2@ZnO by electrodeposition approach by depositing few layered

MoS_2 on the Cu foil followed by the deposition of ZnO. Thereafter, MoS_2@ZnO/PVDF films were obtained by spin coating technique. The addition of ZnO improved both the piezoelectric property of the PENG and the TENG performance. The PVDF/MoS_2@ZnO–based hybrid NG displayed remarkable outcomes of V_{OC} 140 V and I_{SC} 4.6 μA with 256 μW cm^{-2} power density (figures 12.5(C)(a)–(c)). The obtained results outperformed most ultramodern and complicated clean room manufactured NGs. With its high output performance, this hybrid NG can extract energy from mechanical and biological activities such as walking, heel pushing, elbow bending, and machinery vibrations. It can light up 33 LEDs in succession and power electronic equipment such as a calculator and wristwatch (figure 12.5(C)(d)).

Patnam *et al* [48] designed a high-performance hybrid NG using ZnO micro-flowers doped with yttrium (Y-ZnO MFs) incorporated in the PDMS for generating rationally designed composite films. The hybrid NG was built by using cellulose paper as a tribopositive layer and Y-ZnO MFs/PDMS composite film as a tribo-negative layer. For electrode purposes, Al tape was bonded to both cellulose paper and PDMS composite. Wet chemical precipitation was used to produce Y-ZnO MFs. The morphology of the Y-ZnO MFs validated the flower-like shape with a size of 23 μm, and EDS analysis proved the presence of elements Y, Zn, and O, as shown in figures 12.6(A) (a and b). The Y-ZnO MFs/PDMS composite active layer could be able to function in three modes: piezoelectric, triboelectric, and hybrid. Figures 12.6(C)(a) and (b) show the V_{OC} and I_{SC} curves of the PENG, TENG, and hybrid NG, which have voltage values of 15, 75, and 250 V and current values of 1, 5, and 10 μA, respectively. Figures 12.6(C)(c)–(h) show schematic representations of the operational modes of PENG, TENG, and hybrid NG, as well as a single

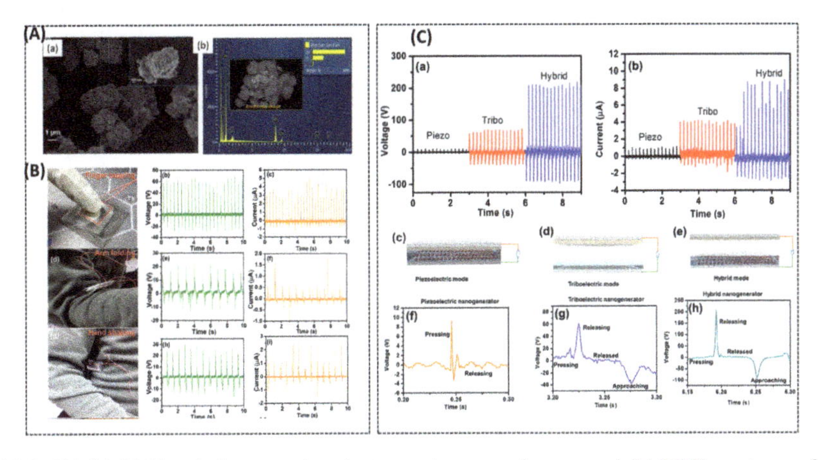

Figure 12.6. (A) (a) Field emission scanning electron microscopy images and (b) EDX spectrum of Y-ZnO micro-flowers. (B) Y-ZnO MFs/PDMS composite hybrid NG to harvest energy from diverse human activities such as (a) finger tapping, (d) arm folding, and (g) hand shaking, as well as the associated output voltage ((b), (e), and (h)) and current signals ((c), (f), and (i)). (C) Comparative (a) V_{OC} and (b) I_{SC} curves at various operation modes (piezoelectric, triboelectric, and hybrid). (c)–(e) Schematic of operation modes of PENG, TENG, and hybrid NG. (f)–(h) Single peak plot of PENG, TENG, and hybrid NGs. Reprinted from [48] with permission the American Chemical Society, Copyright (2021).

signal peak under the various conditions of operation. The findings obtained show that the Hybrid NG outperformed the individual PENG and TENG in terms of output performance. The increased relative permittivity of the composite films was primarily responsible for the improved output performance in the hybrid NG.

The hybrid NG was also used to charge several capacitors ranging from 0.47 to 47 μF, indicating that the manufactured hybrid NG can successfully power small-scale electronic gadgets. The energy held in a 10 μF capacitor was utilized to run a timer on a liquid crystal display. The results show that the Y-ZnO MFs/PDMS–based hybrid NG can be used as a self-powering source for charging and operating portable gadgets. The hybrid NG was further placed on several regions of the human body to harvest and perceive mechanical motion. The hybrid NG's real-time energy harvesting and sensing applications were demonstrated in figure 12.6(B) when subjected to various mechanical motions such as finger tapping, arm folding, and hand shaking (figures 12.6(B)(a), (d), and (g)). During these mechanical motions, the corresponding V_{OC} values were 60, 20, and 25 V (figures 12.6(B)(b), (e), and (h)) and I_{SC} values were 4, 0.5, and 1.5 μA (figures 12.6(B)(c), (f), and (i)). The developed hybrid NG has a potential usage in wearable electronic devices for health monitoring systems.

12.3.2 Implantable devices

Many diseased organs in the medical field must be treated by transplanting micro-medical devices. However, the energy supply of these devices is indeed an immediate matter that needs to be fixed; therefore, employing *in vivo* energy conversion to achieve direct energy supply to these implantable devices is a serious concern. With the exception of lead-based piezoelectric material that affects the living organisms, developing lead-free biocompatible PVDF components is a great option for energy harvesters. Li *et al* [49] reported a soft and flexible implantable NG (IPNG) with a set of specially designed leads and receivers that allow them to monitor the *in vivo* operation of epicardially implanted NGs without opening the chest. A flexible IPNG with a thickness of 40 μm was designed and manufactured using a gold-coated PVDF. *In vitro* testing was used to characterize the performance of packaged IPNG. A computer-controlled actuator bent the IPNG at 1 Hz, resulting in a V_{OC} of 3.8 V and I_{SC} of 60 nA (figure 12.7(b)). The outputs were due to the PVDF film's excellent ferroelectricity. The IPNG system was then implanted in adult swine, and its performance *in vivo* was evaluated. The IPNG component was implanted on the epicardium of the swine heart (left ventricle area), as shown in figure 12.7(a), with the coaxial leads passing through the ribs and tunneling under the skin to the back of the swine. Figure 12.7(b) depicts a schematic of the IPNG operating mechanism.

The heart contracts during systole, causing the PVDF film to bend inward. Due to the deformation, a potential difference is created between the two electrodes deposited on the PVDF films. When the heart contracts during diastole, the PVDF film bends outward, producing an opposite potential and the implantation procedure is shown in the figure 12.7(c). The outputs of the embedded device were examined both during and after implantation. Under open-chest conditions, the implanted IPNG produced a voltage of approximately 2.1 V. (figure 12.7(e)). The

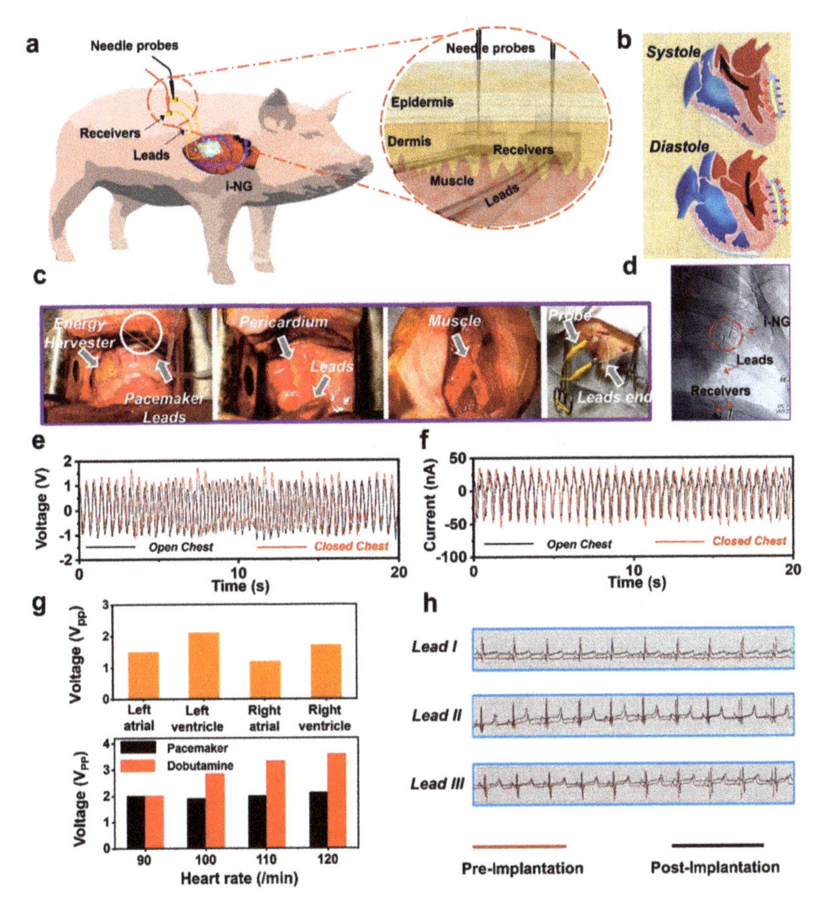

Figure 12.7. Schematic diagrams. (a) IPNG device implanted in swine, (b) IPNG functioning mechanism on heart, (c) IPNG device implantation methods, (d) x-ray image displaying the implanted IPNG system, and (e) and (f) IPNG output voltage and current signals under open- and closed-chest circumstances. (g) The effect of different implantation sites on cardiac contractility on device outputs. The top and bottom figures show voltage outputs at various implantation sites and as a function of heart rate, receptively. (h) Pre- and post-implantation ECG signals in swine. Reprinted from [49] with permission from Elsevier, Copyright (2021).

voltage increased slightly to 2.3 V when the swine chest was closed, possibly due to increased thoracic pressure. Similarly, the current was increased from 68 to 77 nA (open to closed situation). Different implantation sites, including the left atria, left ventricle, right atria, and right ventricle, were tested in separate experiments (figure 12.7(g)). The left ventricle placement produced the highest voltage of 2.1 V, followed by right ventricle of 1.7 V. According to the cardiac enzymes and tissue staining analysis, the device was not subjected to damage or infection to the heart tissues, and the cardiac structure was also unaltered.

Wu *et al* [50] investigated an ultrasound-driven *in vivo* electrical simulation (ES) technique for the repair of peripheral nerve injuries using a biodegradable PENG (BDPENG) and no transcutaneous leads. The BDPENG was made up of biodegradable piezoelectric materials like KNN nanowires, poly (L-lactic acid), and poly

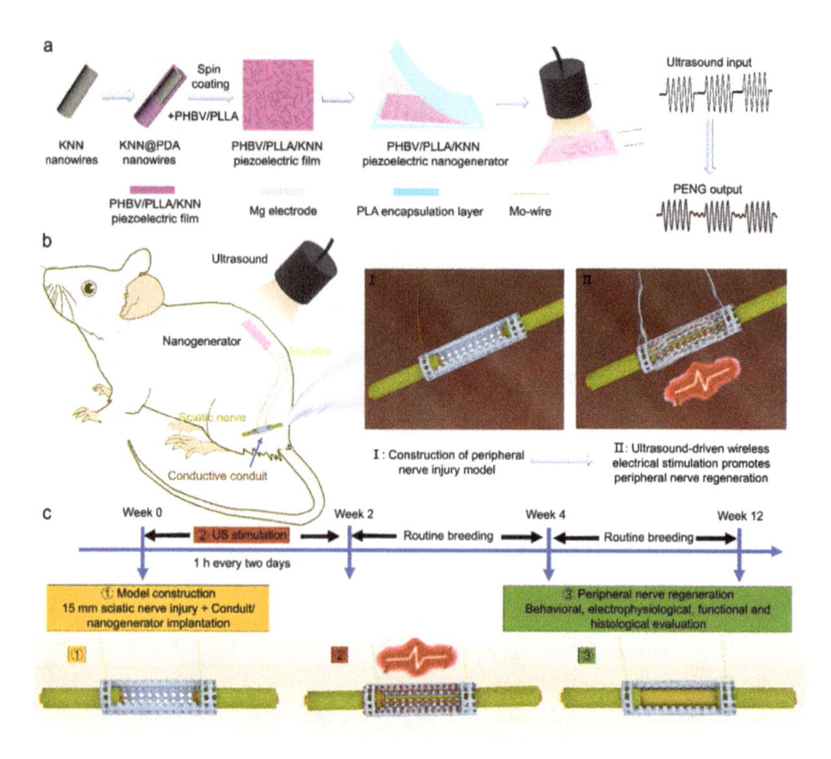

Figure 12.8. Schematics showing (a) BDPENG fabrication, (b) *in vivo* electrical stimulation delivered via implanted BDPENG paired with ultrasound to improve peripheral nerve repair, and (c) fixing damaged peripheral nerves using ultrasound-driven remote electrical stimulation. Reprinted from [50] with permission from Elsevier, Copyright (2022).

(3-hydroxybutyrate co-3-hydroxyvalerate), and the fabrication process is depicted in figure 12.8(a). The implanted BDPENG, as shown in figure 12.8(b), can deliver postoperative *in vivo* ES to enhance nerve regeneration when remotely excited by ultrasound at the appropriate power intensity. Furthermore, *in situ* ES of recovered nerves can be used to monitor nerve repairing dynamics with appropriately recorded muscle electrophysiology response, providing a real-time method to evaluate nerve repair without animal damage and execution (figure 12.8(c)). The BDPENG could also be used as implantable neuro stimulators, providing a real-time, dynamic feedback strategy for evaluating nerve repair without animal sacrifice.

In the medical prospectives, atrial fibrillation (AF) is a silent killer of human life and causes serious illnesses such as myocardial infarction, stroke, and cardiac failure. AF, fortunately, can be detected and treated early. Low-level vagus nerve stimulation (LL-VNS) is an exciting new treatment option for AF. Yet, a few basic hurdles remain in terms of bioelectric stimulation device adaptability, reduction in size, and durability. Sun *et al* [51] constructed a self-powered closed-loop LL-VNS system that can monitor the patient's cardiovascular condition in real time and conduct stimulation impulses automatically as AF develops. The implanted device is flexible, light in weight, and easy to use despite the absence of electronic circuits,

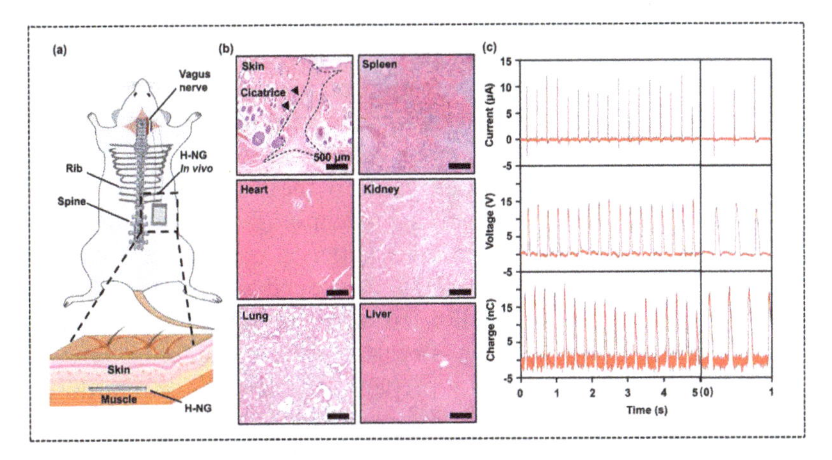

Figure 12.9. (a) The schematic diagram of a hybrid NG implanted in a rat subcutaneously, (b) H&E staining of the rat skin, spleen, liver, heart, kidney, and lung implanted with the hybrid NG subcutaneously after 4 weeks, and (c) the output performance of implanted hybrid NG. Reprinted from [51] with permission from Elsevier, Copyright (2022).

parts, and batteries. The hybrid NG produced an output voltage and current of 14.8 V and 17.8 μA, respectively. The duration of AF was reduced by 90% after LL-VNS therapy in the *in vivo* effect verification investigation, and myocardial fibrosis and atrial connexin levels were substantially improved. To avoid undesired biological responses, excellent biocompatibility is essential. The use of PDMS as the device's encapsulating layer can considerably increase the hybrid NG's flexibility and sealing. Meanwhile, to investigate the device's biocompatibility, the sealed hybrid NG was implanted under the skin in the back of the rat for four weeks (figure 12.9(a)), and rats' H&E staining was observed (figure 12.9(b)). Histological sections of the skin revealed no significant inflammatory response and a few of cicatrices. Meanwhile, no harm was done to the visceral organs. The performance of the hybrid NG implants shown ((figure 12.9(c)) that the human tap could generate a current of 5–15 μA. This corresponds to the device output necessary for AF therapy (figure 12.9(c)). The hybrid NG transverse, sagittal, and coronal planes were visible in the micro-CT scans, and no breaking or considerable movement of the device was observed. Hybrid NG has exceptional sealing qualities and a fixed location *in vivo*, ensuring reliability and precision when a force from the outside drives the internal device.

Implantable medical devices offer a promising treatment option for neurological and cardiovascular illnesses. With the advancement of transient electronics, medical sciences are in desperate need of a new power source that is biocompatible, controllable, and bio-absorbable. Jiang *et al* [52] developed multiple totally bio-absorbable natural-materials–based TENGs (BTNGs) that are produced *in vivo*. The BTNG's device operation time can be adjusted from days to weeks by modifying the natural silk fibroin encapsulating layer. After performing its purpose, the BTNG can be entirely degraded and resorbed in Sprague-Dawley rats, avoiding a second operation and severe side effects.

12.3.3 Artificial intelligence

As artificial intelligence advances, a growing number of smart objects are being created. A significant number of sensing nodes in the sensory network have been needed to power intelligent systems. Commercial batteries presently being used have limited lifespans and seem to be hazardous to the environment, limiting more practical uses in the Internet of Things (IoT) technology. As a result, self-powered piezoelectric materials that do not require batteries are excellent candidates for wireless sensor networks [53]. Tian *et al* [54] developed a self-powered rich lamellar crystal baklava-structured PZT/PVDF piezoelectric sensor for real-time monitoring of table tennis training. A simple two-step, non-solvent induced phase separation and hot-pressing route was used to prepare PZT/PVDF composites. The as-fabricated piezoelectric sensor had a V_{OC} of 2.51 V, I_{SC} of 78.43 nA, a high sensitivity of 6.38 mV N^{-1}, and a superfast response time of 21 ms. Based on the results, a smart table tennis racket equipped with 6×6 sensing units sensory system was designed as shown schematically in figure 12.10(A)(a): PZT/PVDF composites, a signal processor, and a wireless transmitter module. In particular, the sensor array made of PZT/PVDF composites was evenly distributed on the racket, and the signal processor and wireless transmitter module were mounted on the racket handle. The detailed design of the device unit, which was a typical layered sandwich structure prepared using a simple process, is shown schematically in figure 12.10(A)(b). As shown in figure 12.10(A)(d), a sensor array was integrated on the racket to detect the table tennis signal, and the hit location and contact force were quantified to provide individual training guidance for players. The racket could retrieve the hit location and contact force in real time, as well as evaluate the competitive state and scientifically plan individual training guidance for players.

A dynamic piezoelectric tactile sensor was demonstrated by Navaraj *et al* [55]. It had a sensitivity of 2.28 kPa^{-1} when it was interfaced with a metal oxide semiconductor field-effect transistor via an extended gate and a common source configuration. The captured tactile signals were fed into a spiking neural network tempotron classifier system to train the spiking neural network model during a closed-loop tactile scan using an intelligent robotic arm. A classification accuracy of 99.45% was achieved, allowing for real-time binary texture classification, with applications in prosthetics, robotics, medical devices, and wearable sensors (figure 12.10(B)).

As the new era of IoT has emerged, combining artificial intelligence and 5G technologies, a human-machine interface (HMI) is becoming increasingly important in IoT. However, typical HMIs have constraints such as the need for power sources and architectural complexity. Because of the self-powered operation of TENG, TENG-based HMI might be considered an appealing solution. Based on this, Yun *et al* [56] built a self-powered triboelectricity-based touchpad (TTP) that combines with artificial intelligence, consisting of a TENG array (49 pixels) built on a thin, transparent, and flexible substrate, and this TTP operates in two modes: tapping and sliding. The TTP's tracing ability is proved with remarkable precision. Furthermore, using a pre-trained neural network, the TTP identifies digit sequences from '0' to '9'

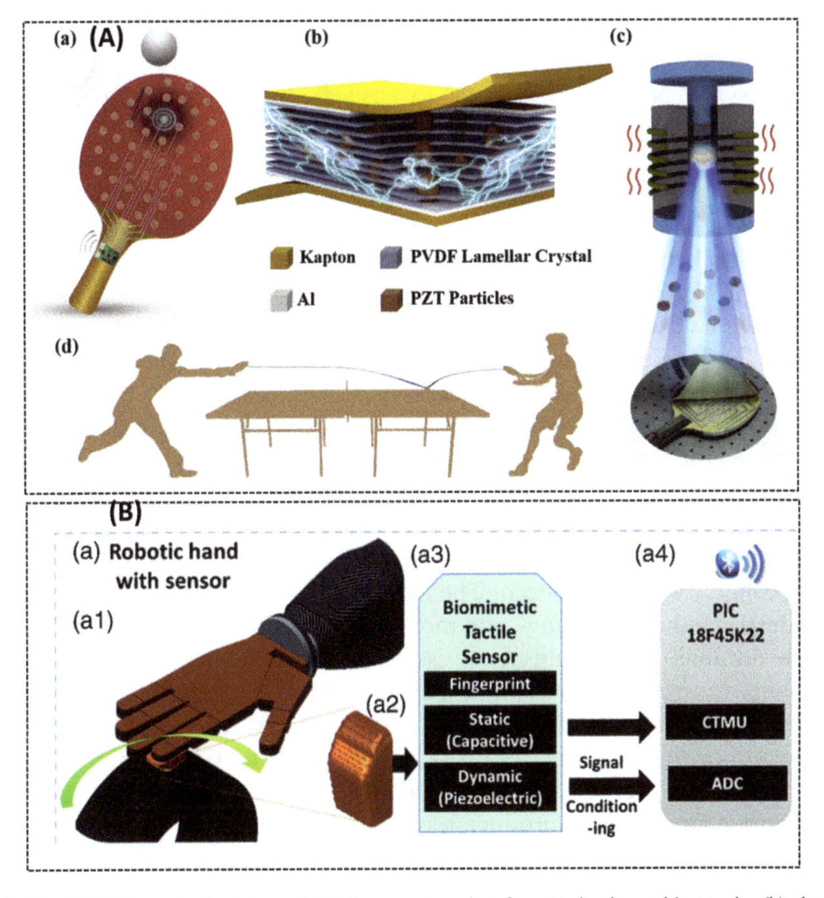

Figure 12.10. (A) Schematic depiction of (a) the smart racket for monitoring table tennis, (b) the precise structure of the developed sensor, (c) the piezoelectric composite construction process and an optical picture of the smart racket, and (d) its use as a smart table tennis racket. Reprinted from [54] with permission from Elsevier, Copyright (2019). (B) A fingerprint-enhanced biomimetic sensor and its implementation in a tactile pattern recognition system are depicted. Reproduced from [55] with permission from John Wiley & Sons, Copyright (2019). CC BY 4.0.

with classification precision of 93.6%, 92.2%, and 91.8% at bending orientations of 0°, 119°, and 165°. Given its high compatibility, it is predicted to be a promising HMI in the near future, with practical IoT leading to artificial intelligence.

12.4 Conclusions and future scope

This chapter provides an overview of the basic structure and various mechanisms of flexible piezoelectric, triboelectric, and hybrid NGs. Each form of mechanical energy harvester has a distinct feature that can be used in a variety of applications. The coupling effect and related theories of NGs for charge generation are examined in the instance of the hybridized NGs. Due to the synergetic boost from two processes, the combination of piezoelectric and triboelectric effects has a better potential in harvesting and sensing performance than an individual unit. In addition, essential elements that improve hybrid NGs performance are detailed in depth. PENG, TENG, hybrid NGs have been extensively discussed in several sectors such as wearable and human monitoring, implementable technologies, and artificial intelligence. Despite the fact that hybrid NGs have potential uses in a variety of industries. For a potentially efficient energy source, some limitations must be overcome before the piezo-tribo hybrid NGs may be used in real-world applications.

- Smart apps used with or without humans have proliferated worldwide as a result of the IoT era's swift growth. Some applications of the self-powered NG demand a very high output voltage; sensitivity and a wide detection range for the sensor are important. As a result, the nanofillers and matrix should be carefully chosen such that they should have high dielectric permittivity, and piezoelectric properties to directly boost power production.
- In order for the triboelectric effect to occur, hybrid NGs often require some degree of contact separation or sliding motion. The surface morphology at nano and micro-scales deteriorates when the NG device is subjected to constant frictional force. As a result, performance may suffer, and life expectancy may be reduced. The hybrid NG's applicability in the healthcare industry and use for long-term monitoring would be hampered by its critical low durability. Durability is an issue, particularly with implantable designs, where recurring surgery should be avoided to the greatest extent possible.
- High-temperature operation of NGs has a deleterious impact on piezoelectric and triboelectric effects. Temperatures beyond the Curie point cause polarization loss in piezoelectric components, while contact-separation processes in triboelectric materials produce less surface charge.
- In general, mechanical energy harvesters generate an alternating current output signal, which creates a barrier for practical applications to operate small electronic appliances. The hybrid NG device could be used to execute strategies for developing a direct current (DC) NG and a p–n junction idea. A built-in electric field generated at the p-n junction with a continuous stimulus can extract charge carriers to provide a DC output current.
- If the pyroelectric effect is combined, the hybrid NG's ability to detect movement, pressure, and temperature makes it possible to track breathing, heart rate, and temperature of the body, among other things. By extending the variety of measures that a hybrid NG can detect, its applications for medical purposes could be significantly expanded.

Acknowledgments

The author D P P acknowledges FONDECYT 2022 Project No. 3220360, Agencia Nacional de Investigación y Desarrollo de Chile (ANID) for the financial support. The author A T acknowledges ANID—SA 77210070 and Universidad de Atacama for their support.

References

[1] Shanmuga Sundar D, Sivanantha Raja A, Sanjeeviraja C and Jeyakumar D 2016 Highly transparent flexible polydimethylsiloxane films—a promising candidate for optoelectronic devices: highly transparent flexible polydimethylsiloxane films *Polym. Int.* **65** 535–43

[2] Dincer I 2000 Renewable energy and sustainable development: a crucial review *Renew. Sustain. Energy Rev.* **4** 157–75

[3] Shanmuga Sundar D, Sivanantharaja A, Sanjeeviraja C and Jeyakumar D 2016 Synthesis and characterization of transparent and flexible polymer clay substrate for OLEDs *Mater. Today: Proc.* **3** 2409–12

[4] Dhanabalan S S, Arun T, Periyasamy G, N D, N C, Avaninathan S R and Carrasco M F 2022 Surface engineering of high-temperature PDMS substrate for flexible optoelectronic applications *Chem. Phys. Lett.* **800** 139692

[5] Dhanabalan S S, Sitharthan R, Madurakavi K, Thirumurugan A, Rajesh M, Avaninathan S R and Carrasco M F 2022 Flexible compact system for wearable health monitoring applications *Comput. Electr. Eng.* **102** 108130

[6] Sriphan S and Vittayakorn N 2022 Hybrid piezoelectric-triboelectric nanogenerators for flexible electronics: recent advances and perspectives *J. Sci.: Adv. Mater. Devices* **7** 100461

[7] Shi Q, Sun Z, Zhang Z and Lee C 2021 Triboelectric nanogenerators and hybridized systems for enabling next-generation IoT applications *Research* **2021** 6849171

[8] Haider Z *et al* 2020 Highly porous polymer cryogel based tribopositive material for high performance triboelectric nanogenerators *Nano Energy* **68** 104294

[9] Haleem A, Haider Z, Ahmad R U S, Claver U P, Shah A, Zhao G and He W 2020 Highly porous and thermally stable tribopositive hybrid bimetallic cryogel to boost up the performance of triboelectric nanogenerators *Int. J. Energy Res.* **44** 8442–54

[10] Graham S A, Dudem B, Mule A R, Patnam H and Yu J S 2019 Engineering squandered cotton into eco-benign microarchitectured triboelectric films for sustainable and highly efficient mechanical energy harvesting *Nano Energy* **61** 505–16

[11] Shanmuga Sundar D and Sivanantharaja A 2013 High efficient plastic substrate polymer white light emitting diode *Opt. Quant. Electron.* **45** 79–85

[12] Melchor-Martínez E M, Macias-Garbett R, Malacara-Becerra A, Iqbal H M N, Sosa-Hernández J E and Parra-Saldívar R 2021 Environmental impact of emerging contaminants from battery waste: a mini review *Case Stud. Chem. Environ. Eng.* **3** 100104

[13] Pabba D P, Rao B V B, Thiam A, Kumar M P, Mangalaraja R V, Udayabhaskar R, Aepuru R and Thirumurugan A 2023 Flexible magnetoelectric PVDF–$CoFe_2O_4$ fiber films for self-powered energy harvesters *Ceram. Int.* **49** 31096–105

[14] Pabba D P *et al* 2023 MXene-based nanocomposites for piezoelectric and triboelectric energy harvesting applications *Micromachines* **14** 1273

[15] Shawon S M A Z *et al* 2021 Piezo-tribo dual effect hybrid nanogenerators for health monitoring *Nano Energy* **82** 105691

[16] Guo H *et al* 2018 A highly sensitive, self-powered triboelectric auditory sensor for social robotics and hearing aids *Sci. Robot.* **3** eaat2516

[17] Fuh Y-K, Li S-C and Chen C-Y 2017 Piezoelectrically and triboelectrically hybridized self-powered sensor with applications to smart window and human motion detection *APL Mater.* **5** 074202

[18] Bhunia S *et al* 2021 Autonomous self-repair in piezoelectric molecular crystals *Science* **373** 321–7

[19] Levitskii R R, Zachek I R, Verkholyak T M and Moina A P 2003 Dielectric, piezoelectric, and elastic properties of the Rochelle salt $NaKC_4\ H_4O_6{\cdot}4H_2O$: a theory *Phys. Rev.* B **67** 174112
[20] Du Q *et al* 2023 Highly -textured $BaTiO_3$ ceramics with high piezoelectric performance prepared by vat photopolymerization *Addit. Manuf.* **66** 103454
[21] Lin J, Cui B, Cheng J, Tan Q and Chen J 2023 Achieving both large transduction coefficient and high Curie temperature of Bi and Fe co-doped PZT piezoelectric ceramics *Ceram. Int.* **49** 474–9
[22] Jaita P, Saenkam K and Rujijanagul G 2023 Improvements in piezoelectric and energy harvesting properties with a slight change in depolarization temperature in modified BNKT ceramics by a simple technique *RSC Adv.* **13** 3743–58
[23] Baraskar B G, Kolekar Y D, Thombare B R, James A R, Kambale R C and Ramana C V 2023 Enhanced piezoelectric, ferroelectric, and electrostrictive properties of lead-free $(1-x)$ BCZT-(x)BCST electroceramics with energy harvesting capability *Small* **19** 2300549
[24] Bairagi S, Shahid-ul-Islam , Shahadat M, Mulvihill D M and Ali W 2023 Mechanical energy harvesting and self-powered electronic applications of textile-based piezoelectric nanogenerators: a systematic review *Nano Energy* **111** 108414
[25] Zou Y *et al* 2022 A high-performance flag-type triboelectric nanogenerator for scavenging wind energy toward self-powered IoTs *Materials* **15** 3696
[26] Zou Y, Raveendran V and Chen J 2020 Wearable triboelectric nanogenerators for biomechanical energy harvesting *Nano Energy* **77** 105303
[27] Javadi M, Heidari A and Darbari S 2018 Realization of enhanced sound-driven CNT-based triboelectric nanogenerator, utilizing sonic array configuration *Curr. Appl Phys.* **18** 361–8
[28] Gong Y, Yang Z, Shan X, Sun Y, Xie T and Zi Y 2019 Capturing flow energy from ocean and wind *Energies* **12** 2184
[29] Singh V and Singh B 2023 MoS_2-PVDF/PDMS based flexible hybrid piezo-triboelectric nanogenerator for harvesting mechanical energy *J. Alloys Compd.* **941** 168850
[30] Liu X, Ma J, Wu X, Lin L and Wang X 2017 Polymeric nanofibers with ultrahigh piezoelectricity via self-orientation of nanocrystals *ACS Nano* **11** 1901–10
[31] Taha E O, Alyousef H A, Dorgham A M, Hemeda O M, Zakaly H M H, Noga P, Abdelhamied M M and Atta M M 2023 Electron beam irradiation and carbon nanotubes influence on PVDF-PZT composites for energy harvesting and storage applications: changes in dynamic-mechanical and dielectric properties *Inorg. Chem. Commun.* **151** 110624
[32] Yan M, Li H, Liu S, Xiao Z, Yuan X, Zhai D, Zhou K, Bowen C R, Zhang Y and Zhang D 2023 3D-printed flexible PVDF-TrFE composites with aligned bczt nanowires and interdigital electrodes for piezoelectric nanogenerator applications *ACS Appl. Polym. Mater.* **5** 4879–88
[33] Joshi B, Seol J, Samuel E, Lim W, Park C, Aldalbahi A, El-Newehy M and Yoon S S 2023 Supersonically sprayed PVDF and ZnO flowers with built-in nanocuboids for wearable piezoelectric nanogenerators *Nano Energy* **112** 108447
[34] Kulkarni N D and Kumari P 2023 Role of rGO on mechanical, thermal, and piezoelectric behaviour of PVDF-BTO nanocomposites for energy harvesting applications *J. Polym. Res.* **30** 79
[35] Prasad P D and Hemalatha J 2021 Multifunctional films of poly(vinylidene fluoride)/ ZnFe2O4 nanofibers for nanogenerator applications *J. Alloys Compd.* **854** 157189

[36] Fan F-R, Tian Z-Q and Lin Wang Z 2012 Flexible triboelectric generator *Nano Energy* **1** 328–34

[37] Liu D, Zhou L, Wang Z L and Wang J 2021 Triboelectric nanogenerator: from alternating current to direct current *iScience* **24** 102018

[38] Minhas J Z, Hasan M A M and Yang Y 2021 Ferroelectric materials based coupled nanogenerators *Nanoenergy Adv.* **1** 131–80

[39] Singh H H and Khare N 2018 Flexible ZnO-PVDF/PTFE based piezo-tribo hybrid nanogenerator *Nano Energy* **51** 216–22

[40] Zhang J, He Y, Boyer C, Kalantar-Zadeh K, Peng S, Chu D and Wang C H 2021 Recent developments of hybrid piezo–triboelectric nanogenerators for flexible sensors and energy harvesters *Nanoscale Adv.* **3** 5465–86

[41] Qian Y and Kang D J 2018 Poly(dimethylsiloxane)/ZnO nanoflakes/three-dimensional graphene heterostructures for high-performance flexible energy harvesters with simultaneous piezoelectric and triboelectric generation *ACS Appl. Mater. Interfaces* **10** 32281–8

[42] Lu L, Ding W, Liu J and Yang B 2020 Flexible PVDF based piezoelectric nanogenerators *Nano Energy* **78** 105251

[43] He Z, Gao B, Li T, Liao J, Liu B, Liu X, Wang C, Feng Z and Gu Z 2019 Piezoelectric-driven self-powered patterned electrochromic supercapacitor for human motion energy harvesting *ACS Sustain. Chem. Eng.* **7** 1745–52

[44] Lee M, Chen C-Y, Wang S, Cha S N, Park Y J, Kim J M, Chou L-J and Wang Z L 2012 A hybrid piezoelectric structure for wearable nanogenerators *Adv. Mater.* **24** 1759–64

[45] Pei H, Shi S, Chen Y, Xiong Y and Lv Q 2022 Combining solid-state shear milling and FFF 3D-printing strategy to fabricate high-performance biomimetic wearable fish-scale PVDF-based piezoelectric energy harvesters *ACS Appl. Mater. Interfaces* **14** 15346–59

[46] Li J, Zhou G, Hong Y, He W, Wang S, Chen Y, Wang C, Tang Y, Sun Y and Zhu Y 2022 Highly sensitive, flexible and wearable piezoelectric motion sensor based on PT promoted β-phase PVDF *Sensors Actuators* A **337** 113415

[47] Ojha S, Bera S, Manna M, Maitra A, Si S K, Halder L, Bera A and Khatua B B 2023 High-performance flexible Piezo–Tribo hybrid nanogenerator based on MoS_2 @ZnO-assisted β-phase-stabilized poly(vinylidene fluoride) nanocomposite *Energy Tech.* **11** 2201086

[48] Patnam H, Graham S A and Yu J S 2021 Y-ZnO microflowers embedded polymeric composite films to enhance the electrical performance of piezo/tribo hybrid nanogenerators for biomechanical energy harvesting and sensing applications *ACS Sustain. Chem. Eng.* **9** 4600–10

[49] Li J, Hacker T A, Wei H, Long Y, Yang F, Ni D, Rodgers A, Cai W and Wang X 2021 Long-term *in vivo* operation of implanted cardiac nanogenerators in swine *Nano Energy* **90** 106507

[50] Wu P, Chen P, Xu C, Wang Q, Zhang F, Yang K, Jiang W, Feng J and Luo Z 2022 Ultrasound-driven *in vivo* electrical stimulation based on biodegradable piezoelectric nano-generators for enhancing and monitoring the nerve tissue repair *Nano Energy* **102** 107707

[51] Sun Y *et al* 2022 Hybrid nanogenerator based closed-loop self-powered low-level vagus nerve stimulation system for atrial fibrillation treatment *Sci. Bull.* **67** 1284–94

[52] Jiang W *et al* 2018 Fully bioabsorbable natural-materials-based triboelectric nanogenerators *Adv. Mater.* **30** 1801895

[53] Cao X, Xiong Y, Sun J, Zhu X, Sun Q and Wang Z L 2021 Piezoelectric Nanogenerators derived self-powered sensors for multifunctional applications and artificial intelligence *Adv Funct Mater.* **31** 2102983

[54] Tian G *et al* 2019 Rich lamellar crystal baklava-structured PZT/PVDF piezoelectric sensor toward individual table tennis training *Nano Energy* **59** 574–81

[55] Navaraj W and Dahiya R 2019 Fingerprint-enhanced capacitive-piezoelectric flexible sensing skin to discriminate static and dynamic tactile stimuli *Adv. Intell. Syst.* **1** 1900051

[56] Yun J, Jayababu N and Kim D 2020 Self-powered transparent and flexible touchpad based on triboelectricity towards artificial intelligence *Nano Energy* **78** 105325